LONDON MATHEMATICAL SOCIETY STUDENT TEXTS

Managing editor: Professor E.B. Davies, Department of Mathematics,
King's College, Strand, London WC2R 2LS

1 Introduction to combinators and λ-calculus, J.R. HINDLEY & J.P. SELDIN

2 Building models by games, WILFRID HODGES

3 Local fields, J.W.S. CASSELS

4 An introduction to twistor theory, S.A. HUGGETT & K.P. TOD

5 Introduction to general relativity, L. HUGHSTON & K.P. TOD

6 Lectures on stochastic analysis: diffusion theory, DANIEL W. STROOCK

7 The theory of evolution and dynamical systems, J. HOFBAUER & K. SIGMUND

8 Summing and nuclear norms in Banach space theory, G.J.O. JAMESON

9 Automorphisms of surfaces after Nielsen and Thurston, A.CASSON & S. BLEILER

10 Nonstandard analysis and its applications, N.CUTLAND (ed)

11 The geometry of spacetime, G. NABER

12 Undergraduate algebraic geometry, MILES REID

LONDON MATHEMATICAL SOCIETY STUDENT TEXTS

Managing editor: Professor E.B. Davies, Department of Mathematics, King's College, Strand, London WC2R 2LS

1 Introduction to combinators and λ-calculus, J.R. HINDLEY & J.P. SELDIN
2 Building models by games, WILFRID HODGES
3 Local fields, J.W.S. CASSELS
4 An introduction to twistor theory, S.A. HUGGETT & K.P. TOD
5 Introduction to general relativity, L.P. HUGHSTON & K.P. TOD
6 Lectures on stochastic analysis: diffusion theory, DANIEL W. STROOCK
7 The theory of evolution and dynamical systems, J. HOFBAUER & K. SIGMUND
8 Summing and nuclear norms in Banach space theory, G.J.O. JAMESON
9 Automorphisms of surfaces after Nielsen and Thurston, A. CASSON & S. BLEILER
10 Nonstandard analysis and its applications, N. CUTLAND (ed)
11 Spacetime and singularities, G. NABER
12 Undergraduate algebraic geometry, MILES REID

London Mathematical Society Student Texts. 10

Nonstandard Analysis and its Applications

Edited by

NIGEL CUTLAND

Department of Mathematics, University of Hull

CAMBRIDGE UNIVERSITY PRESS

Cambridge

New York New Rochelle Melbourne Sydney

CAMBRIDGE UNIVERSITY PRESS
Cambridge, New York, Melbourne, Madrid, Cape Town, Singapore, São Paulo, Delhi

Cambridge University Press
The Edinburgh Building, Cambridge CB2 8RU, UK

Published in the United States of America by Cambridge University Press, New York

www.cambridge.org
Information on this title: www.cambridge.org/9780521351096

First published 1988
Re-issued in this digitally printed version 2008

A catalogue record for this publication is available from the British Library

Library of Congress Cataloguing in Publication data
Nonstandard Analysis and its applications.
(London Mathematical Society student texts ; 10)
Papers presented at a conference held at the University of Hull in 1986.
Includes bibliographical references and index.
1. Mathematical analysis, Nonstandard--Congresses.
I. Cutland, Nigel. II. Series.
QA299.82.I56 1988 519.4 88-16194

ISBN 978-0-521-35109-6 hardback
ISBN 978-0-521-35947-4 paperback

CONTENTS

PREFACE

ex asperis per asteriscos

The methods of Abraham Robinson's Nonstandard (or Infinitesimal) Analysis (NSA) are currently being used across the whole spectrum of mathematics - from 'pure' mathematics through to mathematical physics. This book is designed as an introduction to NSA and to some of its many applications, with the working mathematician or student particularly in mind. It has emerged from a conference with the same title held at the University of Hull in 1986, which had the aim of making NSA more widely known in the mathematical community through a series of introductory lecture courses and lectures on current research. The first part of this book consists of papers based on the introductory lectures given at the conference by Tom Lindstrøm, Ward Henson, Jerry Keisler and Sergio Albeverio. The latter part of the book contains papers that present a sample of recent developments in the more advanced applications of NSA.

Lindstrøm's *An Invitation to Nonstandard Analysis* expounds the foundations of the theory. It is designed to be "a friendly welcome requiring no other background than a smattering of general mathematical culture", offered in the belief that NSA "is of greater interest to the analyst than to the logician". Lindstrøm writes "I have tried to make the subject look the way it would had it been developed by analysts or topologists and not logicians." To this end, his presentation of NSA is somewhat different from others in the literature, in that he builds a nonstandard universe and shows how to practice NSA without any use of logic. Then, in the final chapter of his article, he shows how the language of logic is the natural way to explain and codify in a general way what has been going on in the earlier development.

The choice of topics covered in Lindstrøm's *Invitation* is fairly conventional, and is designed to bring the reader to the point where he can study more specialised nonstandard papers with only an occasional consultation of the literature, and where he can begin to think of making applications in his own field of interest.

One of the most fruitful applications of NSA is in measure theory and probability theory, stemming from the discovery of the Loeb measure construction; this is a simple way to construct a rich class of standard measure spaces from nonstandard spaces, discovered by Peter Loeb in 1975. The article by Keisler discusses applications of Loeb measures to problems in probability theory and the theory of stochastic processes, and explains both how and why it is so successful. Attention is restricted to hyperfinite Loeb spaces, which are particularly easy to work with; it is shown that nothing is lost by working with such spaces, since they have very strong properties (homogeneity and universality) that make them more than adequate for any applications in probability theory.

In functional analysis the construction of nonstandard hulls plays a role similar to that of the Loeb construction in measure theory. Nonstandard hulls are standard topological vector spaces that are constructed in a natural way from nonstandard spaces; they have been used in a variety of ways to solve problems in functional analysis. The article by Henson introduces the nonstandard hull construction for topological vector spaces and operators on them, and is designed to serve as both and introduction and a complement to an earlier survey paper of Henson & Moore[1], so as together to provide a comprehensive discussion of the use of NSA in functional analysis. The earlier survey concentrates on Banach spaces; some recent developments in this area are reported here.

Albeverio's article gives an introduction to the many applications of nonstandard methods in mathematical physics. This field has long been seen as a natural one for such applications, because of the way in which NSA can provide new mathematical models of physical phenomena that are perhaps closer to reality. For example, large finite collections of particles may be more accurately modelled by a hyperfinite set (i.e. a set that is infinite, but finite from the nonstandard point of view, and thus inherits many of the properties of finite sets) than by the continuum. Moreover, the nonstandard framework, with genuine

---

[1] Henson C.W. & Moore, L.C. Jr. (1983). Nonstandard analysis and the theory of Banach spaces; in *Springer Lecture Notes in Mathematics* **983**, 27-112.

infinitesimals and infinite numbers, often allows heuristic reasoning to be made precise in a way that the standard framework prevents. Albeverio's article surveys the different kinds of nonstandard approach that have been productive in mathematical physics, and discusses some specific examples of the kind of results that have been obtained.

The work presented in Loeb's paper is both an alternative approach to the Loeb measure construction and a generalisation of it; he begins with a nonstandard lattice of functions (which could, for example be the measurable functions on a nonstandard measure space) and shows how to construct from it a space of integrable functions. Some recent work extending this approach to vector valued functions is also discussed; here there is an interesting interplay between the Loeb construction and the nonstandard hull construction.

An important but perhaps relatively less well known field of application of nonstandard methods is that of algebra and its interface with the mathematical theory of computation, which is exemplified in the contribution by Benninghofen and Richter. Following a pattern familiar in other applications, the nonstandard approach is used to construct an 'ideal' object (in this case an extension of a free nilpotent group, given as the nonstandard hull of the original group) that is suitably explicit and tractable for the purposes in hand.

The paper of Diener and Stroyan is designed both to introduce *Internal Set Theory* (IST) - an alternative axiomatisation of nonstandard analysis due to Nelson[2] - and to explain the relationship between this and the superstructure approach expounded by Lindstrøm. A slightly restricted version of IST is shown to be valid in a superstructure, and the principal axioms of IST are shown to be equivalent to useful quantifier manipulation rules. It is hoped that the discussion here of the common ground shared between the two approaches will aid the mutual understanding of those familiar with one or other of the dialects[3] of NSA.

A large body of work on the infinitesimal analysis of differential equations has been done over the past ten years or so. The

---

[2]Nelson, E. (1977). Internal set theory, *Bull. Amer. Math. Soc.* **83**, 1165-1193.

[3]This term was used by R. Anderson in his review of Lutz & Goze, *Nonstandard Analysis*, Lecture Notes in Mathematics **881**, Springer, 1981, which appeared in *Bull. London Math. Soc.* **15**(1983), 94-5.

article by the Dieners (written in IST) is an example of the elegance and fruitfulness of nonstandard methods in this area, in particular in the study of singular perturbations. These are naturally represented as nontrivial perturbations by an infinitesimal.

Stroyan's article explores the way in which the theory of superinfinitesimals (due to Benninghofen and Richter) can be used to analyse the detailed structure of the monads of certain topologies arising in functional analysis. This analysis is then applied to obtain new and delicate results for these topologies.

The final paper in the volume, by Arkeryd, surveys the results he has obtained over a number of years on the Boltzmann equation. A nonstandard model of space and time provides the framework for new existence results for this famous equation; here is a further example of one of the themes discussed by Albeverio.

I should like to offer my sincere thanks to each of the authors for their contribution to this book, and also for their key part in the Hull conference from which it has emerged. This is also an opportunity to thank Tom Lindstrøm and Ward Henson, my co-organisers of the conference, and David Ross who was a great help too.

It is a pleasure to acknowledge the generous support for the meeting that was received from the British Logic Colloquium, the Logic Trust, the London Mathematical Society and the SERC.

This volume would not have seen the light of day but for the dedicated services of Eileen Freeman, who battled away with the manuscript on our new T^3 wordprocessor; I am most grateful for all her efforts and patience. David Ross helped us to tame T^3, and read through a copy of the final version of the manuscript, as did Marek Capiński: many thanks are due to both.

Finally, my wife Mary has been most supportive and patient when I have been engrossed in this project: I owe her a big thank you too.

Hull, April 1988

Nigel Cutland

CONTRIBUTORS

Sergio Albeverio	Fakultät für Mathematik, Ruhr-Universität, Bochum, W. Germany.
Leif Arkeryd	Department of Mathematics, Chalmers University of Technology, Göteborg, Sweden.
B. Benninghofen	Fachbereich Informatik, Universität Kaiserslautern, Kaiserslautern, W. Germany.
Francine Diener	UFR S.E.G.M.I., Université de Paris X, France.
Marc Diener	UFR de Mathématiques, Université de Paris VII, France.
Ward C. Henson	Department of Mathematics, University of Illinois, Urbana, Illinois, USA.
H. Jerome Keisler	Department of Mathematics, University of Wisconsin, Madison, Wisconsin, USA.
Tom Lindstrøm	Institute of Mathematics, University of Oslo, Norway.
Peter A. Loeb	Department of Mathematics, University of Illinois, Urbana, Illinois, USA.
M. M. Richter	Fachbereich Informatik, Universität Kaiserslautern, Kaiserslautern, W. Germany.
Keith D. Stroyan	Department of Mathematics, University of Iowa, Iowa City, Iowa, USA.

AN INVITATION TO NONSTANDARD ANALYSIS

TOM LINDSTRØM

INTRODUCTION

Nonstandard Analysis - or the Theory of Infinitesimals as some prefer to call it - is now a little more than 25 years old (see Robinson (1961)). In its early days it was often presented as a surprising solution to the old and - it had seemed - impossible problem of providing infinitesimal methods in analysis with a logical foundation. It soon became clear, however, that the theory was much more than just a reformulation of the Calculus, when Bernstein and Robinson (1966) gave the first indication of its powers as a research tool by proving that all polynomially compact operators on Hilbert spaces have nontrivial invariant subspaces. Since then nonstandard techniques have been used to obtain new results in such diverse fields as Banach spaces, differential equations, probability theory, algebraic number theory, economics, and mathematical physics just to mention a few. Despite the wide variety of topics involved, these applications have enough themes in common that it is natural to regard them as examples of the same general method.

This paper is intended as an exposition of these recurrent themes and the theory uniting them. I have called it "An invititation to nonstandard analysis" because it is meant as an invitation - a friendly welcome requiring no other background than a smattering of general mathematical culture. My point of view is that of applied nonstandard analysis; I'm interested in the theory as a tool for studying and creating standard mathematical structures. As such, I feel that it is of greater interest to the analyst than to the logician, and this attitude is, I hope, reflected in the presentation; put paradoxically, I have tried to make the subject look the way it would had it been developed by analysts or topologists and not logicians. This is the explanation for certain unusual features such as my insistence on working with ultrapower models and my willingness to downplay the importance of first order languages.

Although the presentation may be a little unconventional, the choice of topics is not; there seems to be a fairly general agreement on what are the most important and powerful nonstandard techniques, and I have seen it as my main task to give a full and detailed account of these. The idea has been to bring the reader to the point where he can study more specialized nonstandard papers with only an occasional consultation of the literature, and where he can begin to think of applying nonstandard methods in his own field of interest. Unfortunately, this emphasis on methodology and basic techniques has made it impossible to include convincing examples of new results and at the same time keep the paper within reasonable bounds. But as the other contributions to the present volume contain applications which in depth and variety far exceed anything I could conceivably have put into an introduction of this kind, I do not think that these omissions are of much consequence.

The paper consists of four chapters, each divided into three sections. The first three chapters contain a systematic exposition of nonstandard techniques in different branches of analysis, while the fourth focuses on the underlying logical principles. Not all readers will want or need to read everything; those who are eager to get on to applications may wish to skip Chapter IV at the first reading and only concentrate on the most relevant parts of the other chapters. The chart in Figure 1 traces the dependences between the various sections in detail. Note in particular the sections in the dotted boxes; they are not really part of the systematic development of the theory, but contain examples and applications which add flesh and blood to the bare theoretical bones of the other sections. The paper ends with a comprehensive set of Notes with suggestions for further study.

Acknowledgements. It is a pleasure to thank Nigel Cutland for inviting me to give the lectures on which this paper is based; feedback from many members of the audience both improved the overall quality of the presentation and eliminated some rather embarassing mistakes. Special thanks are due to Keith Stroyan who left me with a heavily annotated copy of the first draft; many of his suggestions have been incorporated into the final version, while some of the more ambitious ones have been left out only for lack of time and space. Through the years a number of people have influenced my view of nonstandard analysis, but none more than Sergio

Albeverio, Jens Erik Fenstad, and Raphael Høegh-Krohn, with whom I spent five years writing a book on the subject. I don't think I want to know how many of my best ideas are really theirs. Finally, I would like to thank the Nansen Fund for generous travel support.

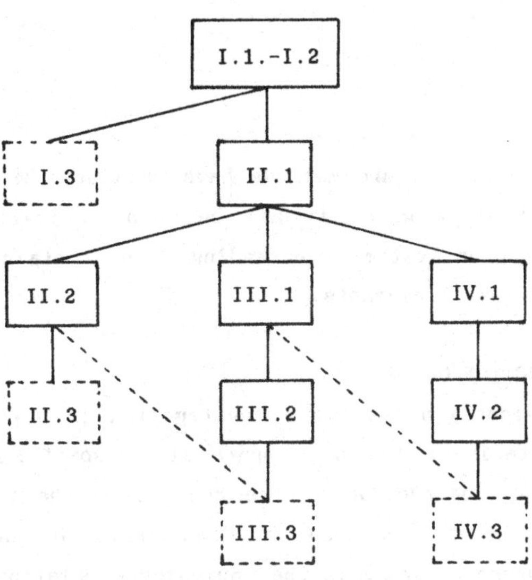

Figure 1

I. A SET OF HYPERREALS

Although nonstandard methods have been used in most parts of mathematics, I will start where it all began historically - with the construction of a number system $^*\mathbb{R}$ extending $\mathbb{R}$ and containing infinitely large and infinitely small elements.

I.1 CONSTRUCTION OF $^*\mathbb{R}$

To convince you that this construction is quite natural and not the least mysterious, let me compare it to something you are all familiar with - the construction of the reals from the rationals using Cauchy-sequences. Recall how this is done: If $\mathcal{C}$ is the set of all rational Cauchy-sequences, and $\equiv$ is the equivalence relation on $\mathcal{C}$ defined by

$$\{a_n\} \equiv \{b_n\} \quad \text{iff} \quad \lim_{n\to\infty} (a_n - b_n) = 0, \tag{1}$$

then the reals are just the set $\mathbb{R} = \mathcal{C}/\equiv$ of all equivalence classes. To define algebraic operations on $\mathbb{R}$, let $\langle a_n \rangle$ denote the equivalence class of the sequence $\{a_n\}$, and define addition and multiplication componentwise

$$\langle a_n \rangle + \langle b_n \rangle = \langle a_n + b_n \rangle; \qquad \langle a_n \rangle \cdot \langle b_n \rangle = \langle a_n \cdot b_n \rangle. \tag{2}$$

The order on $\mathbb{R}$ is defined simply by letting $\langle a_n \rangle < \langle b_n \rangle$ if there is an $\varepsilon \in \mathbb{Q}_+$ such that $a_n < b_n - \varepsilon$ for all sufficiently large n. Finally, we can identify the rationals with a subset of $\mathbb{R}$ through the embedding

$$a \to \langle a, a, a, \ldots \rangle. \tag{3}$$

The construction of $^*\mathbb{R}$ follows exactly the same strategy. Beginning with the set $\mathcal{S}$ of all sequences of real numbers, I shall introduce an equivalence relation $\sim$ on $\mathcal{S}$ and define $^*\mathbb{R}$ as the set $\mathcal{S}/\sim$ of all equivalence classes. If as above $\langle a_n \rangle$ denotes the equivalence class of the sequence $\{a_n\}$, the algebraic operations are defined componentwise as in (2), and I shall also introduce an order on *ℝ which turns it into an ordered field. Finally, ℝ will be identified with a subset of *ℝ through the embedding $a \to \langle a,a,a,\ldots\rangle$.

Before I define the equivalence relation ~, it may be wise to say a few words about the philosophy behind the construction. When we create the reals from the rationals, we are interested in constructing limit points for all "naturally" convergent sequences. Since the limit is all we care about, it is convenient to identify as *many* sequences as possible; i.e. all those which converge to the same "point". No attention is paid to the rate of convergence; hence the two sequences $\{\frac{1}{n}\}$ and $\{\frac{1}{\sqrt{n}}\}$ are identified with the same number 0 although they converge at quite different rates. In creating *ℝ from ℝ, we want to construct a rich and well-organised algebraic structure which encodes not only the *limit* of a sequence but also its *mode of convergence*. To achieve this, we shall reverse the strategy above and identify as *few* sequences as possible.

This sounds silly; to "identify as few sequences as possible" must surely mean the trivial identification $\{a_n\} \sim \{b_n\}$ iff $\{a_n\} = \{b_n\}$. Well, it doesn't if you also want *ℝ to have all the nice algebraic properties of ℝ.

I.1.1 Example

Let $\{a_n\} = \{1,0,1,0,1,\ldots\}$ and $\{b_n\} = \{0,1,0,1,0,\ldots\}$; then $\{a_n\}\cdot\{b_n\} = 0$, although $\{a_n\}$ and $\{b_n\}$ are both non-zero. Thus if we use the trivial identification, we get a structure with zero divisors.

The idea is to make the equivalence relation ~ just strong enough to avoid the problem of zero divisors. Before I can give the definition, I have to fix a finitely additive measure on ℕ with the following properties.

I.1.2 Definition

Throughout this chapter m denotes a (fixed) finitely additive measure on the set ℕ of positive integers such that:

(i) For all $A \subset \mathbb{N}$, $m(A)$ is defined and is either 0 or 1.

(ii) $m(\mathbb{N}) = 1$, and $m(A) = 0$ for all finite A.

That m is a finitely additive measure means, of course, that $m(A \cup B) = m(A) + m(B)$ for all disjoint sets A and B. Note that m divides the subsets of $\mathbb{N}$ into two classes, the "big" ones with measure one and the "small" ones with measure zero, in such a way that all finite sets are "small". The existence of such measures is an exercise in Zorn's lemma (see the Appendix, Proposition A.1).

Observe that for any $A \subset \mathbb{N}$, either $m(A) = 1$ or $m(A^c) = 1$ but not both. Moreover, if $m(A) = 1$ and $m(B) = 1$, then $m(A \cap B) = 1$ since

$$m((A \cap B)^c) = m(A^c \cup B^c) \le m(A^c) + m(B^c) = 0 + 0 = 0.$$

I.1.3 Definition

Let $\sim$ be the equivalence relation on the set $\mathcal{S}$ of all sequences of real numbers defined by

$$\{a_n\} \sim \{b_n\} \text{ iff } m\{n: a_n = b_n\} = 1,$$

i.e. if $\{a_n\}$ equals $\{b_n\}$ almost everywhere.

Having defined the equivalence relation $\sim$, I can now do as promised and let $^*\mathbb{R} = \mathcal{S}/\sim$ be my set of *nonstandard reals* or *hyperreals*. If $\langle a_n \rangle$ denotes the equivalence class of the sequence $\{a_n\}$, define addition and multiplication in $^*\mathbb{R}$ by

$$\langle a_n \rangle + \langle b_n \rangle = \langle a_n + b_n \rangle \quad ; \quad \langle a \rangle \cdot \langle b_n \rangle = \langle a_n \cdot b_n \rangle \tag{4}$$

and order it by

$$\langle a_n \rangle < \langle b_n \rangle \text{ iff } m\{m: a_n < b_n\} = 1. \tag{5}$$

I really ought to check that these definitions are independent of the representatives $\{a_n\}, \{b_n\}$ of the equivalence classes $\langle a_n \rangle, \langle b_n \rangle$, but I shall gladly leave all book-keeping of this sort to you.

To see that the problem of zero divisors has disappeared, assume that $\langle a_n \rangle \cdot \langle b_n \rangle = \langle 0, 0 \ldots \rangle$, i.e. $m\{n: a_n \cdot b_n = 0\} = 1$. Since $\{n: a_n \cdot b_n = 0\} = \{n: a_n = 0\} \cup \{n: b_n = 0\}$, either $\{n: a_n = 0\}$ or $\{n: b_n = 0\}$ has measure one, and thus either $\langle a_n \rangle = \langle 0, 0, \ldots \rangle$ or $\langle b_n \rangle = \langle 0, 0, \ldots \rangle$. Note that the conditions on m are exactly right for this argument to work.

But *ℝ is much more than an algebraic structure without zero divisors; it is an ordered field with zero element $0 = \langle 0,0\ldots\rangle$ and unit $1 = \langle 1,1,\ldots\rangle$. As proving this in detail would just be boring, I'll restrict mystelf to the following typical example.

I.1.4 Example

If $a,b,c \in {}^*\mathbb{R}$ are such that $a > 0$ and $b < c$, how do we prove that $ab < ac$? Well, if $a = \langle a_n\rangle$, $b = \langle b_n\rangle$, and $c = \langle c_n\rangle$, then there are sets $A, B \subset \mathbb{N}$ of m-measure one such that $a_n > 0$ if $n \in A$ and $b_n < c_n$ if $n \in B$. Thus $a_nb_n < a_nc_n$ for all $n \in A\cap B$, and since $m(A\cap B) = 1$, this proves that $ab < ac$.

As already indicated

$$a \to \langle a,a,a,\ldots\rangle \tag{6}$$

is an injective, order preserving homomorphism embedding ℝ in *ℝ, and I shall identify ℝ with its image under this map. Thus all real numbers are elements of *ℝ, but what do its other members look like? In particular, where do the infinitesimal and infinite numbers come from? Let us first agree on the terminology.

I.1.5 Definition

(a) An element $x \in {}^*\mathbb{R}$ is *infinitesimal* if $-a < x < a$ for all positive real numbers a.

(b) An element $x \in {}^*\mathbb{R}$ is *finite* if $-a < x < a$ for some positive real number a. An element in *ℝ which is not finite is called *infinite*.

Three examples of infinitesimals are 0, $\delta_1 = \langle\frac{1}{n}\rangle$, and $\delta_2 = \langle\frac{1}{\sqrt{n}}\rangle$. To check that, say, δ_1 is infinitesimal, note that for any positive $a \in \mathbb{R}$, the set $\{n\colon -a < \frac{1}{n} < a\}$ contains all but a finite number of n's and hence has measure one. Observe also that since $\delta_1 \neq \delta_2$, the two sequences $\{\frac{1}{n}\}$ and $\{\frac{1}{\sqrt{n}}\}$ converging to zero at different rates are represented by different infinitesimals. Finally note that zero is the only infinitesimal real number. Examples of infinite numbers, one positive and one negative, are $\langle n\rangle$ and $\langle -n^2\rangle$.

It is easy to check that the arithmetic rules one would expect really hold; e.g. the sum of two infinitesimals is infinitesimal, and so is the product of a finite number and an infinitesimal one. More interesting is the following observation which shows that the finite part of $^*\mathbb{R}$ has a very simple structure.

I.1.6 **Proposition**

Any finite $x \in {}^*\mathbb{R}$ *can be written uniquely as a sum* $x = a + \varepsilon$, *where* $a \in \mathbb{R}$ *and* ε *is infinitesimal.*

Proof. The uniqueness is obvious since if $x = a_1 + \varepsilon_1 = a_2 + \varepsilon_2$, then $a_1 - a_2 = \varepsilon_2 - \varepsilon_1$; but this quantity is both real and infinitesimal, so it must be zero.

For the existence, let $a = \sup \{b \in \mathbb{R}: b < x\}$; since x is finite, a exists. I must show that x-a is infinitesimal. Assume not, then there is a real number r such that $0 < r < |x-a|$ (absolute values in $^*\mathbb{R}$ are defined exactly as absolute values in $\mathbb{R}$). If $x-a > 0$, this implies that $a+r < x$, contradicting the choice of a. If $x-a < 0$, I get $x < a-r$, also contradicting the choice of a.◂

Let us write $x \approx y$ to mean x and y are *infinitely close*; i.e. x-y is infinitesimal.

I.1.7 **Definition**

For each finite $x \in {}^*\mathbb{R}$, the unique real number a such that $x \approx a$ is called the *standard part* of x and is denoted by $^\circ x$ or st(x). Conversely, for each $a \in \mathbb{R}$, the set of all $x \in {}^*\mathbb{R}$ such that $a = {}^\circ x$ is called the *monad* of a.

The next lemma shows that there is a reasonable relationship between the asymptotic behaviour of $\{a_n\}$ and the value of $\langle a_n \rangle$.

I.1.8 **Lemma**

If the sequence $\{a_n\}$ *has limit* a, *then* $a \approx \langle a_n \rangle$.

***Proof*.** All we have to show is that $a-\varepsilon < \langle a_n \rangle < a+\varepsilon$ for any given $\varepsilon \in \mathbb{R}_+$. But since $\{a_n\}$ converges to a, the set $\{n: a-\varepsilon < a_n < a+\varepsilon\}$ contains all but a finite number of n's and hence has measure one.◂

Let me briefly summarise the contents of this section. We have constructed a set *ℝ of nonstandard reals or *hyperreals* which is an ordered field extension of ℝ and contains infinitely small and infinitely large numbers. A simple but useful picture to have in mind is the one shown in Figure 2; it depicts *ℝ as an ordered structure consisting

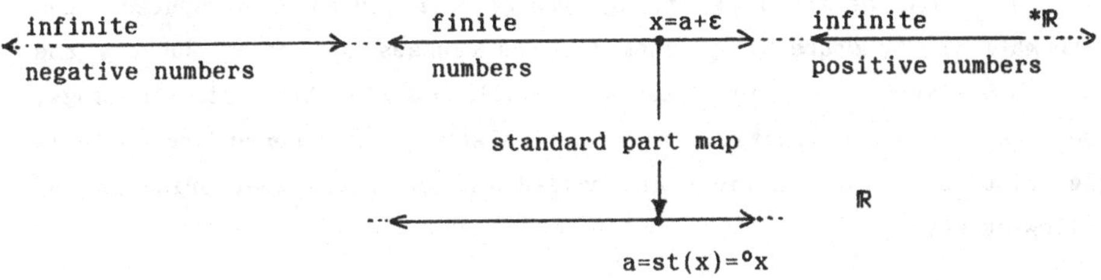

Figure 2

of three parts; the infinite negative numbers, the finite numbers, and the infinite positive numbers. According to Proposition I.1.6, the finite part looks exactly like ℝ except that each point in ℝ has been blown up to become a copy of the set of infinitesimals.

Although the constructions of ℝ and *ℝ are so very similar, there is an important difference between the two sets; the dependence on the measure m makes *ℝ "less canonical" than ℝ. Indeed, if you look back at Example I.1.1, you will see that in *ℝ one of the two sequences {0,1,0,1,0,...}, {1,0,1,0,1,...} is identified with 0 and the other one with 1; and which is which depends on the measure m. If we stick to our philosophy above and consider ℝ and *ℝ as structures constructed to reflect the asymptotic behaviour of sequences, this is not too disconcerting; the difference between the two sets is just that in creating ℝ from the rational Cauchy-sequences we throw out the sequences that do not have a decent asymptotic behaviour at the very beginning, while in creating $^*\mathbb{R}$ we keep them and treat them in an arbitrary but coherent way instead. Mathematically, this point of view is supported by the fact that hyperreals arising from different measures m have the same interesting analytic properties (although they can only be shown to be isomorphic under extra set-theoretic assumptions such as the continuum hypothesis). In Chapter III, I will show that there is occasionally a need for richer sets of hyperreals constructed not from the set of all sequences $\mathbb{R}^{\mathbb{N}}$ but from a larger set $\mathbb{R}^A$, where A is uncountable, and I will continue the present discussion then.

I.2 INTERNAL SETS AND FUNCTIONS

One of the first things you do when you have introduced a new mathematical structure is to look for the classes of "nice" subsets and functions (such as open sets and continuous functions in topology, measurable sets and functions in measure theory). In nonstandard analysis the "nice" sets and functions are called internal, and they arise in the following way.

I.2.1 Definition

(a) A sequence $\{A_n\}$ of subsets of $\mathbb{R}$ defines a subset $\langle A_n\rangle$ of $^*\mathbb{R}$ by

$$\langle x_n\rangle \in \langle A_n\rangle \quad \text{iff} \quad m\{n: x_n \in A_n\} = 1,$$

and a subset of $^*\mathbb{R}$ which can be obtained in this way is called *internal*.

(b) A sequence $\{f_n\}$ of functions $f_n: \mathbb{R} \to \mathbb{R}$ defines a function $\langle f_n\rangle: {}^*\mathbb{R} \to {}^*\mathbb{R}$ by

$$\langle f_n\rangle(\langle x_n\rangle) = \langle f_n(x_n)\rangle,$$

and any function on $^*\mathbb{R}$ which can be obtained in this way is called *internal*.

I.2.2 Example

(a) If $a = \langle a_n\rangle$ and $b = \langle b_n\rangle$ are two elements of $^*\mathbb{R}$, then the interval $[a,b] = \{x\in{}^*\mathbb{R}: a \le x \le b\}$ is internal as it is obtained as $\langle [a_n,b_n]\rangle$.

(b) If $c = \langle c_n\rangle$ is in $^*\mathbb{R}$, the function $\sin(cx)$ is an internal function defined by $\sin(cx) = \langle \sin(c_nx_n)\rangle$.

Note that two internal sets $\langle A_n \rangle$ and $\langle B_n \rangle$ are equal if and only if $A_n = B_n$ for almost all n (and similarly for functions).

What is important about the internal sets and functions is that their product-like structure makes it possible to lift operations and results componentwise from $\mathbb{R}$ to $^*\mathbb{R}$; as an example, we can define the nonstandard integral $\int_A f dx$ (where $A = \langle A_n \rangle$ is an internal set and $f = \langle f_n \rangle$ an internal function) by

$$\int_A f dx = \langle \int_{A_n} f_n \, dx \rangle.$$

This new integral inherits most of the properties of the standard integral; it's easy to check that both general statements such as $\int_A (f+g)dx = \int_A f dx + \int_A g \, dx$ and more specific ones such as

$$\int_a^b \sin(cx)dx = \frac{1}{c}\cos(ca) - \frac{1}{c}\cos(cb)$$

for $a,b,c \in {^*\mathbb{R}}$, remain true.

One of the principles which carry over from $\mathbb{R}$ to the internal sets is the least upper bound formulation of the completeness axiom:

I.2.3 **Proposition**

An internal, non-empty subset of $^*\mathbb{R}$ *which is bounded above has a least upper bound.*

Proof. If the internal set $A = \langle A_i \rangle$ is bounded above by $a = \langle a_i \rangle$, then almost all the A_i's are bounded above by the corresponding a_i's, and without loss of generality we may assume that all the A_i's are bounded above. But then $b = \langle \sup A_i \rangle$ is the least upper bound of A.◄

The least upper bound principle does not hold for all subsets of $^*\mathbb{R}$; if it did, $^*\mathbb{R}$ would satisfy all the axioms for the real numbers and hence be isomorphic to $\mathbb{R}$. That $^*\mathbb{R}$ is not complete often worries beginners in nonstandard analysis, but as you will soon see, the completeness of $\mathbb{R}$ and the least upper bound principle for internal sets always suffice. To complete $^*\mathbb{R}$ by means of, say, Dedekind cuts turns out not to be such a good idea; the completion loses too many of the attractive features of $^*\mathbb{R}$.

A subset of $^*\mathbb{R}$ which is not internal is called *external*. Proposition I.2.3 is an efficient tool for showing that sets are external; as an example, note that the set of infinitesimals does not have a least upper bound and hence must be external. The next result immediately implies that the sets of finite and infinite numbers are both external.

I.2.4 Corollary

Let A *be an internal subset of* $^*\mathbb{R}$

(a) **(Overflow, or overspill)** *If* A *contains arbitrarily large finite elements, then* A *contains an infinite element.*

(b) **(Underflow, or underspill)** *If* A *contains arbitrarily small positive infinite elements, then* A *contains a finite element.*

Proof. (a) If A is unbounded, there is nothing to prove. Thus let a be A's least upper bound; a is clearly infinite, and there must be an $x \in A$ such that $\frac{a}{2} \le x \le a$.

(b) Let be the greatest lower bound of the set A^+ of positive elements in A; then b is finite, and there must be an $x \in A$ such that $b \le x \le b+1$. ◂

Despite their simplicity, "overflow" and "underflow" are quite powerful technical tools and will be used again and again. Less simple but even more powerful is the next result.

I.2.5 Theorem ($\aleph_1$-saturation)

Let $\{A^i\}_{i\in\mathbb{N}}$ *be a sequence of internal sets such that* $\bigcap_{i\le I} A^i \ne \emptyset$ *for all* $I \in \mathbb{N}$*. Then* $\bigcap_{i\in\mathbb{N}} A^i \ne \emptyset$.

Proof. Each A^i is of the form $\langle A^i_n\rangle$, and since $A^i \ne \emptyset$, we can clearly assume that $A^1_n \ne \emptyset$ for all n. It is easy to check that $\langle \bigcap_{i\le I} A^i_n\rangle = \bigcap_{i\le I} \langle A^i_n\rangle = \bigcap_{i\le I} A^i$, and since by assumption $\bigcap_{i\le I} A^i \ne \emptyset$, we see that

$$m\{n: \bigcap_{i\le I} A^i_n \ne \emptyset\} = 1 \tag{1}$$

for all $I \in \mathbb{N}$. For each n, let

$$I_n = \max\{I \in \mathbb{N}: \bigcap_{i\le I} A_n^i \neq \emptyset \text{ and } I \le n\};$$

since $A_n^1 \neq \emptyset$, I_n exists. Choose an element $x_n \in \bigcap_{i\le I_n} A_n^i$ for each n; it suffices to show that $\langle x_n\rangle \in A^I$ for all I. But this follows from (1) since

$$\{n: x_n \in A_n^I\} \supset \{n: I_n \ge I\} = \{n: n \ge I\} \cap \{n: \bigcap_{i\le I} A_n^i \neq \emptyset\},$$

where $\{n: n\ge I\}$ has finite complement and thus measure one.◂

The reason for the mysterious name $\aleph_1$-*saturation* will become clearer in Chapter III.

It is easy to see that the family of internal sets is closed under finite Boolean operations and thus forms an algebra; indeed

$$\langle A_n\rangle \cap \langle B_n\rangle = \langle A_n \cap B_n\rangle; \quad \langle A_n\rangle \cup \langle B_n\rangle = \langle A_n \cup B_n\rangle; \quad \langle A_n\rangle^c = \langle A_n^c\rangle.$$

A quite curious (and, it turns out, extremely useful) consequence of $\aleph_1$-saturation is that this algebra is as far from being a σ-algebra as it could possibly be.

I.2.6 Corollary

If $\{A_n\}_{n\in\mathbb{N}}$ *is a sequence of internal sets, then the union* $\bigcup_{n\in\mathbb{N}} A_n$ *is internal if and only if it equals* $\bigcup_{n\le N} A_n$ *for some* $N \in \mathbb{N}$.

Proof. Assume that $A = \bigcup_{n\in\mathbb{N}} A_n$ is internal. Then all the sets $A\backslash A_n$ are internal, and clearly $\bigcap_{n\in\mathbb{N}} (A\backslash A_n) = \emptyset$. By $\aleph_1$-saturation there is an $N \in \mathbb{N}$ such that $\bigcap_{n\le N} (A\backslash A_n) = \emptyset$, and consequently $A = \bigcup_{n\le N} A_n$. ◂

$\aleph_1$-saturation makes the internal sets look a little like compacts; the next result connects them more directly with closed sets. If $A \subset {}^*\mathbb{R}$, let

$$st(A) = \{st(x): x\in A\} \tag{2}$$

be the *standard part* of A (recall Definition I.1.7).

I.2.7 Proposition

If $A \subset {}^*\mathbb{R}$ *is internal, then* $st(A)$ *is closed.*

Proof. Pick a point $x \in \overline{st(A)}$. For each $n \in \mathbb{N}$, the set

$$A_n = A \cap \{y \in {}^*\mathbb{R}: \ |x-y| < \frac{1}{n}\}$$

is internal and - since $x \in \overline{st(A)}$ - nonempty. By $\aleph_1$-saturation, $\bigcap_{n\in\mathbb{N}} A_n \neq \emptyset$. Choose a point $y \in \bigcap_{n\in\mathbb{N}} A_n$; clearly $y \in A$ and $y \approx x$, and thus $x \in st\,(A)$. ◄

So far we have been studying internal sets in general, but time has now come to take a look at two important subclasses; the standard and the hyperfinite sets.

I.2.8 Definition

For each $A \subseteq \mathbb{R}$, the internal set $^*A = \langle A,A,A,\ldots\rangle$ is called the *nonstandard version of* A. Similarly, if $f: \mathbb{R} \to \mathbb{R}$, the internal function $^*f = \langle f,f,f,\ldots\rangle$ is called the *nonstandard version of* f. An internal set or function is called *standard* if it is of the form $^*A$ or $^*f$.

Note that $^*A$ is usually a much richer set than A; e.g. the nonstandard interval $^*(a,b)$ contains not only all real numbers between a and b, but also all nonstandard numbers with the same property. In fact, we have

I.2.9 Proposition

For all $A \subseteq \mathbb{R}$, $A \subseteq {}^*A$, *with equality if and only if* A *is finite.*

Proof. The inclusion is trivial; if $a \in A$, then

$$a = \langle a,a,a,\ldots\rangle \in \langle A,A,A,\ldots\rangle = {}^*A.$$

Assume next that A is infinite. To produce an element in $^*A$ that is not in A, choose a sequence $\{a_1,a_2,a_3\ldots\}$ of distinct elements from A. Clearly $\langle a_n\rangle$ is in $^*A$, but is different from all elements in A.

Finally, if $A = \{b_1, b_2, \ldots b_k\}$ is finite and $\langle a_n \rangle \in {}^*A$, note that since

$$\{n\colon a_n \in A\} = \{n\colon a_n = b_1\} \cup \{n\colon a_n = b_2\} \cup \ldots \cup \{n\colon a_n = b_k\}$$

and the set on the left has measure one, there must be an i such that $\{n\colon a_m = b_i\}$ has measure one. Consequently, $\langle a_n \rangle = b_i \in A$.◂

The corresponding result for functions says that *f is always an extension of f; this is because for all $a \in \mathbb{R}$,

$${}^*f(a) = \langle f \rangle(\langle a \rangle) = \langle f(a) \rangle = f(a).$$

Important examples of standard sets are $^*\mathbb{Z}$, $^*\mathbb{N}$, and $^*\mathbb{Q}$ - the sets of nonstandard integers, natural numbers, and rationals. The set $^*\mathbb{N}$ consists of all elements of $^*\mathbb{R}$ of the form $\langle N_n \rangle$, where $N_n \in \mathbb{N}$ for (almost) all n. In additon to the ordinary natural numbers 1,2,3,4..., $^*\mathbb{N}$ also contains infinite integers such as $\langle 1,2,3,\ldots \rangle$. Concepts and operations carry over from $\mathbb{N}$ to $^*\mathbb{N}$ in the usual manner; for example, if $N = \langle N_n \rangle$ and $M = \langle M_n \rangle$ are elements of $^*\mathbb{N}$, then N is *divisible by* M if $N/M \in {}^*\mathbb{N}$ or, equivalently, N_n is divisible by M_n for almost all n.

The standard sets are important but not very exciting; they are just the appropriate nonstandard versions of the sets occurring in standard analysis. *Hyperfinite* sets, on the other hand, are new and interesting; they are infinite sets with all the combinatorial structure of finite sets.

I.2.10 Definition

An internal set $A = \langle A_n \rangle$ is called *hyperfinite* if (almost) all the A_n's are finite. The *internal cardinality* of A is the nonstandard integer $|A| = \langle |A_n| \rangle$, where $|A_n|$ is the number of elements in A_n.

I.2.11 Example

If $N \in {}^*\mathbb{N}$, the set

$$T = \{0, \frac{1}{N}, \frac{2}{N}, \ldots, \frac{N-1}{N}, 1\}$$

ought to be a hyperfinite set with internal cardinaltiy N+1. To see that this is actually the case, note that if $N = \langle N_n \rangle$, then $T = \langle T_n \rangle$, where

$$T_n = \{0, \frac{1}{N_n}, \frac{2}{N_n}, \dots, 1\}.$$

Hence $|T| = \langle |T_n| \rangle = \langle N_n + 1 \rangle = N+1$.

As a first illustration of how finite notions can be extended to hyperfinite sets, take an internal function $f = \langle f_n \rangle$ and a hyperfinite set $A = \langle A_n \rangle$, and define the *sum of* f *over* A by

$$\sum_{a \in A} f(a) = \langle \sum_{a_n \in A_n} f_n(a_n) \rangle$$

If T is as in the example above and $g\colon \mathbb{R} \to \mathbb{R}$ is a function, then according to this definition

$$\sum_{t \in T} {}^*g(t) \frac{1}{N} = \langle \sum_{t \in T_n} g(t_n) \frac{1}{N_n} \rangle.$$

If g is continuous and N is infinite, the sequence on the right "converges" to $\int_0^1 g(t)dt$, and thus

$$\int_0^1 g(t)dt = st(\sum_{t \in T} {}^*g(t) \frac{1}{N}); \tag{3}$$

the Riemann integral is nothing but a hyperfinite sum!

It is easy to check that the hyperfinite sets inherit most of the properties of finite sets; e.g. a hyperfinite set always has a smallest and largest element, and the union of two hyperfinite sets is again hyperfinite. The following alternative definition of hyperfinite sets is often useful: I leave the proof to you.

I.2.12 **Proposition**

An internal set A *is hyperfinite with internal cardinality* N *if and only if there is an internal bijection* $f\colon \{1,2,3,\dots,N\} \to A$.

In my opinion, hyperfinite sets are one of the most important and interesting discoveries of nonstandard analysis, and we shall meet them again in a variety of contexts, such as hyperfinite difference equations, hyperfinite probability spaces, and hyperfinite dimensional linear spaces.

I.3 INFINITESIMAL CALCULUS

Although their importance and relevance are probably not evident at this stage, the results presented in the two previous sections form much of the theoretical basis for nonstandard methods in analysis. As a first example of how to use them, I'll give you a very brief introduction to nonstandard calculus - not because this is of such great interest in itself, but because it illustrates in a simple and elegant way themes and techniques that will become important in more complex settings later. Note in particular the way in which these arguments exemplify our basic philosophy - that nonstandard numbers are convenient representations for the various asymptotic behaviours of real sequences.

Our strategy will be simple and always the same; first translate standard conditions into nonstandard terms, and then use these nonstandard reformulations to prove standard theorems. I'll first show you how this works with continuity; recall that $a \approx b$ means that a and b are infinitely close, i.e. $a-b$ is infinitesimal.

I.3.1 Proposition

The function $f\colon \mathbb{R} \to \mathbb{R}$ *is continuous at the point* $a \in \mathbb{R}$ *if and only if* $^*f(x) \approx f(a)$ *for all* $x \approx a$.

Proof. Assume that f is continuous at a and that $x = \langle x_n \rangle$ is infinitely close to a. Given $\varepsilon \in \mathbb{R}_+$ we must show that $|^*f(x)-f(a)| < \varepsilon$. Choose $\delta \in \mathbb{R}_+$ such that for all $y \in \mathbb{R}$, $|y-a| < \delta$ implies $|f(y)-f(a)| < \varepsilon$. Then

$$\{n\colon |f(x_n)-f(a)| < \varepsilon\} \supset \{n\colon |x_n-a| < \delta\},$$

and since $x \approx a$, the set on the right has measure one. Consequently, $|^*f(x)-f(a)| < \varepsilon$.

If f is not continuous at a, there exist an $\varepsilon \in \mathbb{R}_+$ and a sequence $\{x_n\}$ of reals converging to a such that $|f(x_n)-f(a)| > \varepsilon$ for all n. But then $x = \langle x_n \rangle$ is infinitely close to a and $|^*f(x)-f(a)| > \varepsilon$.◂

This characterization of continuity has certain pedagogical advantages over the usual definition. On the very elementary level, note that the continuity of the composition $f\circ g$ of two continuous functions f and g now is obvious; if $x \approx a$, then $^*g(x) \approx g(a)$ and hence $^*f(^*g(x)) \approx f(g(a))$. Another example is the Maximum Value Theorem:

I.3.2 Proposition

A continuous function on a compact interval attains a maximal value.

Proof. Choose an infinite integer N and divide the interval [a,b] into N pieces $a, a + \frac{b-a}{N}, a + 2\frac{(b-a)}{N}, \ldots$. The set $\{{}^*f(a), {}^*f(a + \frac{b-a}{N}), \ldots\}$ of values of $^*f$ is hyperfinite and hence it must have a largest element $^*f(a + i\frac{(b-a)}{N})$. Put $c = st(a + i\frac{(b-a)}{N})$; then by Proposition I.3.1, f attains its maximal value at c.◂

The description of continuity in Proposition I.3.1 seems so canonical that one may begin to wonder how one can distinguish other notions such as uniform continuity.

I.3.3 Proposition

A function $f: \mathbb{R} \to \mathbb{R}$ *is uniformly continuous on a set* A *if and only if* $^*f(x) \approx {}^*f(y)$ *for all infinitely close* $x,y \in {}^*A$.

Proof. Assume that f is uniformly continuous on A and that $x = \langle x_n \rangle$, $y = \langle y_n \rangle$ are infinitely close elements in $^*A$. Give $\varepsilon \in \mathbb{R}_+$, we must show that $|{}^*f(x) - {}^*f(y)| < \varepsilon$. Choose $\delta \in \mathbb{R}_+$ such that $|f(u)-f(v)| < \varepsilon$ whenever $u,v \in A$ and $|u-v| < \delta$. Then

$$\{n: |f(x_n)-f(y_n)| < \varepsilon\} \supset \{n: |x_n - y_n| < \delta\},$$

and since $x \approx y$, the set on the right has measure one. Consequently, $|{}^*f(x) - {}^*f(y)| < \varepsilon$.

If f is not uniformly continuous on A, there exist an $\varepsilon \in \mathbb{R}_+$ and sequences $\{x_n\}$, $\{y_n\}$ from A such that $x_n - y_n \to 0$, but $|f(x_n)-f(y_n)| > \varepsilon$ for all n. Hence $x = \langle x_n \rangle$, $y = \langle y_n \rangle$ are two infinitely close elements in $^*A$ such that $|{}^*f(x) - {}^*f(y)| > \varepsilon$.◂

I.3.4 Corollary

A continuous function on a compact interval is uniformly continuous.

Proof. Let I = [a,b] be the interval and pick two infinitely close elements $x, y \in {}^*I$. Since I is compact, c = st(x) = st(y) belongs to I, and since f is continuous,

$$^*f(x) \approx f(c) \approx {}^*f(y).◂$$

We now turn to derivatives.

I.3.5 Proposition

A function $f\colon \mathbb{R} \to \mathbb{R}$ *is differentiable at* $a \in \mathbb{R}$ *if and only if there is a number* $b \in \mathbb{R}$ *such that*

$$\frac{^*f(x)-{}^*f(a)}{x-a} \approx b$$

for all $x \approx a$, $x \neq a$. *Moreover, if such a* b *exists, it equals* f'(a).

The proof is almost identical to the proof of Proposition I.3.1. and is left to you. As a benefit you can now prove the chain rule the way you always wanted to:

I.3.6 Corollary

If g *is differentiable at* a *and* f *at* g(a), *then* $f \circ g$ *is differentiable at* a, *and* $(f \circ g)'(a) = f'(g(a))g'(a)$.

Proof. Let $x \approx a$; all we have to prove is that

$$\frac{^*f(^*g(x))-{}^*f(^*g(a))}{x-a} \approx f'(g(a))g'(a). \tag{1}$$

But if $^*g(x) = {}^*g(a)$, then both sides of (1) are zero, and if $^*g(x) \neq {}^*g(a)$, then

$$\frac{^*f(^*g(x))-{}^*f(^*g(a))}{x-a} = \frac{^*f(^*g(x))-{}^*f(^*g(a))}{^*g(x)-{}^*g(a)} \cdot \frac{^*g(x)-{}^*g(a)}{x-a} \approx f'(g(a))\cdot g'(a)$$

by Proposition I.3.5.◂

Let us finally take a look at a slightly more sophisticated result - Peano's existence theorem for ordinary differential equations.

I.3.7 **Theorem**

Let f: $\mathbb{R} \times [0,1] \to \mathbb{R}$ *be a bounded, continuous function. Then the initial value problem*

$$y'(t) = f(y(t),t);\ y(0) = y_0 \tag{2}$$

has a solution for all $y_0 \in \mathbb{R}$.

Proof. The idea is as follows. Choose an infinite integer N and let T be the hyperfinite set $\{0, \frac{1}{N}, \frac{2}{N}, \ldots, 1\}$. Define an internal function $Y: T \to {}^*\mathbb{R}$ inductively by

$$Y(k/N) = y_0 + \sum_{i=0}^{k-1} {}^*f(Y(i/N), i/N)\frac{1}{N} \tag{3}$$

and let $y: [0,1] \to \mathbb{R}$ be the function defined by $y(t) = st(Y(\tilde{t}))$, where $\tilde{t}$ is the element in T to the immediate left of t (this is necessary since t itself need not be an element of T). Then (3) implies

$$y(t) = y_0 + \int_0^t f(y(s),s)\,ds, \tag{4}$$

and hence y is a solution of (2).

In order to make this sketch rigorous, observe first that there really exists an internal function Y satisfying (3); if $N = \langle N_n \rangle$, let $Y = \langle Y_n \rangle$, where

$$Y_n(k/N_n) = y_0 + \sum_{i=0}^{k-1} f(Y_n(i/N_n), i/N_n)\frac{1}{N_n}.$$

Next, note that since f is bounded by some real number M,

$$|Y(t) - Y(s)| \le M|t-s|$$

for all $s,t, \in T$. Consequently, y is continuous and so is Y in the sense that $Y(s) \approx Y(t)$ whenever $s \approx t$.

It remains to show that (3) implies (4), and to do so it clearly suffices to show that

$$\int_0^t f(y(s),s)ds \approx \sum_{i=0}^{\tilde{t}\cdot N} {}^*f(Y(i/N), i/N)\frac{i}{N} \tag{5}$$

Since $s \to f(y(s),s)$ is continuous, the argument leading up to formula (3) at the end of the last section tells us that

$$\int_0^t f(y(s),s)ds \approx \sum_{i=0}^{\tilde{t}.N} {}^*f({}^*y(i/N),i/N)\frac{1}{N} . \qquad (6)$$

Moreover, by the continuity of f, y, and Y,

$${}^*f({}^*y(i/N),i/N) \approx {}^*f(Y(i/N),i/N), \qquad (7)$$

and combining (6) and (7), we prove (5) and hence the theorem.◂

This technique of solving differential equations by reducing them to hyperfinite difference equations is quite powerful; in the next chapter I'll indicate how it can be applied to the much more complicated theory of stochastic differential equations. One noteworthy aspect of the proof above is that Ascoli's theorem - which plays such an important part in the usual standard proof of Peano's theorem - is not mentioned. As I'll show you in Chapter III, the reason for this is that in nonstandard topology Ascoli's theorem is a triviality (in fact, its proof is hidden in the observation that the function $y(t) = st(Y(\tilde{t}))$ is bounded and continuous).

II. SUPERSTRUCTURES AND LOEB MEASURES

In the early 1970's, Peter Loeb (1975) introduced the measure theoretic construction that leads to what have since become known as *Loeb measures*. Combined with other nonstandard techniques, this construction has played an increasingly important role in applications to analysis, probability, and mathematical physics. The purpose of this chapter is to develop the basic theory for Loeb measures and to take a brief look at a few selected applications. But first I have to extend the framework of the theory slightly.

II.1 SUPERSTRUCTURES

In the first chapter I explained how to construct a nonstandard version $^*\mathbb{R}$ of the set $\mathbb{R}$ of real numbers, but it is clear that the same method can be used to construct a nonstandard version $^*S$ of any given set S; just let $\mathcal{S} = S^{\mathbb{N}}$ be the set of all S-valued sequences, and put

$$^*S = \mathcal{S}/\sim$$

where $\sim$ is the equivalence relation

$$\{a_n\} \sim \{b_n\} \quad \text{iff} \quad m\{n: a_n = b_n\} = 1$$

(m, of course, is still the finitely additive measure in Definition I.1.2). As before, I can introduce internal sets and internal functions over S. Note that the notion of internal can, in fact, be pushed much further; if, for instance, $\{F_n\}$ is a sequence of functionals over S, I can define an internal functional $F = \langle F_n\rangle$ acting on internal functions $f = \langle f_n\rangle$ by $F(f) = \langle F_n(f_n)\rangle$, and I can go on to talk about internal sets of internal functionals etc. To be able to treat all generalisations of this sort at one stroke, it is convenient to introduce the *superstructure* over S.

If T is a set, I'll write $\mathcal{P}(T)$ for the *power set* of T; i.e. the set of all subsets of T.

II.1.1 Definition

Given a set S, define a sequence $\{V_n(S)\}$ of sets inductively by

$$V_0(S) = S, \quad V_{n+1} = V_n(S) \cup \mathcal{P}(V_n(S)).$$

The *superstructure* over S is the union $V(S) = \bigcup_{n\in\mathbb{N}} V_n(S)$. If $x \in V(S)$, the *rank* of x is the smallest n such that $x \in V_n(S)$.

In this construction, we shall always consider elements of S as *urelements* and not as *sets*; thus if $x \in S$, we shall simply pretend that x does not have any elements. This is just a convenient way of avoiding certain trivial, but irritating ambiguities which arise if one element of S is allowed to be an element of another; recall for instance that in the usual set-theoretic definition of the natural numbers, $0 \in 1$, while this is certainly not the case in the Cauchy-sequence approach to the reals. If we treat S as a collection of urelements, we ensure that the superstructure V(S) only depends on the structure of the set S and not on the accidental construction of its elements.

Using the usual set theoretic definition of a function (a set of ordered pairs where each first component only occurs once), it is easy to check that all functions, functionals, functions of functionals etc. over S, live in V(S). Indeed, since an ordered pair $\langle x,y\rangle = \{\{x\}, \{x,y\}\}$ of elements from S belongs to $V_2(S)$, a function $f: S \to S$ is a subset of $V_2(S)$ and hence belongs to $V_3(S)$; by a similar argument, a functional belongs to $V_6(S)$, a function of functionals to $V_9(S)$ etc.

Our generalized internal entities will be elements of $V({}^*S)$ arising from sequences $\{A_n\}$ of elements in V(S). Such a sequence $\{A_n\}$ is called *bounded* if there is a $p \in \mathbb{N}$ such that $A_n \in V_p(S)$ for all n. If $\{A_n\}$ is bounded,

$$\mathbb{N} = \{n: A_n \text{ has rank } 0\} \cup \dots\dots \cup \{n: A_n \text{ has rank } p\},$$

and thus there is an $i \le p$ such that the set $\{n: A_n \text{ has rank } i\}$ has m-measure one. This i is called the *rank* of $\{A_n\}$. Note that if $\{A_n\}$ has rank i, then we can assume that all the A_n's have rank i by changing $\{A_n\}$ on a set of measure zero.

To each bounded sequence $\{A_n\}$ we associate an element $\langle A_n\rangle$ in V(*S) by induction on the rank. If the rank of $\{A_n\}$ is zero, then $\langle A_n\rangle$ is just the element $\langle A_n\rangle$ in *S. If $\langle B_n\rangle$ has been defined for all $\{B_n\}$ with rank less than i and $\{A_n\}$ has rank i, then

$$\langle A_n\rangle = \{\langle B_n\rangle: \{B_n\} \text{ has rank less than } i \text{ and } B_n \in A_n \text{ for almost all } n\}.$$

As in section I.2, we can now make the following definitions.

II.1.2 Definition

An element of V(*S) of the form $A = \langle A_n\rangle$ is called *internal.* If (almost) all the A_n's are finite, A is *hyperfinite.* An internal set of the form $^*B = \langle B,B,B,\ldots\rangle$ is called *standard.*

Note that if F is an internal function from one internal set $A = \langle A_n\rangle$ to another $B = \langle B_n\rangle$, then there is a sequence of functions $\{F_n\}$, with $F_n: A_n \to B_n$, such that for all $a = \langle a_n\rangle \in A$,

$$F(a) = \langle F_n(a_n)\rangle. \tag{1}$$

Conversely, any such sequence $\{F_n\}$ of functions defines an internal function F through (1). Thus the definition above extends the definition of an internal function given in section I.2.

Let me state here for later reference a very simple observation which will be of some importance in Chapter IV.

II.1.3 Lemma

A set in V(*S) *is internal if and only if it is an element of some standard set.*

Proof. By the construction of the internal sets, any element of an internal set is internal. In particular, any element of a standard set must be internal. For the converse, note that if A is an internal set of rank p, then $A \in {}^*V_p(S)$. ◂

All general properties of internal sets (such as $\aleph_1$-saturation and its corollary I.2.6) remain true in V(*S) with exactly the same proofs as before. The following extension principle will be useful in the next section.

II.1.4 **Proposition**

Let $\{A^i\}_{i\in\mathbb{N}}$ *be a sequence of internal sets all with rank less than some* $p \in \mathbb{N}$. *Then there is an internal sequence* $\{A^i\}_{i\in{}^*\mathbb{N}}$ *extending* $\{A^i\}_{i\in\mathbb{N}}$.

Proof. That $\{A^i\}_{i\in{}^*\mathbb{N}}$ is an internal sequence means that there is an internal function $A: {}^*\mathbb{N} \to V({}^*S)$ such that $A(i) = A^i$ for all $i\in{}^*\mathbb{N}$. The original sequence $\{A^i\}_{i\in\mathbb{N}}$ is not internal since an internal function always has an internal domain, and $\mathbb{N}$ is external.

To construct A, note that all the A^i are of the form $\langle A_n^i\rangle$ with $A_n^i \in V_p(S)$, and define a function $A_n: \mathbb{N} \to V_p(S)$ by $A_n(i) = A_n^i$ for each $n \in \mathbb{N}$. Letting $A = \langle A_n\rangle$, we get $A(i) = \langle A_n(i)\rangle = \langle A_n^i\rangle = A^i$ for all $i \in \mathbb{N}$, and the proposition is proved.◂

If the internal sets are the most important objects in nonstandard analysis, a natural question is how to prove that a given set is internal. This is quite easy; one only has to produce a sequence $\{A_n\}$ of sets such that $A = \langle A_n\rangle$. If A is defined in terms of other internal sets $B = \langle B_n\rangle$, $C^{(1)} = \langle C_n^{(1)}\rangle, \ldots, C^{(k)} = \langle C_n^{(k)}\rangle$ by means of a statement φ, i.e.,

$$A = \{x\in B:\ \varphi(x, B, C^{(1)}, \ldots, C^{(k)})\},$$

one just lets

$$A_n = \{x\in B_n:\ \varphi(x, B_n, C_n^{(1)}, \ldots, C_n^{(k)})\}.$$

This method is usualy referred to as the *Internal Definition Principle*, and if one gives precise meaning to what is meant by a "statement" φ, it can actually be proved to work (see Theorem IV.2.5 below). As a simple example, assume that $B = \langle B_n \rangle$ and $C = \langle C_n \rangle$ are internal sets and that $F = \langle F_n \rangle$ is an internal function. If A consists of those elements in B that belong to the image of C under F, i.e.,

$$A = \{x \in B\colon \exists y \in C(F(y) = x)\},$$

then $A = \langle A_n \rangle$, where

$$A_n = \{x \in B_n\colon \exists y \in C_n(F_n(y) = x)\},$$

as is easily checked. I shall usually leave it to the reader to prove that the sets I claim are internal really are.

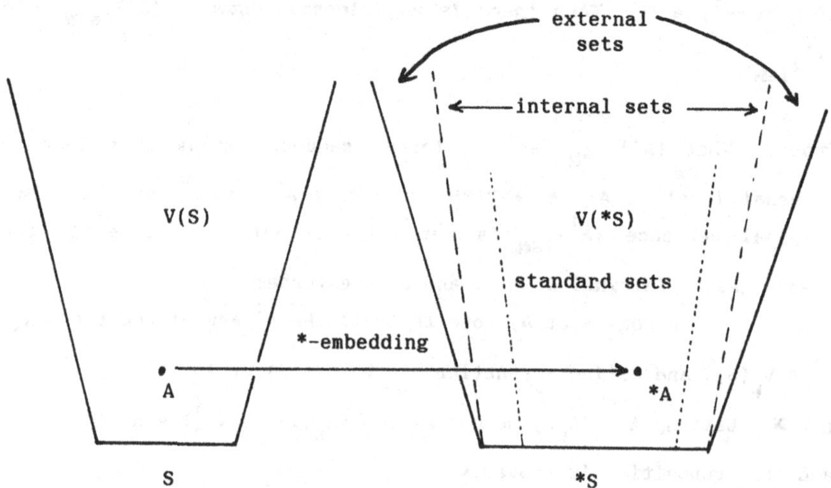

Figure 3

Figure 3 shows the two superstructures V(S) and V(*S) and the *-embedding mapping each set A in V(S) to its nonstandard version *A. If in an application one wants to study a specific mathematical structure M, it is often convenient to let S contain not only M but also all the auxiliary structures one may need such as $\mathbb{R}$, $\mathbb{C}$, certain topological spaces and measure spaces etc. - hence $S = M \cup \mathbb{C} \cup \ldots$. As a matter of fact, one usually doesn't specify S at all, just assumes that it is large enough to contain all standard entities one runs into.

Let me end this section with a brief remark on terminology.

Given some abstract property P of standard sets, there is a corresponding property *P of internal sets defined componentwise; e.g., an internal subset $A = \langle A_n \rangle$ of $^*\mathbb{R}$ is *-open if (almost) all the A_n's are open, and an internal function $f = \langle f_n \rangle$ is *-continuous if (almost) all the f_n's are continuous. There is always an intrinsic definition of these *-properties obtained by translating the definition of the original property into nonstandard terms; e.g., an internal function f is *-continuous if for all $x \in {}^*\mathbb{R}$ and all positive $\varepsilon \in {}^*\mathbb{R}$, there is a positive $\delta \in {}^*\mathbb{R}$ such that if $|x-y| < \delta$, then $|f(x) - f(y)| < \varepsilon$.

There is also another procedure for turning standard concepts into nonstandard ones where the appropriate adjective is decorated with a capital S instead of an asterisk * - thus continuous becomes S-continuous and not *-continuous. The idea here is much looser than in the asterisk case, and all I can say is that one in general keeps the standard definition, but applies it to nonstandard objects; as an example, an internal function f is S-continuous if for all $x \in \mathbb{R}$ and all $\varepsilon \in \mathbb{R}_+$, there is a $\delta \in \mathbb{R}_+$ such that if $|x-y| < \delta$ then $|f(x) - f(y)| < \varepsilon$ (but, mind you, this should hold for all $y \in {}^*\mathbb{R}$).

II.2 LOEB MEASURES

As I have tried to convince you, the internal sets are the "nice" sets of nonstandard analysis. But however nice and interesting they are individually, the class of internal sets does not have very nice closure properties; in particular, Corollary I.2.6 tells us that it is not closed under countable Boolean operations. Since analysis abounds in countable Boolean operations, you may find this rather discouraging and discomforting. The good news of this section is that if there is a measure around, there is also a complete σ-algebra extending the algebra of internal sets such that the sets in the σ-algebra can be approximated up to null-sets by internal sets. Thus we can have the best of two worlds; using the approximation property we can pass back and forth between the internal sets with their nice individual structure and the σ-algebra with its nice closure properties. This method is particularly powerful in hyperfinite probability spaces where we can within the same structure combine discrete, combinatorial ideas with a well-developed analytical machinery.

The technical basis for these developments is the Loeb construction which takes a finitely additive, internal measure and turns it into a real-valued, σ-additive measure. To make this more precise, consider an internal set Ω in some superstructure $V(^*S)$. An *internal algebra* on Ω is an internal set $\mathcal{A}$ of subsets of Ω which contains $\emptyset$ and Ω, and is closed under complements and finite unions. Since $\mathcal{A}$ is internal, it automatically satisfies a stronger property; if $\mathcal{F}$ is a hyperfinite subset of $\mathcal{A}$, then the union $\cup\{A: A\in\mathcal{F}\}$ is an element of $\mathcal{A}$. To see this, notice that $\mathcal{A}$ can be written as $\langle\mathcal{A}_n\rangle$, where each $\mathcal{A}_n$ is an algebra of sets, and $\mathcal{F}$ as $\langle\mathcal{F}_n\rangle$, where each $\mathcal{F}_n$ is a finite subset of $\mathcal{A}_n$. Since for all n, $\cup\{A_n: A_n\in\mathcal{F}_n\} \in \mathcal{A}_n$ it follows that $\cup\{A: A\in\mathcal{F}\} = \langle\cup\{A_n: A_n\in\mathcal{F}_n\}\rangle \in \mathcal{A}$.

An example of an internal algebra is the set of all internal subsets of Ω.

Let $^*\overline{\mathbb{R}}_+ = {}^*\mathbb{R}_+ \cup \{0,\infty\}$ be the set of extended, nonnegative hyperreals. An *internal, finitely additive measure* on the internal algebra $\mathcal{A}$ is an internal function μ: $\mathcal{A} \to {}^*\overline{\mathbb{R}}_+$ such that $\mu(\emptyset) = 0$ and

$$\mu(A \cup B) = \mu(A) + \mu(B)$$

for all disjoint $A,B \in \mathcal{A}$. Since μ is internal, the additivity extends automatically to hyperfinite unions;

$$\mu(\bigcup_{A\in\mathcal{F}} A) = \sum_{A\in\mathcal{F}} \mu(A)$$

for all disjoint, hyperfinite subsets $\mathcal{F}$ of $\mathcal{A}$.

It is measures of this kind I want to turn into σ-additive, real-valued measures. But before I explain how this is done, it may be helpful to take a look at an example.

II.2.1 **Example**

Let Ω be a hyperfinite set and $\mathcal{A}$ the set of all internal subsets of Ω. The internal, finitely additive measure μ: $\mathcal{A} \to {}^*\overline{\mathbb{R}}_+$ defined by

$$\mu(A) = \frac{|A|}{|\Omega|},$$

is often referred to as the *uniform probability measure* or the *normalised counting measure* on Ω. It is a typical example of the kind of measures I will have in mind in this chapter.

An internal, finitely additive measure $\mu: \mathcal{A} \to {}^*\overline{\mathbb{R}}_+$ can be turned into a finitely additive, real-valued measure ${}^\circ\mu: \mathcal{A} \to \overline{\mathbb{R}}_+$ simply by taking standard parts:

$$ {}^\circ\mu(A) = {}^\circ(\mu(A)) $$

(where ${}^\circ r = \infty$ when r is an infinitely large element of ${}^*\overline{\mathbb{R}}_+$). The Loeb measure of μ is basically the extension of ${}^\circ\mu$ to a σ-additive measure. There are several ways of constructing this extension, and I shall show you two of them. The first one is extremely simple and elegant, but requires some knowledge of measure theory - more precisely Caratheodory's Extension Theorem.

Recall that this theorem says that a finitely additive measure ${}^\circ\mu$ on an algebra $\mathcal{A}$ has a σ-additive extension $\widetilde{{}^\circ\mu}$ if whenever $\{A_n\}$ is a disjoint, countable sequence of elements from $\mathcal{A}$ such that $\bigcup_{n\in\mathbb{N}} A_n \in \mathcal{A}$, then

$$ {}^\circ\mu\left(\bigcup_{n\in\mathbb{N}} A_n\right) = \sum_{n=1}^{\infty} {}^\circ\mu(A_n) $$

In our case, this condition is trivially satisfied since Corollary I.2.6 tells us that $\bigcup_{n\in\mathbb{N}} A_n \in \mathcal{A}$ if and only if all but a finite number of A_n's are empty. Thus ${}^\circ\mu$ has a σ-additive extension $\widetilde{{}^\circ\mu}$ and the completion $L(\mu)$ of $\widetilde{{}^\circ\mu}$ is the Loeb-measure of μ.

Easy and elegant as this construction is, I still prefer to base my exposition on the more pedestrian approach I shall now present. The gist of the method should be clear from the following definitions.

II.2.2 **Definition**

(i) A subset B of Ω is *μ-approximable* if for each $\varepsilon \in \mathbb{R}_+$, there are sets $A,C \in \mathcal{A}$ such that $A \subset B \subset C$ and $\mu(C) - \mu(A) < \varepsilon$.

(ii) The *Loeb algebra* $L(\mathcal{A})$ consists of those $B \subset \Omega$ such that $B \cap F$ is μ-approximable for all $F \in \mathcal{A}$ with finite μ-measure.

(iii) The *Loeb measure* of μ is the map $L(\mu): L(\mathcal{A}) \to \overline{\mathbb{R}}_+$ defined by

$$ L(\mu)(B) = \inf\{{}^\circ\mu(C): C \in \mathcal{A} \text{ and } C \supset B\}. $$

There are two obvious lemmas to be proved - that $L(\mathcal{A})$ is a

σ-algebra, and that $L(\mu)$ is a measure on $L(\mathcal{A})$. Before I do this, allow me the following "methodological" remark: If you have ever worked seriously with measure extensions, your first reaction to the set-up in Definition II.2.2 is probably that it won't work; to obtain a measure by an outer measure construction of this kind, a more sophisticated, two step approximation procedure is usually needed. That the naive approach above works for Loeb measures is due to the excellent control we have over internal sets and, in particular, to the extension principle II.1.4 and "overflow" I.2.4.

II.2.3 **Lemma**

$L(\mathcal{A})$ is a σ-algebra extending $\mathcal{A}$.

Proof. That $\mathcal{A}$ is a subset of $L(\mathcal{A})$ follows immediately from the definition. In particular, $\emptyset$ and Ω are elements of $L(\mathcal{A})$, and to prove that $L(\mathcal{A})$ is a σ-algebra, it suffices to show that it is closed under complements and countable unions.

Assume that $B \in L(\mathcal{A})$. To prove that $B^c \in L(\mathcal{A})$, assume that $\varepsilon \in \mathbb{R}_+$ and $F \in \mathcal{A}$ with finite measure are given, and choose $A,C \in \mathcal{A}$ such that $A \subset B \cap F \subset C$ and $\mu(C) - \mu(A) < \varepsilon$. Then

$$C^c \cap F \subset B^c \cap F \subset A^c \cap F,$$

and since $\mu(A^c \cap F) - \mu(C^c \cap F) < \varepsilon$, this shows that $B^c \in L(\mathcal{A})$.

Next, let $\{B_n\}_{n\in\mathbb{N}}$ be a sequence of elements of $L(\mathcal{A})$, and assume that $\varepsilon \in \mathbb{R}_+$ and $F \in \mathcal{A}$ of finite measure are given. To prove that $B = \bigcup_{n\in\mathbb{N}} B_n$ belongs to $L(\mathcal{A})$, I shall construct inner and outer approximations $A,C, \in \mathcal{A}$ such that $A \subset B\cap F \subset C$ and $\mu(C) - \mu(A) < \varepsilon$. Since each B_n is in $L(\mathcal{A})$, there are $A_n,C_n \in \mathcal{A}$ such that $A_n \subset B_n \cap F \subset C_n$ and $\mu(C_n) - \mu(A_n) < \varepsilon\cdot 2^{-(n+1)}$. Let

$$\alpha = \lim_{n\to\infty} {}^\circ\mu\Big(\bigcup_{k=1}^{n} A_k\Big),$$

and note that since F has finite measure, α is finite. Hence there is an $N \in \mathbb{N}$ such that $\bigcup_{n=1}^{N} A_n$ has measure greater than $\alpha - \frac{\varepsilon}{2}$, and I let

$$A = \bigcup_{n=1}^{N} A_n.$$

To find the outer approximation C, use Proposition II.I.4 to extend $\{C_n\}_{n\in\mathbb{N}}$ to an internal sequence $\{C_n\}_{n\in{}^*\mathbb{N}}$. Since $\mathcal{A}$, μ, and $\{C_n\}_{n\in{}^*\mathbb{N}}$ are internal, so is the set

$$E = \{n \in {}^*\mathbb{N}: \bigcup_{k=1}^{n} C_k \in \mathcal{A} \text{ and } \mu(\bigcup_{k=1}^{n} C_k) < \alpha + \frac{\varepsilon}{2}\}.$$

If $n \in \mathbb{N}$,

$$\mu(\bigcup_{k=1}^{n} C_k) < \mu(\bigcup_{k=1}^{n} A_k) + \sum_{k=1}^{n} \varepsilon\cdot 2^{-(k+1)} < \alpha + \frac{\varepsilon}{2},$$

and hence $\mathbb{N} \subset E$. By "overflow" (Corollary I.2.4), E has an infinite element M. Let $C = \bigcup_{k=1}^{M} C_k$; then $A \subset B\cap F \subset C$, and since $\mu(C) - \mu(A) < \varepsilon$, $B \in L(\mathcal{A})$ and the lemma is proved.◂

II.2.4 Lemma

$L(\mu)$ is a complete measure on $L(\mathcal{A})$.

Proof. That $L(\mu)$ is complete just means that if $L(\mu)(B) = 0$ and $B'\subset B$, then $L(\mu)(B') = 0$. This is obviously satisfied, and thus it suffices to prove that if $\{B_n\}_{n\in\mathbb{N}}$ is a disjoint sequence of elements from $L(\mathcal{A})$, then

$$L(\mu)(\bigcup_{n\in\mathbb{N}} B_n) = \sum_{n=1}^{\infty} L(\mu)(B_n). \tag{1}$$

I can assume that each $L(\mu)(B_n)$ is finite, since otherwise both sides of (1) would be infinite. But if $L(\mu)(B_n)$ is finite, B_n is μ-approximable, and given $\varepsilon \in \mathbb{R}_+$, there are A_n, $C_n \in \mathcal{A}$ such that $A_n \subset B_n \subset C_n$ and

$$°\mu(C_n) - \varepsilon\cdot 2^{-(n+1)} < L(\mu)(B_n) < °\mu(A_n) + \varepsilon\cdot 2^{-(n+1)}. \tag{2}$$

Assume first that $\sum_{n=1}^{\infty} L(\mu)(B_n) = \infty$. To prove (1) in this case, just observe that by (2), the sum $\sum_{n=1}^{\infty} °\mu(A_n)$ is also infinite, and hence $\bigcup_{n\in\mathbb{N}} B_n$ can be approximated from the inside by internal sets $\bigcup_{n=1}^{N} A_n$ of arbitrarily large finite measure.

Assume next that $\sum_{n=1}^{\infty} L(\mu)(B_n) = \alpha < \infty$. Then by (2)

$$\sum_{n=1}^{\infty} {}^{\circ}\mu(C_n) - \frac{\varepsilon}{2} < \alpha < \sum_{n=1}^{\infty} {}^{\circ}\mu(A_n) + \frac{\varepsilon}{2}.$$

Thus if N is a sufficiently large element of $\mathbb{N}$,

$\mu(\bigcup_{n=1}^{N} A_n) = \sum_{n=1}^{N} \mu(A_n) > \alpha-\varepsilon$. To get an approximation from the outside, argue as in the proof of the previous lemma: let $\{C_n\}_{n\in{}^*\mathbb{N}}$ be an internal extension of $\{C_n\}_{n\in\mathbb{N}}$, and choose an infinite element M of the internal set

$$\{n \in {}^*\mathbb{N}\colon \bigcup_{k=1}^{n} C_k \in \mathcal{A} \text{ and } \mu(\bigcup_{k=1}^{n} C_k) < \alpha+\varepsilon\}.$$

Since $\bigcup_{n=1}^{N} A_n \subset \bigcup_{n\in\mathbb{N}} B_n \subset \bigcup_{k=1}^{M} C_k$,

$$\alpha-\varepsilon \le L(\mu)(\bigcup_{n\in\mathbb{N}} B_n) < \alpha+\varepsilon,$$

and as $\varepsilon \in \mathbb{R}_+$ is arbitrary, $L(\mu)(\bigcup_{n\in\mathbb{N}} B_n) = \alpha$.◂

The most important and useful properties of the Loeb-measure are summarised in the following theorem.

II.2.5 **Theorem.**

$L(\mu)$ is a complete measure on the σ-algebra $L(\mathcal{A})$ satisfying:

(i) $L(\mu)(A) = {}^{\circ}\mu(A)$ *for all* $A \in \mathcal{A}$.

(ii) *If* $B \in L(\mathcal{A})$ *with* $L(\mu)(B) < \infty$, *then for each* $\varepsilon \in \mathbb{R}_+$ *there are sets* $A, C, \in \mathcal{A}$ *such that* $A \subset B \subset C$ *and the numbers* $L(\mu)(C\backslash B)$ *and* $L(\mu)(B\backslash A)$ *are both less than* ε.

(iii) *If* $B \in L(\mathcal{A})$ *and* $L(\mu)(B) < \infty$, *then there is a set* $D \in \mathcal{A}$ *such that* $L(\mu)(B\Delta D) = 0$ *(where* $B\Delta D$ *is the symmetric difference* $(B\backslash D)\cup(D\backslash B)$*).*

Proof. We have already checked that $L(\mu)$ is a complete measure and $L(\mathcal{A})$ a σ-algebra. Moreover, (i) is an immediate consequence of the definition of $L(\mu)$, and (ii) is a slight reformulation of μ-approximability. To prove (iii), choose one increasing sequence $\{A_n\}_{n\in\mathbb{N}}$ and one decreasing sequence $\{C_n\}_{n\in\mathbb{N}}$ of elements of $\mathcal{A}$ such that $A_n \subset B \subset C_n$ and $\mu(C_n)-\mu(A_n) < \frac{1}{n}$ for all $n \in \mathbb{N}$. Extend $\{A_n\}_{n\in\mathbb{N}}$ and $\{C_n\}_{n\in\mathbb{N}}$ to two internal sequences $\{A_n\}_{n\in{}^*\mathbb{N}}$ and $\{C_n\}_{n\in{}^*\mathbb{N}}$. The internal set

$$\{n\in{}^*\mathbb{N}\colon \{A_k\}_{k\le n}, \{C_k\}_{k\le n} \text{ are monotone, } A_n \in \mathcal{A}, \text{ and } A_n \subset C_n\}$$

contains all finite n and hence has a infinite element N. Put $D = A_N$; then $A_n \subset D \subset C_n$ for all $n \in \mathbb{N}$, and thus $L(\mu)(B\Delta D) = 0$.◂

Part (iii) of this theorem is the approximation result that I was referring to in the introduction to this section and which makes it possible to pass back and forth between $\mathcal{A}$ and $L(\mathcal{A})$. I'll show you a similar result for $\mathcal{A}$-measurable and $L(\mathcal{A})$-measurable functions later, but first I want to take a look at the following example.

II.2.6 **Example**

Choose an infinite integer $N \in {}^*\mathbb{N}\backslash\mathbb{N}$, and let

$$\Omega = \{\tfrac{k}{N}\colon k \in {}^*\mathbb{Z} \text{ and } -N^2 \le k \le N^2\}.$$

If $\mathcal{A}$ is the family of all internal subsets of Ω, define the internal measure $\mu\colon \mathcal{A} \to {}^*\overline{\mathbb{R}}_+$ by

$$\mu(A) = \frac{|A|}{N}.$$

Let $L(\mu)$ be the Loeb-measure of μ and $L(\mathcal{A})$ the associated Loeb algebra. If $\mathcal{B}$ denotes the σ-algebra of those subsets B of $\mathbb{R}$ such that $st^{-1}(B) \cap \Omega \in L(\mathcal{A})$, define a measure λ on $\mathcal{B}$ by $\lambda(B) = L(\mu)(st^{-1}(B))$. Then λ is the Lebesgue measure.

This simple example illustrates a much used technique in nonstandard measure theory: in order to construct a standard measure, first find a discrete version of it on a hyperfinite space, then apply the Loeb construction to get a σ-additive measure, and finally push the Loeb measure down to a standard space by means of the standard part map. A more exciting application of this method is the construction of Brownian motion which I will show you in the next section.

It's quite instructive to check that Example II.2.6 keeps what it promises; i.e. that λ really is the Lebesgue measure. First note that the interval (a,b) is λ-measurable with measure b-a since

$$st^{-1}(a,b) \cap \Omega = \bigcup_{n\in\mathbb{N}} \{x\in\Omega: a + \frac{1}{n} < x < b - \frac{1}{n}\}$$

is in $L(\mathcal{A})$ and ${}^{\circ}\mu\{x\in\Omega: a + \frac{1}{n} < x < b - \frac{1}{n}\} = b - a - \frac{2}{n}$. Since λ is complete, this means that it is either the Lebesgue measure or some σ-additive extension of it. To see that the latter can't be the case, it suffices to show that if $B \in \mathcal{B}$ has finite measure $\lambda(B) = \alpha$, then B is Lebesgue measurable. By definition of λ, we have $st^{-1}(B) \in L(\mathcal{A})$ and $L(\mu)(st^{-1}(B)) = \alpha$. Given $\varepsilon \in \mathbb{R}_+$, choose internal sets $A,C \subset \Omega$ such that $A \subset st^{-1}(B) \subset C$, $L(\mu)(A) > \alpha-\varepsilon$, and $L(\mu)(C) < \alpha+\varepsilon$. By Proposition I.2.7, $A' = st(A)$ is a closed subset of B, and since $st^{-1}(A') \supset A$, we see that $\lambda(A') = L(\mu)(st^{-1}(A')) \geq L(\mu)(A) > \alpha-\varepsilon$. By a symmetric argument, $C' = \mathbb{R}\setminus st(\Omega\setminus C)$ is an open set containing B as a subset, and $\lambda(C') < \alpha+\varepsilon$. Since $A' \subset B \subset C'$, and the Lebesgue measure is complete and agrees with λ on Borel sets, B must be Lebesgue measurable with Lebesgue measure α.

There is a natural extension of the Loeb theory from measures to integrals. Let $\overline{\mathbb{R}} = \mathbb{R} \cup \{\pm\infty\}$ be the set of extended real numbers with the usual topology, and let ${}^*\overline{\mathbb{R}}$ be its nonstandard version. The standard part ${}^{\circ}x$ of an element x in ${}^*\overline{\mathbb{R}}$ is interpreted in $\overline{\mathbb{R}}$ in the obvious way. Given a finitely additive internal measure space $(\Omega,\mathcal{A},\mu)$, an internal function $F: \Omega \to {}^*\overline{\mathbb{R}}$ is *$\mathcal{A}$-measurable* if $F^{-1}(O) \in \mathcal{A}$ for all *-open sets $O \subset {}^*\overline{\mathbb{R}}$ (a set is *-open if it is of the form $\langle O_n\rangle$ with each O_n open). It is trivial to check that if F is $\mathcal{A}$-measurable, then its standard part $\omega \to {}^{\circ}F(\omega)$ is $L(\mathcal{A})$-measurable.

If F is an internal, $\mathcal{A}$-measurable function, define

$$\int F d\mu = \langle \int F_n d\mu_n \rangle \tag{3}$$

where $F = \langle F_n\rangle$ and $\mu = \langle \mu_n\rangle$. Notice that if F is a hyperfinite simple function $\sum_{k=1}^{K} a_k 1_{A_k}$, $K \in {}^*\mathbb{N}$, then $\int F d\mu = \sum_{k=1}^{K} a_k \mu(A_k)$, and that the general integral (3) can also be defined by approximating F by hyperfinite simple functions in the usual way.

The next two results concern the relationship between the two integrals $\int F d\mu$ and $\int {}^{\circ}F dL(\mu)$. It's too optimistic to hope that they will always be equal:

II.2.7 **Example**

Let Ω be as in Example II.2.6, and define F and G by

$$F(\omega) = \begin{cases} N \text{ if } \omega = 0 \\ 0 \text{ otherwise} \end{cases} ; \qquad G(\omega) = \frac{N}{2N^2+1} \text{ for all } \omega.$$

Then $\int F d\mu = \int G d\mu = 1$ and $\int {}^{\circ}F dL(\mu) = \int {}^{\circ}G dL(\mu) = 0$.

The example shows us exactly what to avoid. Define an internal, $\mathcal{A}$-measurable function to be *finite* if $\mu\{\omega\colon F(\omega) \neq 0\}$ and $\sup|F(\omega)|$ are both finite.

II.2.8 ***Lemma***.

If F is finite, ${}^{\circ}\int F d\mu = \int {}^{\circ}F dL(\mu)$.

Proof. Fix $\varepsilon \in \mathbb{R}_+$ and let $\Omega_0 = \{\omega\colon F(\omega) \neq 0\}$. For each $k \in \mathbf{Z}$, let

$$A_k = \{\omega \in \Omega_0\colon k\varepsilon \le F(\omega) < (k+1)\varepsilon\}.$$

Then $F_\varepsilon^+(\omega) = \sum_{k\in\mathbf{Z}} (k+2)\varepsilon 1_{A_k}(\omega)$ and $F_\varepsilon^-(\omega) = \sum_{k\in\mathbf{Z}} (k-1)\varepsilon 1_{A_k}(\omega)$ are upper and lower approximations of both F and ${}^{\circ}F$, and since $\sup|F(\omega)|$ is finite, F_ε^+, F_ε^- are internal and $\mathcal{A}$-measurable. Thus

$$\left|{}^{\circ}\int F d\mu - \int {}^{\circ}F dL(\mu)\right| \le {}^{\circ}\int |F_\varepsilon^+ - F_\varepsilon^-| d\mu$$

$$= {}^{\circ}\sum_{k\in\mathbf{Z}} 3\varepsilon\mu(A_k) = 3\varepsilon{}^{\circ}\mu(\Omega_0) \to 0$$

as $\varepsilon \to 0$.◂

This result can be extended to a larger class of functions.

II.2.9 **Definition**.

A function $F\colon \Omega \to {}^*\overline{\mathbb{R}}$ is S-*integrable* if it is internal and $\mathcal{A}$-measurable and satisfies:

(i) $\int |F| d\mu$ is finite

(ii) If $\mu(A) \approx 0$, then $\int_A |F| d\mu \approx 0$

(iii) If $F(\omega) \approx 0$ for all $\omega \in A$, then $\int_A |F| d\mu \approx 0$.

II.2.10 **Theorem.**

If F is S-integrable, $°\int F d\mu = \int °F dL(\mu)$.

Proof. It suffices to prove the theorem when F is nonnegative. For each $n \in {}^*\mathbb{N}$, define $F_n: \Omega \to {}^*\mathbb{R}$ by

$$F_n(\omega) = \begin{cases} F(\omega) & \text{if } \frac{1}{n} \le F(\omega) \le n \\ 0 & \text{otherwise.} \end{cases}$$

If n is finite, F_n is a finite function, and thus $°\int F_n d\mu = \int °F_n dL(\mu)$. By the monotone convergence theorem,

$$\int °F dL(\mu) = \lim_{n\to\infty} \int °F_n dL(\mu) = \lim_{n\to\infty} °\int F_n d\mu \le °\int F d\mu,$$

where the limits are in the standard sense (i.e. n runs through $\mathbb{N}$). Assume for contradiction that we do not have equality; say, $°\int F d\mu = \alpha$ and $\int °F dL(\mu) = \beta$, where $\alpha > \beta$. Since the internal set

$$\{n \in {}^*\mathbb{N}: \int F_n d\mu < \frac{\alpha+\beta}{2}\}$$

contains all finite n, it has an infinite element N. But then by S-integrability

$$\alpha = °\int F d\mu = °\int F_N d\mu + °\int_{\{F>N\}} F d\mu + °\int_{\{F<\frac{1}{N}\}} F d\mu = °\int F_N d\mu + 0 + 0 \le \frac{\alpha+\beta}{2} < \alpha$$

and the theorem follows.◂

This result is the best possible in the sense that if F is nonnegative and $°\int F d\mu = \int °F dL(\mu)$, then F is S-integrable. The proof suggests an alternative description of S-integrability which is occasionally convenient: a function F is S-*integrable* if and only if

$$\int_{\{|F|>N\}} |F| d\mu + \int_{\{|F|<\frac{1}{N}\}} |F| d\mu \approx 0$$

for all infinite N.

Given an $L(\mathcal{A})$-measurable function $f\colon \Omega \to \overline{\mathbb{R}}$, a *lifting* of f is an internal, $\mathcal{A}$-measurable function $F\colon \Omega \to {}^*\overline{\mathbb{R}}$ such that ${}^\circ F(\omega) = f(\omega)$ $L(\mu)$-a.e. Liftings are extremely important as they allow us to replace external functions by internal ones. But do they always exist?

II.2.11 Theorem.

Assume that $\mu(\Omega)$ *is finite. Then all* $L(\mathcal{A})$*-measurable functions have liftings.*

Proof. Let $\{G_n\}_{n\in\mathbb{N}}$ be a countable basis for the topology of $\overline{\mathbb{R}}$. Since $f^{-1}(G_n)$ is $L(\mathcal{A})$-measurable and has finite measure, I can find an increasing sequence $\{U_{n,m}\}_{m\in\mathbb{N}}$ of internal subsets of $f^{-1}(G_n)$ such that

$$L(\mu)(f^{-1}(G_n) \setminus \bigcup_{m\in\mathbb{N}} U_{n,m}) = 0.$$

If $\mathcal{F}_m$ is the set of all internal, $\mathcal{A}$-measurable functions $\tilde{F}$ such that for all $n \le m$, $\tilde{F}(U_{n,m}) \subset {}^*G_n$, then each $\mathcal{F}_m$ is non-empty and the sequence $\{\mathcal{F}_m\}_{m\in\mathbb{N}}$ is decreasing. By $\aleph_1$-saturation, $\bigcap_{m\in\mathbb{N}} \mathcal{F}_m \neq \emptyset$. If F is an element of the intersection, it is easy to check that ${}^\circ F(\omega) = f(\omega)$ for all ω outside the null set

$$\Omega' = \bigcup_{n\in\mathbb{N}} (f^{-1}(G_n) \setminus \bigcup_{m\in\mathbb{N}} U_{n,m}),$$

and thus F is a lifting of f.◄

If we remove the assumption that $\mu(\Omega)$ is finite, the result no longer holds; to see this, let Ω be as in Example II.2.6, and let $f(\omega)$ be 1 if ω is finite and 0 if ω is infinite. However, by putting a simple finiteness condition on f, we get the following lifting theorem for infinite measure spaces:

II.2.12 Corollary

Assume that $f\colon \Omega \to \overline{\mathbb{R}}$ *is* $L(\mathcal{A})$*-measurable and that*

$$L(\mu)\{x\colon |f(x)| \ge \frac{1}{n}\} < \infty$$

for all $n \in \mathbb{N}$. *Then* f *has an* $\mathcal{A}$*-measurable lifting.*

Proof. Let $\{\Omega_n\}_{n\in\mathbb{N}}$ be a disjoint family of elements of $\mathcal{A}$ such that

$$L(\mu)(\Omega_n \Delta \{x: \frac{1}{n} < |f(x)| \le \frac{1}{n-1}\}) = 0$$

for all n. By the theorem there is an internal, $\mathcal{A}$-measurable function $F_n: \Omega_n \to {}^*\overline{\mathbb{R}}$ such that ${}^\circ F_n(\omega) = f(\omega)$ for almost all $\omega \in \Omega_n$. If $n \ge 2$, we can clearly assume that $|F_n(\omega)| \le \frac{1}{n-1}$ for all $\omega \in \Omega_n$.

Extend $\{F_n\}_{n\in\mathbb{N}}$ to an internal sequence $\{F_n\}_{n\in{}^*\mathbb{N}}$, and consider the internal set

$$\{n\in{}^*\mathbb{N}: \text{For all } k,\ 1<k\le n,\ F_k \text{ is } \mathcal{A}\text{-measurable and } |F_k(\omega)| < \frac{1}{k-1}\}.$$

By overflow, this set contains an infinite element N. Define $F:\Omega \to {}^*\overline{\mathbb{R}}$ as follows: if there is an $n \le N$ such that $F_n(\omega)$ is defined, put $F(\omega) = F_n(\omega)$ for the smallest such n, and let $F(\omega) = 0$ otherwise. Then F is a lifting of f.◂

I'll end my discussion of general nonstandard measure theory here, but to give you a little more of the flavour of typical applications of the theory, the next section gives an account of a hyperfinite approach to Brownian motion.

II.3 BROWNIAN MOTION

Small particles suspended in a stationary liquid move about in a highly irregular and seemingly random way. This movement is caused by collisions with the molecules of the liquid and is known to physicists, chemists, and biologists as Brownian motion. When mathematicians speak of Brownian motion, they usually have in mind a specific mathematical model of this physical phenomenon due to Norbert Wiener.

Wiener's model is of a statistical nature; it consists of a probability space $(\Omega,\mathcal{F},P)$ and a stochastic process $b: \Omega \times [0,\infty) \to \mathbb{R}^3$ with certain properties which I shall soon specify. The idea is that each path $t \to b(\omega,t)$ represents a possible trajectory and that the measure P contains all necessary information about the statistical behaviour of the process; e.g. $P\{\omega: b(\omega,t) \in A\}$ is the probability that the particle is in the set A at time t.

As a physical phenomenon, Brownian motion is no longer of any great research interest, but the mathematical model has found an impressive number of applications in pure and applied mathematics as well as in related areas such as physics and electrical engineering. The purpose of this section is to show that nonstandard analysis - and, in particular, Loeb measure on a hyperfinite probability space - is a well-adapted tool for the study of stochastic models of this kind. The approach I shall follow is due to R.M. Anderson (1976).

Let me begin by a crash course in probabilistic jargon. A *probability* space $(\Omega, \mathcal{A}, P)$ is just a measure space where $P(\Omega) = 1$, and a *random variable* on Ω is a measurable function $X: \Omega \to \overline{\mathbb{R}}$. The random variables $X_1, \ldots, X_n$ are *independent* if for all Borel sets $A_1, \ldots, A_n$,

$$P\{\omega: X_1(\omega) \in A_1, \ldots, X_n(\omega) \in A_n\} = \prod_{i=1}^{n} P\{\omega: X_i(\omega) \in A_i\};$$

this means that if we observe the outcome of some of the variables $X_{i_1}, \ldots, X_{i_k}$, we get no information whatsoever about the distribution of the remaining ones. A random variable X is *Gaussian distributed with mean* a *and variance* t if for all Borel sets A,

$$P\{\omega: X(\omega) \in A\} = \int_A \frac{1}{\sqrt{(2\pi t)}} \exp\left(-\frac{(x-a)^2}{2t}\right)dx$$

The *expectation* E(X) of a random variable is just its integral with respect to P:

$$E(X) = \int X dP.$$

Finally, a *stochastic process* is a map

$$Y: \Omega \times [0,\infty) \to \overline{\mathbb{R}}$$

such that $\omega \to Y(\omega,t)$ is measurable for each t. The functions $t \to Y(\omega,t)$ are called the *paths* of the process.

II.3.1 **Definition.**

A *Brownian motion* is a stochastic process $b: \Omega \times [0,\infty) \to \overline{\mathbb{R}}$ such that $b(\omega,0) = 0$ for all ω and:

(i) If $s_1 < t_1 \le s_2 < t_2 \le \ldots \le s_n < t_n$, then the random variables $b(\cdot,t_1)-b(\cdot,s_1), \ldots, b(\cdot,t_n)-b(\cdot,s_n)$ are independent.

(ii) If $s < t$, the random variable $b(\cdot,t)-b(\cdot,s)$ is Gaussian distributed with mean zero and variance t-s.

(iii) For almost all ω, the path $t \to b(\omega,t)$ is continuous.

This is a one-dimensional Brownian motion; higher dimensional versions $\vec{b}$ are obtained by letting independent one-dimensional copies $b_1,\dots,b_n$ run along orthogonal axes; i.e. $\vec{b}(\omega,t) = (b_1(\omega,t),\dots,b_n(\omega,t))$.

The nonstandard construction of Brownian motion is quite easy and intuitive. Choose an infinite integer $N \in {}^*\mathbb{N}$, and let T be the *hyperfinite* timeline

$$T = \{0, \frac{1}{N}, \frac{2}{N}, \dots, \frac{N^2-1}{N}, N\}.$$

The collection Ω of all internal maps $\omega : T \to \{-1,1\}$ is a hyperfinite set with 2^{N^2+1} elements. Think of each $\omega \in \Omega$ as a sequence $\omega(0)$, $\omega(\frac{1}{N})$, $\omega(\frac{2}{N})$,... of coin tosses where the value 1 means heads and -1 means tails. The *hyperfinite random walk* $B\colon \Omega \times T \to {}^*\mathbb{R}$ defined by

$$B(\omega,\frac{k}{N}) = \sum_{j=0}^{k-1} \frac{\omega(j/N)}{\sqrt{N}}$$

starts at the origin and walks along the hyperreal axis with steps of length $\frac{1}{\sqrt{N}}$; the direction of the jth step is decided by the outcome of the $(j-1)^{th}$ coin toss. Figure 4 shows the path corresponding to a sequence beginning $\omega(0) = -1$, $\omega(\frac{1}{N}) = -1$, $\omega(\frac{2}{N}) = 1$, $\omega(\frac{3}{N}) = -1$, ...

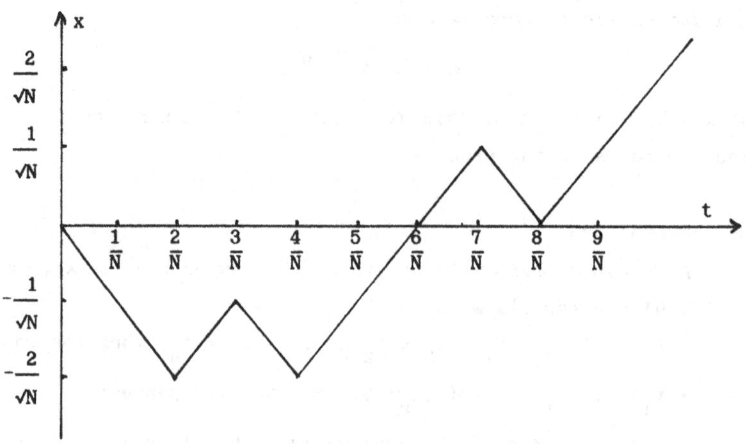

Figure 4

Define a standard map $b: \Omega \times [0,\infty) \to \overline{\mathbb{R}}$ by

$$b(\omega,t) = {}^{\circ}B(\omega,t^{+}),$$

where t^{+} is the element in T to the immediate right of t (note that t itself need not be an element of T). I claim that b is a Brownian motion.

So far the claim doesn't even make sense since I haven't got a measure on Ω. But this is easily remedied: let $\mathcal{A}$ be the algebra of all internal subsets of Ω, and let P be the normalised counting measure;

$$P(A) = \frac{|A|}{2^{N^2+1}}$$

for all $A \in \mathcal{A}$. P just says that all sequences of coin tosses are equally probable. Applying the Loeb construction, I get the measure space $(\Omega, L(\mathcal{A}), L(P))$ I shall be using.

The plan is to prove that the three parts of Definition II.3.1 are satisfied one by one. The first part is easy:

II.3.2 **Lemma**

If $s_1 < t_1 \le s_2 < t_2 \le \dots \le s_n < t_n$, *then the increments* $b(\cdot,t_1)-b(\cdot,s_1), \dots, b(\cdot,t_n)-b(\cdot,s_n)$ *are independent.*

Proof. Observe first that $B(\cdot,t_1^{+})-B(\cdot,s_1^{+}), \dots, B(\cdot t_n^{+})-B(\cdot,s_n^{+})$ are $*$-independent in the sense that if $\tilde{A}_1,\dots,\tilde{A}_n$ are internal subsets of ${}^*\mathbb{R}$, then

$$P\{\omega: B(\omega,t_1^{+})-B(\omega,s_i^{+}) \in \tilde{A}_1, \dots, B(\omega,t_n^{+})-B(\omega,s_n^{+}) \in \tilde{A}_n\} =$$

$$= \prod_{i=1}^{n} P\{\omega: B(\omega,t_i^{+})-B(\omega,s_i^{+}) \in \tilde{A}_i\}.$$

For each i, let P_i be the internal measure on ${}^*\mathbb{R}$ defined by

$$P_i(C) = P\{\omega: B(\omega,t_i^{+})-B(\omega,s_i^{+}) \in C\},$$

and note that if $A \subset \mathbb{R}$ is a Borel set, then

$$L(P_i)(st^{-1}(A)) = L(P)\{\omega: b(\omega,t_i)-b(\omega,s_i) \in A\}.$$

Choose internal sets $\tilde{A}_i$ such that

$$L(P_i)(\tilde{A}_i \,\Delta\, st^{-1}(A_i)) = 0$$

for all i. Then

$$L(P)\{\omega\colon b(\omega,t_1)-b(\omega,s_1)\in A_1,\ \dots\ ,\ b(\omega,t_n)-b(\omega,s_n)\in A_n\}$$

$$\approx P\{\omega\colon B(\omega,t_1^+)-B(\omega,s_1^+)\in \tilde{A}_1,\ \dots\ ,\ B(\omega,t_n^+)-B(\omega,s_n^+)\in \tilde{A}_n\}$$

$$= \prod_{i=1}^{n} P\{\omega\colon B(\omega,t_i^+)-B(\omega,s_i^+)\in \tilde{A}_i\}$$

$$\approx \prod_{i=1}^{n} L(P)\{\omega\colon b(\omega,t_i)-b(\omega,s_i)\in A_i\},$$

and the lemma is proved.◂

Before turning to the second condition of Definition II.3.1, it's convenient to introduce the following notation: if $s,t\in T$, $s<t$, and $X\colon T\to {}^*\mathbb{R}$ is an internal function, I shall write

$$\sum_{r=s}^{t} X(r) \quad \text{for} \quad X(s)+X(s+\tfrac{1}{N})+\ \dots\ +X(t-\tfrac{1}{N})$$

and

$$\prod_{r=s}^{t} X(r) \quad \text{for} \quad X(s)\cdot X(s+\tfrac{1}{N})\ \dots\ \cdot X(t-\tfrac{1}{N}).$$

Note that $X(t)$ is *not* included in these expressions.

II.3.3 **Lemma**

If $s<t$, *then* $b(\cdot,t)-b(\cdot,s)$ *is Gaussian distributed with mean zero and variance* $t-s$.

Proof. Since the Fourier transform of a Gaussian distribution with mean zero and variance $t-s$ is $\exp(-\tfrac{1}{2}y^2(t-s))$, it suffices to show that

$$\int \exp[iy(b(t)-b(s))]dL(P) = \exp\left(-\tfrac{1}{2}y^2(t-s)\right).$$

But

$$\int \exp[iy(b(t)-b(s))]dL(P)$$

$$\approx \int {}^*\exp[iy(B(t^+)-B(s^+))]dP \qquad \text{(by lemma II.2.8)}$$

$$= \int {}^*\exp\left[iy\sum_{r=s^+}^{t^+}\frac{\omega(r)}{\sqrt{N}}\right]dP \qquad \text{(by definition of B)}$$

$$= \prod_{r=s^+}^{t^+} \int {}^*\exp\left[iy\,\frac{\omega(r)}{\sqrt{N}}\right] dP \qquad \text{(by independence)}$$

$$= \prod_{r=s^+}^{t^+} {}^*\cos\left[\frac{y}{\sqrt{N}}\right] \qquad \text{(since } \omega(r) = \pm 1 \text{ with probability } \tfrac{1}{2}\text{)}$$

$$= \prod_{r=s^+}^{t^+} \left[1 - \frac{y^2}{2N} - O\left(\frac{y^4}{N^2}\right)\right] \qquad \text{(Taylor expansion)}$$

$$\approx {}^*\exp[-\tfrac{1}{2}y^2(t^+-s^+)] \qquad \text{(by definition of exp)}$$

$$\approx \exp[-\tfrac{1}{2}y^2(t-s)], \qquad \text{(by continuity of exp)}$$

and the lemma is proved.◂

Let $\Delta t = \frac{1}{N}$ be the time increment and let $\Delta B(\omega,t)$ be the forward increment $B(\omega,t+\Delta t)-B(\omega,t)$ of the process B. I shall write E for the expectation with respect to the internal measure P, hence $E(X) = \int X(\omega)dP(\omega)$. The following simple estimates are all that is needed to prove the continuity of b.

II.3.4 Lemma

For all $s,t \in T$, $s < t$:

(i) $E((B(t)-B(s))^2) = t-s$

(ii) $E((B(t)-B(s))^4) = 3(t-s)^2 - 2(t-s)\Delta t < 3(t-s)^2$

Proof. (i) Note that

$$E[(B(t)-B(s))^2] = E[(\sum_{r=s}^{t} \Delta B(r))^2]$$

$$= E[\sum_{r=s}^{t} \Delta B(r)^2 + \sum_{r\neq q} \Delta B(r)\Delta B(q)].$$

Since $\Delta B(r)^2 = \Delta t$ and $E(\Delta B(r)\Delta B(q)) = 0$ for $r \neq q$, this implies

$$E[(B(t)-B(s))^2] = \sum_{r=s}^{t} \Delta t = t-s.$$

(ii) It is convenient and no loss of generality to prove this part for s = 0. Notice that

$$E[B(t)^4] = E[\sum_{r=0}^{t} (B(r+\Delta r)^4 - B(r)^4)] = \sum_{r=0}^{t} E[(B(r)+\Delta B(r))^4 - B(r)^4]$$

$$= \sum_{r=0}^{t} E[4B(r)^3\Delta B(r) + 6B(r)^2\Delta B(r)^2 + 4B(r)\Delta B(r)^3 + \Delta B(r)^4]$$

$$= \sum_{r=0}^{t} E[6B(r)^2\Delta B(r)^2 + \Delta B(r)^4],$$

where the last step uses that $E(B(r)^3\Delta B(r))$ and $E(B(r)\Delta B(r)^3)$ are both zero. Since $\Delta B(r)^2 = \Delta t$ and (by part (i)) $E(B(r)^2) = r$, we get

$$E[B(t)^4] = \sum_{r=0}^{t} 6r\Delta t + \sum_{r=0}^{t} \Delta t^2.$$

Summing the arithmetic series $\sum_{r=0}^{t} r\Delta t = \frac{1}{2}t(t-\Delta t)$, this becomes

$$E[B(t)^4] = 3t(t-\Delta t) + \Delta t\cdot t = 3t^2 - 2t\Delta t < 3t^2,$$

and the proof is complete.◂

II.3.5 **Lemma**

For $L(P)$*-almost all* ω*, the path* $t \to b(\omega,t)$ *is continuous.*

Proof. For each $k \in \mathbb{N}$, let T_k be the truncated timeline $T_k = \{t \in T: t \le k\}$. Given a triple $(m,n,k) \in \mathbb{N}^3$, define a "bad" set $\Omega_{m,n,k}$ by

$$\Omega_{m,n,k} = \{\omega: \exists i\in\mathbb{N}\ \exists s\in T_k\cap[\tfrac{i}{m}, \tfrac{i+1}{m}] \text{ with } |B(\omega,\tfrac{i}{m})-B(w,s)|^4 \ge \tfrac{1}{n}\}.$$

To show that b is continuous, it suffices to prove that for each pair $(n,k) \in \mathbb{N}^2$, ${}^\circ P(\Omega_{m,n,k}) \to 0$ as $m \to \infty$. Now

$$P(\Omega_{m,n,k}) \le \sum_{i=0}^{km-1} P\{\omega: \exists s \in T\cap[\tfrac{i}{m}, \tfrac{i+1}{m}] \text{ with } |B(w,\tfrac{i}{m})-B(\omega,s)|^4 \ge \tfrac{1}{n}\}$$

$$\le 2\sum_{i=0}^{km-1} P\{\omega: |B(\omega,\tfrac{i}{m})-B(\omega, \tfrac{i+1}{m})|^4 \ge \tfrac{1}{n}\},$$

where the last inequality comes from the following reflection argument: If $|B(\omega, \frac{i}{m})-B(\omega, \frac{i+1}{m})|^4 < \frac{1}{n}$, but there is an $s \in [\frac{i}{m}, \frac{i+1}{m}]$ such that $|B(\omega, \frac{i}{m})-B(\omega,s)|^4 \ge \frac{1}{n}$, let s_ω be the smallest such s. Define $\tilde{\omega}$ to be the reflected path

$$\tilde{\omega}(r) = \begin{cases} \omega(r) & \text{if } r < s_\omega \\ -\omega(r) & \text{if } r \geq s_\omega. \end{cases}$$

Then $|B(\tilde{\omega}, \frac{i}{m})-B(\tilde{\omega}, \frac{i+1}{m})|^4 \geq \frac{1}{n}$ (see Figure 5). Since there is an internal one-to-one correspondence between reflected and unreflected paths, (1) follows.

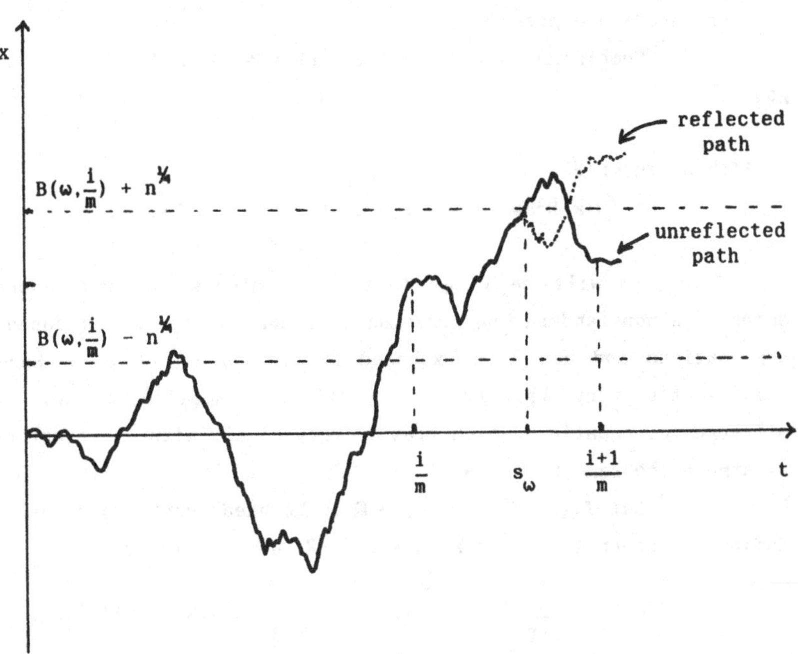

Figure 5

The rest is just an easy computation using the previous lemma:

$$P(\Omega_{m,n,k}) \leq 2\sum_{i=0}^{km-1} P\{\omega\colon |B(\omega,\tfrac{i}{m})-B(\omega, \tfrac{i+1}{m})|^4 \geq \tfrac{1}{n}\}$$

$$\leq 2\sum_{i=0}^{km-1} nE(|B(\omega, \tfrac{i}{m})-B(\omega, \tfrac{i+1}{m})|^4)$$

$$\leq 6n\sum_{i=0}^{km-1} \frac{1}{m^2} \leq \frac{6kn}{m} \to 0 \text{ as } m \to 0.$$

This proof illustrates one of the most interesting features of the nonstandard construction of Brownian motion; the way in which combinatorial principles (such as the reflection argument above) can be exploited in an infinite setting. In a certain sense, b is at the same time both a random walk and a Brownian motion, and the combination of a hyperfinite probability space and the Loeb-construction makes it possible to move freely back and forth between a discrete and a continuous point of view. A probabilist might say that b has a built-in version of the Donsker invariance principle.

Combining lemmas II.3.2, II.3.3, and II.3.5, we reach our goal:

II.3.6 **Theorem.**

b is a Brownian motion on $(\Omega, L(\mathcal{A}), L(P))$.

As will be clear from the comments and references in the Notes, the nonstandard construction of Brownian motion has found numerous applications and has been extended in various directions. Here I shall only sketch very briefly one particular application to stochastic differential equations which ties in very nicely with the proof of Peano's existence theorem in Section I.3.

Let $f,g : \mathbb{R} \times [0,\infty) \to \mathbb{R}$ be bounded, continuous functions, and define an internal process $X: \Omega \times T \to {}^*\mathbb{R}$ inductively by

$$X(\omega,t) = \sum_{s=0}^{t} {}^*f(X(\omega,s),s)\Delta t + \sum_{s=0}^{t} {}^*g(X(\omega,s),s)\Delta B(\omega,s)$$

Then the standard process $x(\omega,t) = {}^\circ X(\omega,t^+)$ is a solution of the stochastic differential equation

$$x(\omega,t) = \int_0^t f(x(\omega,s),s)ds + \int_0^t g(X(\omega,s),s)db(\omega,s).$$

This is proved by checking that

$$\int_0^t f(x(\omega,s),s)ds = {}^\circ\sum_{s=0}^{t^+} {}^*f(X(\omega,s),s)\Delta t$$

and

$$\int_0^t g(x(\omega,s),s)db(s) = {}^\circ\sum_{s=0}^{t^+} {}^*g(X(\omega,s),s)\Delta B(\omega,s).$$

The proof of the first equality is very similar to the proof of Peano's existence theorem I.3.7, while the proof of the second equality is a little more technical as it involves a stochastic integral. This method for proving existence results for stochastic differential equations is due to H.J. Keisler (1984), and can be extended to more complicated situations where f and g are no longer continuous and where the processes take values in higher dimensional spaces; see the Notes for references to this and related subjects.

III. SATURATION AND TOPOLOGY

If, as I claimed in Chapter I, the hyperreals provide a richer and more flexible setting for the study of asymptotic behaviour than what is found within the usual topological framework of Cauchy-sequences, a natural question is what happens to nonstandard theory in topological spaces with uncountable bases. In such spaces even the simple notion of convergence cannot be phrased in terms of sequences, and it is natural to suspect that the finer nuances of asymptotic behaviour which the nonstandard construction is meant to capture, will fare even worse. The goal of this chapter is to show that by generalising the nonstandard approach slightly, we can obtain a theory that works equally well in all spaces.

III.1 BEYOND $\aleph_1$-SATURATION

The basic idea is the same as in topology; if ordinary sequences won't do, we will work with generalised "sequences" with richer index sets. Thus fix an infinite set I and a finitely additive, {0,1}-valued measure m defined on all subsets of I such that

(i) $m(A) = 0$ if A is finite

(ii) $m(I) = 1$

(recall Definition I.1.2). Given a set S, let *S be the set of all sequences $\langle a_i \rangle_{i\in I}$ of elements of S modulo the equivalence relation

$$\{a_i\}_{i\in I} \sim \{b_i\}_{i\in I} \quad \text{iff} \quad a_i = b_i \text{ for m-almost all } i.$$

If V(S) and V(*S) are the superstructures over S and *S as defined in Section II.1, we can as in that section identify equivalence classes of bounded sequences $\{A_i\}_{i\in I}$ from V(S) with subsets $\langle A_i \rangle$ of V(*S). As before, subsets of V(*S) arising in this manner will be called *internal*, and we can obtain a nonstandard theory for V(*S) by just copying the definitions and arguments in Section II.1. I'll leave this to you.

One of our strongest tools in the first two chapters has been $\aleph_1$-saturation. If we choose the index set I and the measure m cleverly, this principle can be strengthened to deal with larger families of internal sets. Recall that a family $\{A_\gamma\}_{\gamma\in\Gamma}$ has the *finite intersection property* if for all finite subsets $\{\gamma_1, \gamma_2, \ldots, \gamma_n\}$ of Γ,

$$A_{\gamma_1} \cap A_{\gamma_2} \cap \ldots \cap A_{\gamma_n} \neq \emptyset.$$

III.1.1 **Definition**

The nonstandard model V(*S) is *κ-saturated* (for a cardinal κ) if whenever card $\Gamma < \kappa$ and $\{A_\gamma\}_{\gamma\in\Gamma}$ is a family of internal sets with the finite intersection property, then

$$\bigcap_{\gamma\in\Gamma} A_\gamma \neq \emptyset$$

As I'll show in the next section, κ-saturated models give the right setting for the study of topological spaces with bases of cardinality less than κ. If you wish, you can skip the remainder of this section and proceed directly to these applications.

To construct κ-saturated models we need an index set I of cardinality κ and a special kind of finitely additive measure m on I. First of all, we need m to be *countably incomplete* in the sense that there is a decreasing sequence $I = I_1 \supset I_2 \supset \ldots$ of sets of measure one such that $\bigcap_{n\in\mathbb{N}} I_n = \emptyset$. In addition, m must be *κ-good*, which is a more technical condition I shall now explain.

Given a set Γ, let $\mathcal{P}_\omega(\Gamma)$ be the collection of all finite subsets of Γ. If $\mathcal{M}$ is the set of all subsets of I of measure one, a *reversal* is a map $f: \mathcal{P}_\omega(\Gamma) \to \mathcal{M}$ such that

$$\text{if } \Gamma_1 \supset \Gamma_2, \text{ then } f(\Gamma_1) \subset f(\Gamma_2). \qquad (1)$$

A *strict reversal* is a map $g: \mathcal{P}_\omega(\Gamma) \to \mathcal{M}$ such that

$$g(\Gamma_1 \cup \Gamma_2) = g(\Gamma_1) \cap g(\Gamma_2). \qquad (2)$$

The measure m is *κ-good* if for all sets Γ of cardinality less than κ, there is for each reversal $f: \mathcal{P}_\omega(\Gamma) \to \mathcal{M}$ a strict reversal $g: \mathcal{P}_\omega(\Gamma) \to \mathcal{M}$ such that $g(\Gamma') \subseteq f(\Gamma')$ for all $\Gamma' \in \mathcal{P}_\omega(\Gamma)$.

III.1.2 Theorem

If I *is a set of infinite cardinality* κ, *there is a countably incomplete,* κ^+*-good,* $\{0,1\}$*-valued measure* m *on* I *(where* κ^+ *is the successor of* κ*).*

The proof is an interesting but quite technical piece of infinitary combinatorics, and is relegated to the Appendix.

III.1.3 Theorem

A nonstandard model $V({}^*S)$ *constructed from a countably incomplete,* κ*-good measure is* κ*-saturated.*

Proof. Let $\{A^\gamma\}_{\gamma\in\Gamma}$ be a family of internal sets with the finite intersection property, and assume that card $\Gamma < \kappa$. We must show that $\bigcup_{\gamma\in\Gamma} A^\gamma \neq \emptyset$. Since each A^γ is of the form $\langle A_i^\gamma\rangle$, it suffices to find an element $x = \langle x_i\rangle$ in $V({}^*S)$ such that $\{i\in I \mid x_i\in A_i^\gamma\}$ has measure one for all γ.

By the countable incompleteness, there is a decreasing sequence $I = I_1 \supset I_2 \supset \ldots$ of sets of measure one such that $\bigcap_{n\in\mathbb{N}} I_n = \emptyset$. Define a map $f: \mathcal{P}_\omega(\Gamma) \to \mathcal{M}$ by

$$f(\{\gamma_1,\ldots,\gamma_n\}) = I_n \cap \{i\in I:\ A_i^{\gamma_1}\cap\ldots\cap A_i^{\gamma_n} \neq \emptyset\}$$

whenever $\gamma_1,\ldots,\gamma_n$ are distinct. Since f is a reversal and m is κ-good, f is minorized by a strict reversal g.

For each $i\in I$ define a subset Γ_i of Γ by

$$\Gamma_i = \{\gamma:\ i \in g(\{\gamma\})\}.$$

Note that if Γ_i has n distinct elements $\gamma_1,\ldots,\gamma_n$, then $i \in I_n$; this is because

$$i \in g(\{\gamma_1\}) \cap \ldots \cap g(\{\gamma_n\}) = g\{\gamma_1,\ldots,\gamma_n\} \subset f\{\gamma_1,\ldots,\gamma_n\} \subset I_n,$$

where I have used that g is a strict reversal. Since $\bigcap_{n\in\mathbb{N}} I_n = \emptyset$, it follows that Γ_i must be finite.

Let $\Gamma_i = \{\gamma_1^i,\ldots,\gamma_{n_1}^i\}$. Since $i \in g(\Gamma_i) \subset f(\Gamma_i)$, the intersection $A_i^{\gamma_1^i} \cap \ldots \cap A_i^{\gamma_{n_i}^i}$ is nonempty, and we can choose

$$x_i \in A_i^{\gamma_1^i} \cap \ldots A_i^{\gamma_{n_i}^i}$$

Fix $\gamma \in \Gamma$; it only remains to check that $\{i \in I: x_i \in A_i^\gamma\}$ has measure one. But

$$\{i \in I: x_i \in A_i^\gamma\} \supset \{i: \gamma \in \Gamma_i\} = g(\{\gamma\}) \in \mathcal{M}$$

and the proof is complete.◂

By combining theorems III.1.2 and III.1.3, we get the existence of κ^+-saturated models for all infinite cardinals κ. It is often convenient to choose $\kappa \geq \text{card } V(S)$ as we can then use saturation on all families indexed by sets in V(S) - models of this kind are called *polysaturated*. Since the interesting nonstandard sets are the internal ones, and the next lemma tells us that we can never expect to apply saturation to families indexed by internal sets, little is usually gained by requiring saturation beyond polysaturation.

III.1.4 **Lemma**

An infinite, internal set in a κ-saturated model has cardinality at least κ.

***Proof*.** If A were an internal set of cardinality less than κ, then by using saturation on the family $\{A\backslash\{a\}\}_{a\in A}$, we would get

$$\bigcap_{a\in A} (A\backslash\{a\}) \neq \emptyset,$$

which is absurd.◂

This lemma has some interesting consequences for a discussion I started at the end of Section I.1 and will continue here. Assume that we need to work in a superstructure V(S) where S contains both the reals $\mathbb{R}$ and a topological space X with a base of cardinality κ. The appropriate nonstandard model V(*S) is κ-saturated, and hence $^*\mathbb{R}$ has cardinality at least κ. Thus it is impossible to fix a canonical set $^*\mathbb{R}$ of hyperreals once and for all; we need models of arbitrary large cardinality.

From a technical point of view, this lack of uniqueness is of little or no consequence, but if one is more philosophically inclined and wants to interpret $^*\mathbb{R}$ as an analytic model of the geometric line, it's a little disquieting. I shall not pursue the matter here; a possible solution is suggested by Fenstad (1985, 198?) who describes the geometric line as a basic geometric object which contains points, but which is not to be identified with any collection of points. Let me also remind you that the reals themselves are not as canonical as we like to believe; once we have fixed a model of set-theory, $\mathbb{R}$ is determined up to isomorphism, but different set-theoretical universes have quite different copies of $\mathbb{R}$ inside them.

III.2 GENERAL TOPOLOGY

Fix a topological space (X,τ). Let S be a set which contains X, $\mathbb{R}$, and all other standard entities we shall come across in this section. We shall work with a polysaturated model $V(^*S)$ of $V(S)$, and the goal is to see what the basic notions of topology look like in a nonstandard outfit.

The two most important concepts will be "*monad*" and "*nearstandard*". Given a point a in X, the *monad* of a is the subset of $^*X$ defined by

$$\mu(a) = \cap \{^*O\colon a \in O \text{ and } O \in \tau\};$$

this is a straightforward generalisation of the monads defined in Definition I.1.7. An element $x \in {}^*X$ is *nearstandard* if it belongs to $\mu(a)$ for some $a \in X$; I shall say that it is *nearstandard to* a. The set of all nearstandard points is denoted by $ns(^*X)$. Here are the nonstandard characterisations of open, closed, and compact sets:

III.2.1 Proposition

Let A be a subset of X. *Then*

(i) A *is open iff for all* $a \in A$, $\mu(a) \subset {}^*A$.

(ii) A *is closed iff whenever* $x \in {}^*A$ *is nearstandard to some* $a \in X$, *then* $a \in A$.

(iii) A *is compact iff all* $x \in {}^*A$ *is nearstandard to some* $a \in A$.

Proof. (i) If A is open and $a \in A$, then $\mu(a) \subset {}^*A$ by the defintion of monads. On the other hand, if A is not open there is an $a \in A$ such that $O \cap A^c$ is nonempty for all open neighbourhoods O of a. By polysaturation

$$\bigcap\{{}^*O \cap {}^*A^c : a \in O \text{ and } O \in \tau\} \neq \emptyset,$$

and any element of this set belongs to $\mu(a)$ but not to $^*A$.

(ii) Just apply (i) to A^c.

(iii) Let A be compact, and assume for contradiction that $x \in {}^*A$ is not nearstandard to any element in A. For each $a \in A$, there is then a neighbourhood O_a such that $x \notin {}^*O_a$. By compactness, the covering $\{O_a\}_{a\in A}$ has a finite subcovering $\{O_{a_1},\ldots,O_{a_n}\}$, and since the *-operation commutes with finite unions, $^*A \subset {}^*O_{a_1} \cup\ldots\cup {}^*O_{a_n}$. This contradicts our assumptions $x \in {}^*A$, $x \notin {}^*O_{a_i}$.

Assume next that A is not compact; then there is a family $\{F_\gamma\}_{\gamma\in\Gamma}$ of closed sets such that $\{F_\gamma \cap A\}_{\gamma\in\Gamma}$ has the finite intersection property, but

$$\bigcap_{\gamma\in\Gamma} (F_\gamma \cap A) = \emptyset. \tag{1}$$

Obviously, $\{{}^*F_\gamma \cap {}^*A\}_{\gamma\in\Gamma}$ also has the finite intersection property, and by polysaturation $\bigcap_{\gamma\in\Gamma} ({}^*F_\gamma \cap {}^*A)$ must have an element x. Assume that x is nearstandard to an element a in A. Since $x \in {}^*F_\gamma$ for all $\gamma \in \Gamma$, and F_γ is closed, it follows from (ii) that $a \in F_\gamma$. But then $a \in \bigcap_{\gamma\in\Gamma} (F_\gamma \cap A)$, contradicting (1). Hence x cannot be nearstandard to an element of A, and the proof is complete.◄

If X is Hausdorff, each element $x \in ns({}^*X)$ is nearstandard to exactly one element a in X. I shall call a the *standard part* of x and denote it by $^\circ x$ or $st(x)$. It's often useful to think of the standard part operation as a map

$$st: ns({}^*X) \to X.$$

I shall say that x and y are *infinitely close* if they are nearstandard and have the same standard part.

In the Hausdorff case the three parts of Proposition III.2.1 can be rephrased as follows:

(i)' *A is open iff* $st^{-1}(A) \subset {}^*A$

(ii)' *A is closed iff* ${}^*A \cap ns({}^*X) \subset st^{-1}(A)$

(iii)' *A is compact iff* ${}^*A \subset st^{-1}(A)$

The next result is a generalisation of Proposition I.2.7.

III.2.2 **Proposition**

If X *is Hausdorff and* A *is an internal subset of* $^*X$, *then* st(A) *is closed.*

Proof. Choose a point $a \in \overline{st(A)}$. The family

$$\{^*O \cap A\colon a \in O \text{ and } O \in \tau\}$$

has the finite intersection property (since $a \in \overline{st(A)}$), and hence by polysaturation, the set

$$\bigcap \{^*O \cap A\colon a \in O \text{ and } O \in \tau\}$$

has an element x. Clearly, $x \in A$ and $st(x) = a$, and thus $a \in st(A)$.◂

To study continuous functions, introduce another topological space Y which is also supposed to be subset of S.

III.2.3 **Proposition**

A function $f\colon X \to Y$ *is continuous* at $a \in X$ *if and only if* $^*f(\mu(a)) \subset \mu(f(a))$.

Proof. Assume that f is continuous at a and that $x \in \mu(a)$. Given a neighbourhood G of f(a), I must show that $^*f(x) \in {}^*G$. By continuity, there is a neighbourhood O of a such that $f(O) \subset G$, and it is trivial to check that this implies $^*f(^*O) \subset {}^*G$. Since $x \in {}^*O$, it follows that $^*f(x) \in {}^*G$.

Assume that f is *not* continuous at a; then there is a neighbourhood G of f(a) such that

$$A_O = \{x \in O\colon f(x) \notin G\}$$

is nonempty for all neighbourhoods O of a. The family $\{^*A_O\}$ has the finite intersection property, and thus by polysaturation the sets $^*A_O$ have a common element x. Clearly, $x \in \mu(A)$ and $^*f(x) \notin {}^*G$.◂

If you are interested in studying a specific topological space, it is often necessary to understand in detail how the standard part map operates. As an example, let me show you what happens with the space

$C(X,Y)$ of continuous functions $f: X \to Y$ given the compact-open topology:

III.2.4 Proposition

Let X *and* Y *be Hausdorff spaces and assume that* X *is locally compact. If* $f \in C(X,Y)$ *and* $F \in {}^*C(X,Y)$, *then* f *is the standard part of* F *if and only if whenever* x *is nearstandard to* a, *then* $F(x)$ *is nearstandard to* $f(a)$.

Proof. Since the compact-open topology has basic open sets of the form

$$C(K,G) = \{f \in C(X,Y): f(K) \subset G\},$$

where K is a compact neighbourhood in X and G an open neighbourhood in Y, f is the standard part of F iff $F \in {}^*C(K,G)$ whenever $f \in C(K,G)$. Assume first that f is the standard part of F, that $a \in X$, and that x is nearstandard to a. To show that $F(x)$ is nearstandard to $f(a)$, let G be a neighbourhood of $f(a)$ and choose a compact neighbourhood K of a such that $f \in C(K,G)$. Then $F \in {}^*C(K,G)$, and since $x \in {}^*K$, $F(x) \in {}^*G$. But G is an arbitrary neighbourhood of $f(a)$, and thus $F(x)$ is nearstandard to $f(a)$.

For the converse, let $C(K,G)$ be a neighbourhood of f. If $x \in {}^*K$, then by Proposition III.3.1 (iii) x is nearstandard to an element b in K. Thus $F(x)$ is nearstandard to $f(b)$ which is an element of G. By Proposition III.3.1 (i), $F(x) \in {}^*G$, and thus $F \in {}^*C(K,G)$.◂

Proposition III.2.4 is often interpreted as saying that if F maps infinitely close elements to infinitely close elements, then the function

$$f(x) = {}^\circ F(x)$$

is continuous. But this is not always true.

III.2.5 Example

Let X be the real line $\mathbb{R}$ with the usual topology. The space Y will be the plane $\mathbb{R}^2$ with a topology that is slightly finer than the usual one; in addition to the ordinary open sets, I shall assume that sets of the form $O \backslash \Gamma$, where O is an ordinary open set and $\Gamma = \{(x,0) \in \mathbb{R}^2: x \neq 0\}$, are open. Pick a non zero infinitesimal ε and define $F: {}^*X \to {}^*Y$ by $F(x) = (x,\varepsilon)$. Clearly, F is an element in ${}^*C(X, Y)$ which maps infinitely close elements to infinitely close elements but the function

$$f(x) = {}^\circ F(x) = (x,0)$$

nevertheless fails to be continuous at the origin.

As the next proposition shows, what goes wrong in this example is that Y is not a regular space; the point (0,0) and the closed set Γ can not be separated by open sets. You may skip the proof if you want to.

III.2.6 **Proposition**

Let X be a locally compact Hausdorff space and Y a regular space. If $F: {}^*X \to {}^*Y$ *is an internal function mapping infinitely close elements in* $^*X$ *to infinitely close elements in* $^*Y$*, then the function* $f: X \to Y$ *defined by* $f(x) = {}^\circ F(x)$ *is continuous.*

Proof. Assume not; then there is a point $a \in X$ and a neighbourhood G of f(a) such that $f^{-1}(G)$ is *not* a neighbourhood of a. Since Y is regular, there are disjoint open sets, O_1, O_2 such that $x \in O_1$ and $Y\backslash G \subset O_2$. For each neighbourhood O of a, let

$$A_O = \{x \in {}^*O: F(x) \notin {}^*O_1\}.$$

It is easy to see that A_O is nonempty: by definition of G, there is an $x \in O$ such that $f(x) \in Y\backslash G$, and hence $F(x) \in {}^*O_2$. Since O_1 and O_2 are disjoint, this means that $F(x) \notin {}^*O_1$.

The family $\{A_O\}$ is closed under finite intersections and thus has the finite intersection property. By saturation we can find an $x \in \bigcap A_O$. Obviously, x is infinitely close to a, but F(x) can not be infinitely close to F(a) since $F(x) \notin {}^*O_1$ and $F(a) \in {}^*O_1$. This contradicts the assumption in the theorem, and the proof is complete.◂

I'll end my discussion of C(X,Y) here; if you have found it unnecessarily detailed for an elementary introduction of this kind, my only excuse is that I wanted to give you a realistic impression of how nonstandard topology works in a concrete setting.

The results above are primarily important as technical tools in nonstandard contexts, but they can also with advantage be used to give simple proofs of classical theorems of topology. In this respect, the characterisation of compact sets given in Proposition III.2.1 (iii) is particularly useful. Here are a few examples.

III.2.7 **Tychonov's Theorem**

A product $X = \prod_{\alpha\in A} X_\alpha$ *of compact spaces is itself compact.*

Proof. An element x in $^*X$ is of the form $(x_\alpha)_{\alpha\in {}^*A}$. For each $\alpha\in A$, x_α is nearstandard to some element $y_\alpha \in X_\alpha$, and it is easy to check that x is nearstandard to $y = (y_\alpha)_{\alpha\in A}$.◂

III.2.8 **Alaoglu's Theorem**

Let $\hat{B}$ *be the dual space of a Banach space B. Then the unit ball in* $\hat{B}$ *is compact in the weak*-topology.*

Proof. If K is the unit ball in $\hat{B}$, an element u in $^*K$ is a linear functional on $^*B$ with norm less than or equal to one. Define an element $v \in K$ by

$$v(x) = {}^\circ u(x) \qquad \text{for all } x \in B;$$

clearly, v is the standard part of u in the weak*-topology.◂

Let X be a topological space and $\mathcal{F}$ a family of functions $f:X \to \mathbb{R}$. I shall say that $\mathcal{F}$ is *equicontinuous* if for each $\varepsilon \in \mathbb{R}_+$ and each $a \in X$, there is a neighbourhood O of a such that $|f(x)-f(a)| < \varepsilon$ for all $x \in O$ and all $f \in \mathcal{F}$. The family $\mathcal{F}$ is *bounded* if there is a constant K such that $|f(x)| \le K$ for all $f \in \mathcal{F}$ and $x \in X$.

III.2.9 **Ascoli's Theorem**

Let $\mathcal{F}$ *be a bounded and equicontinuous family of functions on a compact space X. Then* $\mathcal{F}$ *is pre-compact in the compact-open topology (i.e. the closure of* $\mathcal{F}$ *is compact).*

Proof. It suffices to show that any element $f \in {}^*\mathcal{F}$ is nearstandard. Define ${}^\circ f:X \to \mathbb{R}$ by ${}^\circ f(x) = \mathrm{st}\, f(x)$; since $|f(x)| \le K$, and $|f(x)-f(a)| < \varepsilon$ whenever $x \in {}^*O$ (where K and O are as in the definitions above), ${}^\circ f$ is a well-defined function in $C(X,\mathbb{R})$, and it is easily seen to be the standard part of f.◂

Despite the ease and elegance of these nonstandard proofs of classical topological results, they are not, in my opinion, what makes nonstandard topology worthwhile. Rather, I feel that the strength of nonstandard methods in topology is that they often make such abstract results unnecessary. A point in case is the proof of Peano's existence theorem in Section I.3, where Ascoli's theorem (which is an important ingredient in the standard proof) was replaced by a simple standard part operation. Something similar is at play in the construction of Brownian motion in Section II.3; a standard approach based on random walks involves a study of weak convergence of measures (see, e.g., Billingsley (1968)), which in the nonstandard approach is replaced by a standard part argument. The reason for these simplifications seems to be that a very general limit construction is built, once and for all, into the existence of sufficiently saturated, nonstandard models.

Throughout this section I have made systematic use of polysaturation. I'll end with an example which shows that I couldn't have obtained the same results by sticking to the $\aleph_1$-saturated models of the first two chapters. If you don't like ordinals, you should probably skip this example.

III.2.10 **Example**

Let X be the first uncountable ordinal (considered as the set $X = \{0,1,\dots,\omega,\omega+1,\dots\}$ of all countable ordinals). For each countable ordinal α, let $O_\alpha = \{\beta: \beta < \alpha\}$ be the set of all smaller ordinals. The family $\tau = \{O_\alpha\}$ is a topology on X, and (X,τ) is noncompact since $\{O_\alpha\}$ is a covering of X with no finite subcovering.

If *X is a nonstandard model of X of the kind studied in the first two chapters, an element α in *X is of the form $\langle\alpha_n\rangle$, where each α_n is a countable ordinal. The sequence $\{\alpha_n\}_{n\in\mathbb{N}}$ is bounded in X, and thus the set

$$A_\alpha = \{\beta \in X: \alpha \le \beta\}$$

is nonempty. Let $\gamma \in A_\alpha$, then α is nearstandard to γ. Consequently, *X is a nonstandard model of a noncompact space, but all the elements of *X are nearstandard.

III.3 COMPLETIONS, COMPACTIFICATIONS, AND NONSTANDARD HULLS

Nonstandard analysis may be thought of as a general method of producing ideal elements. Examples are legion; an infinitesimal is an idealised "extremely small number"; a hyperfinite set is an ideal version of a "very large, finite set"; the function $(2\pi\varepsilon)^{-1/2}\exp(-x^2/2\varepsilon)$ with $\varepsilon \approx 0$ is a nonstandard realization of a "C^∞ delta function", and so on. General topology is another area of mathematics which abounds in ideal elements - just think of the ways in which we adjoin ideal points to make a space complete or compact. The purpose of this section is to show by a few examples how the ideal elements of topology can be constructed using nonstandard methods, and how these constructions can be used in functional analysis.

Let me begin with completions of metric spaces. Given a standard metric space (X,d), call an element $x \in {}^*X$ *finite* if ${}^*d(x,a)$ is finite for some (hence all) $a \in X$, and let $\mathrm{fin}({}^*X)$ be the set of all finite elements. Introduce an equivalence relation $\approx$ on $\mathrm{fin}({}^*X)$ by

$$x \approx y \quad \text{if} \quad {}^*d(x,y) \approx 0,$$

and let the *nonstandard* or *infinitesimal hull* $\hat{X}$ of ${}^*X$ be the set

$$\hat{X} = \mathrm{fin}({}^*X)/\approx$$

of all equivalence classes. Note that $\hat{X}$ carries a natural metric $\hat{d}$ defined by

$$\hat{d}(\hat{x},\hat{y}) = \mathrm{st}({}^*d(x,y)),$$

where x and y are arbitrary elements of the equivalence classes $\hat{x}$ and $\hat{y}$, respectively.

III.3.1 Proposition

The nonstandard hull is a complete metric space.

Proof. The only nontrivial part is the completeness. Let $\{\hat{x}_n\}_{n\in\mathbb{N}}$ be a Cauchy-sequence, and pick an element x_n in each equivalence class $\hat{x}_n$. By Proposition II.1.4, there is an internal sequence $\{x_n\}_{n\in{}^*\mathbb{N}}$ extending $\{x_n\}_{n\in\mathbb{N}}$. The idea is to show that $\{x_n\}_{n\in\mathbb{N}}$ converges to $\hat{x}_H$ for all sufficiently small, infinite H.

Since $\{\hat{x}_n\}$ is Cauchy, there is for each $m \in \mathbb{N}$ an element $N_m \in \mathbb{N}$ such that

$$*d(x_{N_m}, x_n) < \frac{1}{m} \tag{1}$$

for all finite $n \geq N_m$. By "overflow" (I.2.4), there is an infinite H_m such that (1) holds for all n, $N_m \leq n \leq H_m$. The family $\{A_m\}$, where

$$A_m = \{k \in *\mathbb{N}: m \leq k \leq H_m\},$$

has the finite intersection property, and hence there is an infinite H less than all H_m. But then $\{x_n\}_{n \in \mathbb{N}}$ converges to $\hat{x}_H$.◂

The nonstandard hull $(\hat{X}, \hat{d})$ is a complete extension of (X,d), but it is usually too large to be the completion. Call an element $x \in *X$ *pre-nearstandard* if for all $\varepsilon \in \mathbb{R}_+$ there is an $a \in X$ such that $d(x,a) < \varepsilon$; the set of all pre-nearstandard elements is denoted by pns(*X).

III.3.2 Corollary

The space $(\text{pns}(*X)/\approx, \hat{d})$ *is the completion of* (X,d).

Note in particular that the pre-nearstandard points and the nearstandard ones (as defined in the last section) coincide if and only if X is complete.

This example is typical; the nonstandard extension *X provides us with all the ideal elements we need, and our task is just to find the equivalence relation on *X suitable for our purpose. I'll show you how to obtain the Stone-Čech compactification by the same procedure.

This time X is a completely regular topological space, and $C_b(X)$ is the set of all bounded, continuous function from X to $\mathbb{R}$. Define an equivalence relation ~ on *X by

$$x \sim y \text{ iff } *f(x) \approx *f(y) \text{ for all } f \in C_b(X),$$

and let

$$\beta X = *X/\sim$$

be the set of all equivalence classes. Each function $f \in C_b(X)$ can be extended to a function $\bar{f}: \beta X \to \mathbb{R}$ by

$$\bar{f}(\tilde{x}) = \text{st}(*f(x)),$$

where x is an arbitrary element of the equivalence class $\tilde{x}$. Let βX have the weakest topology making all the functions $\bar{f}$ continuous.

III.3.3 **Proposition**

βX is the Stone-Čech compactification of X, *and $\bar{f}$ is the canonical extension of* f.

I'll leave the proof to you with the warning that it does take some work. (Hurd and Loeb (1985) is a convenient reference if you get stuck.) There are many other constructions based on the same general philosophy as the two above - the Notes give a few references.

The nonstandard hulls constructed above were derived from the nonstandard versions of standard metric spaces, but the same method applies to a much larger family of internal spaces. A *nonstandard metric space* (X,d) is simply an internal set X and an internal function $d: X \to {}^*\mathbb{R}$ satisfying

(i) $d(x,y) \geq 0$ with equality if and only if $x = y$,

(ii) $d(x,y) = d(y,x)$ for all $x,y \in X$,

(iii) $d(x,y) \leq d(x,z) + d(z,y)$ for all $x,y,z \in X$.

The nonstandard version $({}^*X, {}^*d)$ of a standard metric space is clearly a nonstandard metric space; here are few other examples.

III.3.4 **Example**

If $n \in {}^*\mathbb{N}$ and $p \in {}^*\mathbb{R}$, $p \geq 1$, let $\ell_p(n)$ be the space of all internal sequences $x_1, x_2, \ldots, x_n$ of hyperreals given the metric

$$d(x,y) = \|x-y\|_p = \Big(\sum_{i=1}^{n} |x_i - y_i|^p\Big)^{1/p}.$$

These spaces are natural nonstandard generalizations of finite ℓ_p spaces.

III.3.5 **Example**

There is a similar way of generalising Sobolev spaces. Given an internal domain Ω, define a norm $\|\cdot\|_{m,p,\Omega}$ for each $m \in {}^*\mathbb{N}$, $p \geq 1$, by

$$\|u\|_{m,p,\Omega} = \Big(\sum_{0 \leq |\alpha| \leq m} \int_\Omega |D^\alpha u(x)|^p dx\Big)^{1/p},$$

and let $H_{m,p,\Omega}$ be the set of all internal functions u with *-finite norm. Notice that these spaces have some strange elements; e.g.

$$f(x) = \delta \sin \frac{x}{\delta}, \qquad \delta \approx 0$$

is a noninfinitesimal element of $H_{1,1,[0,1]}$.

To define the nonstandard hull, assume that a nonstandard metric space is given, and fix a *point of reference* $a_0 \in X$ (if, as in the examples above, X is a normed space, I shall always let $a_0 = 0$). A point $x \in X$ is *finite* if $d(x,a_0)$ is finite, and the set of all finite points is denoted by fin(X). As before, $\approx$ is the equivalence relation

$$x \approx y \text{ iff } d(x,y) \approx 0,$$

and the *nonstandard hull* is the set

$$\hat{X} = \text{fin}(X)/\approx$$

of all equivalence classes given the metric

$$\hat{d}(\hat{x},\hat{y}) = \text{st}(d(x,y)),$$

where x,y are arbitrary elements of $\hat{x}$, $\hat{y}$. With exactly the same proof as above we have

III.3.6 **Proposition.**

$(\hat{X},\hat{d})$ *is a complete metric space.*

There are several reasons for considering hulls of nonstandard metric spaces. The most obviously appealing is, perhaps, that it is an excellent device for constructing new and interesting classes of spaces. To see this, return for a moment to Example III.3.4: if p is finite, or p is infinite and n finite, nothing very exciting happens; in the first case the nonstandard hull is an abstract L_r-space with $r = \text{st}(p)$, and in the second it is isomorphic to $\ell_\infty(n)$. But if n and p are both infinite, a whole new family of space $\hat{\ell}_p(n)$ appears (with certain repetitions; Henson and Moore (1983) have shown that $\hat{\ell}_p(n)$ and $\hat{\ell}_q(m)$ are isomorphic if and only if $n^{1/p}$ and $m^{1/q}$ have the same standard part (possibly ∞)).

Another reason for studying nonstandard hulls is that complicated conditions on standard spaces X frequently translate into simpler conditions on their nonstandard hulls $\hat{X}$; as an example I mention that X is superreflexive if and only if $\hat{X}$ is reflexive.

But perhaps the most common reason for taking nonstandard hulls is simply a desire to come back to a standard setting - either because of a wish to apply a standard theorem or simply because of a feeling that the results are best presented in standard terms. In many such situations one needs to impose a certain regularity; some of the elements in the nonstandard hull may just be too wild for the applications one has in mind (such as, for example, the function f in Example III.3.5). A useful technical tool in formulating and proving various forms of regularity is the following generalized notion of pre-nearstandard.

Fix a subset $S \subset fin(X)$. An element $x \in X$ is *pre-nearstandard* to S if for all $\varepsilon \in \mathbb{R}_+$, there is an $s \in S$ such that $d(s,x) < \varepsilon$. The collection of all elements which are pre-nearstandard to S is denoted by $pns(X,S)$. As an immediate consequence of Proposition III.3.6, we have:

III.3.7 **Corollary**

$(pns(X,S)/\approx, \hat{d})$ *is a complete metric space.*

III.3.8 **Example.**

In the sequence space $\ell_p(n)$ of Example III.3.4 it is often natural to let S consist of all sequences $x_1, x_2, \ldots, x_m, 0, 0 \ldots, 0$ where both m and all the x_i's are finite. If n is infinite and $r = st(p)$ is finite, $pns(\ell_p(n),S)/\approx$ is then isomorphic to ℓ_r.

The final reason I will give for studying nonstandard hulls is embeddings in hyperfinite dimensional spaces. An instance of this phenomenon can be seen in the example above; the space $\ell_p(n)$ is (in an obvious sense) an n-dimensional linear space over $^*\mathbb{R}$ and yet it contains a copy of the infinite dimensional space ℓ_r. Since hyperfinite dimensional spaces have all the algebraic properties of finite dimensional spaces, embeddings of this kind are extremely useful tools in extending finite dimensional results to infinite dimensional spaces. I'll show you how by sketching a proof of the spectral theorem for bounded, symmetric operators on Hilbert spaces.

Let me first fix the terminology. An (internal) *linear space over* $^*\mathbb{R}$ is an internal set X with internal addition $(x,y) \to x+y$ and scalar multiplication $(\alpha,x) \to \alpha x$ operations satisfying the usual axioms. It is *N-dimensional* if there is an internal set $\{e_1,e_2,\ldots,e_N\}$ with N elements such that each element $x \in X$ can be written as an internal sum $x = \sum_{i=1}^{N} x_i e_i$ in exactly one way, and it is *hyperfinite dimensional* if it is N-dimensional for some $N \in {}^*\mathbb{N}$. The basic embedding result is an immediate consequence of polysaturation.

III.3.9 **Proposition**

Let E be an infinite dimensional linear space over $\mathbb{R}$. Then there is a hyperfinite dimensional linear space X over $^*\mathbb{R}$ such that $E \subset X \subset {}^*E$.

Proof. For each finite subset S of E, let

$$A_S = \{X\colon X \text{ is a hyperfinite dimensional subspace of } {}^*E, \text{ and } S \subset X\}.$$

The family $\{A_S\}$ has the finite intersection property, and by polysaturation has a common element X which satisfies the proposition.◂

Linear algebra in a hyperfinite dimensional space X over $^*\mathbb{R}$ is exactly like linear algebra in a finite dimensional space over $\mathbb{R}$. As an example, I'll show that the spectral theory is the same. Let $\langle\cdot,\cdot\rangle$ be an internal inner product on X, and let $T\colon X \to X$ be an internal symmetric operator. Since X is a hyperfinite dimensional linear space, it's of the form $\langle X_i\rangle$ where each X_i is a finite dimensional space. Similarly, we have $T = \langle T_i\rangle$ where T_i is a symmetric operator from X_i to X_i. By the finite dimensional spectral theorem, each T_i has an orthonormal basis $B_i = \{e^i_1,\ldots,e^i_{n_i}\}$ of eigenvectors with a corresponding set of eigenvalues $\Lambda_i = \{\lambda^i_1,\ldots,\lambda^i_{n_i}\}$. The set $B = \langle B_i\rangle$ is an orthonormal basis of eigenvectors for T, and $\Lambda = \langle\Lambda_i\rangle$ is the corresponding set of eigenvalues.

If $B = \{e_1,\dots,e_N\}$ and $\Lambda = \{\lambda_1,\dots,\lambda_N\}$, each element $x \in X$ can be written uniquely as an internal sum

$$x = \sum_{i=1}^{N} x_i e_i \tag{1}$$

and

$$Tx = \sum_{i=1}^{N} \lambda_i x_i e_i, \tag{2}$$

exactly as in the finite dimensional case.

Turning to spectral theory in Hilbert spaces, assume that $A:H \to H$ is a bounded symmetric operator on a Hilbert space, and choose a hyperfinite dimensional space X such that $H \subset X \subset {}^*H$. If $P_X: {}^*H \to X$ is the orthogonal projection, define an internal, symmetric operator

$$T: X \to X$$

by $T = P_X {}^*A$. The idea is to use (1) and (2) in combination with a nonstandard hull argument to obtain the spectral decomposition of A.

Let $\hat{X}$ be the nonstandard hull of X (with respect to the norm inherited from *H), and define an operator $\hat{A}:\hat{X} \to \hat{X}$ by

$$\hat{A}(\hat{x}) = \widehat{T(x)}$$

where $\hat{}$ denotes the equivalence class. If we identify H with a subspace of $\hat{X}$ through the embedding $h \to \widehat{{}^*h}$, $\hat{A}$ becomes an extension of A. Note that if $H^\perp$ is the orthogonal complement of H, then for all $u \in H$, $v \in H^\perp$

$$\langle \hat{A}v,u\rangle = \langle v,\hat{A}u\rangle = 0$$

by the symmetry of $\hat{A}$ and the fact that $\hat{A}$ maps H into itself. Hence for all $\hat{x} \in \hat{X}$

$$\hat{A}P_H(\hat{x}) = P_H\hat{A}(\hat{x}), \tag{3}$$

where $P_H: \hat{X} \to H$ is the projection. This simple equality is quite useful; note, for instance, that if $x \in X$ is a finite eigenvector of T with eigenvalue λ, then

$$\hat{A}(P_H(x)) = P_H(\hat{A}(\hat{x})) = {}^\circ\lambda P_H(\hat{x}) \tag{4}$$

(but be careful; since $P_H(x)$ may be zero, this doesn't mean that ${}^\circ\lambda$ is necessarily an eigenvalue of $\hat{A}$).

We are now ready to study the spectral decomposition of A. Let e_i and λ_i be the eigenvectors and eigenvalues of T as above. For each Borel set $\Omega \subset \mathbb{R}$, let

$$\hat{X}_\Omega = \text{Span}\,\{\hat{e}_i \in \hat{X}: {}^\circ\lambda_i \in \Omega\}.$$

Clearly, $\hat{X}_\Omega$ is a closed subspace of $\hat{X}$. Let

$$\overline{P}_\Omega: H \to H$$

be the operator defined by $\overline{P}_\Omega(\hat{x}) = P_H P_\Omega(x)$ and note that by (4), $(P_H P_\Omega)^2 = P_H P_\Omega P_H P_\Omega = P_H P_\Omega$, which means that $\overline{P}_\Omega$ is a projection. The idea is that $\Omega \to \overline{P}_\Omega$ is the projection valued measure associated with A.

In order to prove this, I have to show that for each $\hat{x} \in H$, there is a measure $\mu_{\hat{x}}$ on $\mathbb{R}$ such that

$$\mu_{\hat{x}}(\Omega) = \langle \overline{P}_\Omega \hat{x}, \hat{x}\rangle \tag{5}$$

and

$$\langle A\hat{x}, \hat{x}\rangle = \int \lambda\, d\mu_{\hat{x}}(\lambda), \tag{6}$$

and I shall do this by using the Loeb measure techniques of Section II.2.

For each $x = \sum x_i e_i \in X$ define an internal measure ν_x on $^*\mathbb{R}$ by

$$\nu_x(\tilde{\Omega}) = \sum_{\lambda_i \in \tilde{\Omega}} x_i^2.$$

If $L(\nu_x)$ is the Loeb measure of ν_x, then by the arguments following Example II.2.6, $\mu_x = L(\nu_x)\circ st^{-1}$ is a completed Borel measure on $\mathbb{R}$. Given a Borel set $\Omega \subset \mathbb{R}$, use Theorem II.2.5 (iii) to pick an internal subset $\tilde{\Omega}$ of $^*\mathbb{R}$ such that $L(\nu_x)(\tilde{\Omega}\Delta st^{-1}(\Omega)) = 0$. If $P_{\tilde{\Omega}}: X \to X_{\tilde{\Omega}}$ is the internal projection onto the space generated by all eigenvectors of T with eigenvalues belonging to $\tilde{\Omega}$, then for all $\hat{x} \in H$

$$\langle \overline{P}_\Omega \hat{x}, \hat{x}\rangle = {}^\circ\langle P_{\tilde{\Omega}} x, x\rangle = {}^\circ\sum_{\lambda_i \in \tilde{\Omega}} x_i^2 = {}^\circ\nu_x(\tilde{\Omega}) = \mu_x(\Omega)$$

which is (5). To get (6), note that by lemma II.2.8:

$$\langle A\hat{x}, \hat{x}\rangle = {}^\circ\langle Tx, x\rangle = {}^\circ\sum \lambda_i x_i^2 = \int_{^*\mathbb{R}} \lambda\, d\nu_x(\lambda) =$$

$$= \int_{^*\mathbb{R}} {}^\circ\lambda\, dL(\nu_x)(\lambda) = \int_{\mathbb{R}} \lambda\, d\mu_x(\lambda).$$

Defining spectral integrals by

$$\int f(\lambda)d\langle P_\lambda\hat{x},\hat{x}\rangle = \int f(\lambda)d\mu_x(\lambda),$$

we get the well-known formula

$$\langle A\hat{x},\hat{x}\rangle = \int \lambda d\langle P_\lambda\hat{x},\hat{x}\rangle$$

(compare, e.g., Reed & Simon (1972)).◂

This proof is rather typical of hyperfinite dimensional techniques in functional analysis; the starting point is a hyperfinite analog of the desired result - which is often easy, or even trivial, to establish - but some honest, technical work is required to push this hyperfinite result down to the embedded, infinite dimensional space.

IV. THE TRANSFER PRINCIPLE

Introducing the internal sets and functions back in Section I.2, I commented that what made them so important was that their product-like structure $A = \langle A_i \rangle$, $f = \langle f_i \rangle$ made it possible to lift standard definitions and results componentwise; recall, for example, how I could define the internal integral $\int_A f dx$ as $\langle \int_{A_i} f_i dx \rangle$. Since then I've used this trick time and time again in a variety of contexts, and it has become natural to ask whether there is a general principle at play here; is it possibly to classify, once and for all, what statements can be lifted in this way and with what consequences? There is both a theoretical and practical side to this question; an affirmative answer would not only provide us with a better understanding of nonstandard models in general, but it would also relieve us of the burden of having to carry out essentially the same argument in each individual case.

IV.1 THE LANGUAGES L(V(S)) AND L*(V(S))

To answer the question above, I first of all need to give precise meaning to the word *statement*. I shall interpret it as a *grammatically correct formula* in a certain language L(V(S)), which I shall use this section to describe. The general principle - aptly named the *Transfer Principle* - which I shall prove in the next section, says that a formula φ in L(V(S)) is a true statement about V(S) if and only if a corresponding statement $^*\varphi$ is true about V(*S).

Throughout this section I will keep fixed a superstructure V(S) and its nonstandard companion V(*S). A function $F\colon V(S)^k \to V(S)$ is *tame* if for each $n \in \mathbb{N}$, there is an $m \in \mathbb{N}$ such that $F(a_1,\dots,a_k) \in V_m(S)$ whenever $a_1,\dots,a_k \in V_n(S)$ (recall the notation and terminology of Section II.1). Note that if F is tame and $a^{(1)} = \langle a_i^{(1)} \rangle, \dots, a^{(k)} = \langle a_i^{(k)} \rangle$ are internal sets, then

$$*F(a^{(1)},\ldots,a^{(k)}) = \langle F(a_i^{(1)},\ldots,a_i^{(k)})\rangle$$

defines an internal set. Extend *F to a function $*F:V(*S)^k \to V(*S)$ by assigning arbitrary values to $*F(b_1,\ldots,b_k)$ when some of the b_i's are external. This is not as silly as it may look; the point is that I shall never want to apply *F to external sets, but it is technically inconvenient to work with partially defined functions in what follows.

To define the language L(V(S)), I must first specify its *alphabet* $\mathcal{A}$(V(S)). It consists of the following symbols:

variables: $v_1, v_2, v_3, \ldots$,

constant symbols: one symbol $\underline{a}$ for each element $a \in V(S)$,

relation symbols: $=$, $\in$,

function symbols: one symbol $\underline{F}$ for each tame function $F: V(S)^k \to V(S)$,

connectives: $\neg$ (not), $\wedge$ (and), $\vee$ (or), $\Rightarrow$ (implies),
$\Leftrightarrow$ (if and only if),

quantifiers: $\exists$ (there exists), $\forall$ (for each),

parentheses: (,).

A *string* over $\mathcal{A}$(V(S)) is just a finite sequence $s_1s_2s_3\ldots s_n$ of symbols. Arbitrary strings make no sense, but it is possible to single out subclasses which can be interpreted in a natural way.

IV.1.1 **Definition**

The class $\mathcal{T}$ of *terms* is the smallest class Γ of strings such that:

(i) If a string t consists of a single variable or of a single constant symbol, then $t \in \Gamma$.

(ii) If F is a tame function of k variables and $t_1,\dots,t_k \in \Gamma$, then $\underline{F}(t_1,\dots,t_k) \in \Gamma$.

The class $\mathcal{F}$ of *formulas* is the smallest set Φ of strings such that:

(iii) If t_1 and t_2 are terms, then the strings $t_1 = t_2$ and $t_1 \in t_2$ belong to Φ.

(iv) if $\varphi \in \Phi$, then $\neg\varphi \in \Phi$.

(v) If $\varphi,\psi \in \Phi$, then $(\varphi \wedge \psi)$, $(\varphi \vee \psi)$, $\varphi \Rightarrow \psi$, $(\varphi \Longleftrightarrow \psi)$ all belong Φ.

(vi) If $\varphi \in \Phi$, x is a variable, and t is a term which doesn't contain x, then $\exists x{\in}t\ \varphi$ and $\forall x{\in}t\ \varphi$ belong Φ.

There is no problem with the existence of $\mathcal{T}$ and $\mathcal{F}$; e.g. $\mathcal{F}$ can be obtained as the intersection of all sets Φ satisfying (iii)-(vi).

IV.1.2 **Example**

The terms and formulas are built up inductively. To check, for example, that

$$\exists v_1{\in}\underline{b}\ (\underline{F}(v_1,\underline{a},v_2) = v_1 \vee \neg\ v_1 = \underline{a}) \tag{1}$$

(where $\underline{a}$ and $\underline{b}$ are constant symbols, and $\underline{F}$ is a function symbol taking three variables) is a formula, I would proceed as follows:

By (i), v_1, $\underline{a}$, and v_2 are terms, and thus $\underline{F}(v_1,\underline{a},v_2)$ is a term by (ii). Combining this with (iii), I get that $\underline{F}(v_1,\underline{a},v_2) = v_1$ and $v_1 = \underline{a}$ are formulas. By (iv), $\neg v_1 = \underline{a}$ is a formula, and by applying (v), I see that

$$(\underline{F}(v_1,\underline{a},v_2) = v_1 \vee \neg\ v_1 = \underline{a}) \tag{2}$$

must be a formula. Finally, $\underline{b}$ is a term by (i), and hence I can apply (vi) to (2) to conclude that (1) is a formula.

Definition IV.1.1 suggests the following general proof strategy known as *induction on the complexity of formulas*. To prove that all formulas have a certain property P, let Φ be the set of all formulas having P and show that it satisfies IV.1.1 (iii)-(vi). One of the results that can be proved by this method is the unique readability of formulas;

each formula can be decomposed as $t_1 = t_2$, $t_1 \in t_2$, $\neg\varphi$, $(\varphi \wedge \psi)$, $(\varphi \vee \psi)$, $(\varphi \Rightarrow \psi)$, $(\varphi \Leftrightarrow \psi)$, $\exists x \in t\ \varphi$, or $\forall x \in t\ \varphi$ (where t_1, t_2, t are terms and φ, ψ formulas) in exactly one way. Hence there is no ambiguity; a formula cannot be interpreted in different ways by different parsings. There is a similar result for terms.

Since the constant symbol $\underline{a}$ and function symbols $\underline{F}$ are derived from elements $a \in V(S)$ and functions $F\colon V(S)^k \to V(S)$, all formulas have natural interpretations as statements about V(S); e.g., the formula

$$\exists v_2 \in \underline{a}\ \exists v_1 \in \underline{b}\ \underline{F}(v_1) = v_2 \tag{3}$$

says that for every v_2 in a, there is an element v_1 in b such that $F(v_1) = v_2$, i.e., a is a subset of the image of b under F. There is a companion language L*(V(S)) which in a similar way expresses statements about V(*S).

The *alphabet* $\mathcal{A}$*(V(S)) of L*(V(S)) is identical to the alphabet $\mathcal{A}$(V(S)) above except that each constant symbol $\underline{a}$ in $\mathcal{A}$(V(S)) has been replaced by a constant symbol $^*\underline{a}$ and each function symbol $\underline{F}$ by a function symbol $^*\underline{F}$. Terms and formulas are formed as before. To each term t or formula φ in L(V(S)), there is a term *t or formula $^*\varphi$ in L*(V(S)) obtained by replacing all $\underline{a}$'s by $^*\underline{a}$'s and all $\underline{F}$'s by $^*\underline{F}$'s. I shall refer to *t and $^*\varphi$ as the **-transforms* of t and φ, respectively.

The formulas in L*(V(S)) are easily interpreted as statements about V(*S); for example the *-transform

$$\forall v_2 \in {}^*\underline{a}\ \exists v_1 \in {}^*\underline{b}\ {}^*\underline{F}(v_1) = v_2$$

of formula (3) says that *a is a subset of the image of *b under *F.

The main result I am aiming at - the Transfer Principle - will basically say that a formula φ in L(V(S)) is true if and only if its *-transform $^*\varphi$ is. Before I prove this, I'll try to illustrate by a few examples how ordinary mathematical statements can be expressed in L(V(S)) and what happens to them when we apply the *-transformation.

It's convenient to begin by naming a few tame functions. I shall be sloppy and refer to the *pairing function*

$$P(a,b) = \{a,b\}$$

and *ordered pairing function*

$$Q(a,b) = \langle a,b \rangle$$

simply as $\{\underline{a},\underline{b}\}$ and $\langle \underline{a},\underline{b} \rangle$ instead of the more correct $\underline{P}(\underline{a},\underline{b})$ and $\underline{Q}(\underline{a},\underline{b})$.

I shall also drop the stars on the transformed version of these two function symbols. Given integers $n > 0$, $m \geq 0$, let me also introduce the functions

$$V_m^{(n)}(a_1,\dots,a_n) = V_k(S)$$

where $k = m + \max(\text{rank}(a_1),\dots,\text{rank}(a_n))$ (recall that the rank of an element $a \in V(S)$ is the smallest p such that $a \in V_p(S)$).

Assume that $a,b,c \in V(S)$ and that we want to express that c is a function from a to b (i.e., c is a set of ordered pairs $\langle x,y\rangle$ where $x \in a$, $y \in b$ and where each $x \in a$ appears as a first component exactly once). Since

$$\varphi_1(\underline{a},\underline{b},\underline{c}) = \forall v_1 \in \underline{c}\ \exists v_2 \in \underline{a}\ \exists v_3 \in \underline{b}\ \ v_1 = \langle v_2,v_3\rangle$$

says that c is a set of ordered pairs from $a \times b$ and

$$\varphi_2(\underline{a},\underline{b},\underline{c}) = \forall v_1 \in \underline{a}\ \exists v_2 \in \underline{b}\ (\langle v_1,v_2\rangle \in \underline{c} \wedge \forall v_3 \in \underline{b}\ (\langle v_1,v_3\rangle \in c \Rightarrow v_2 = v_3))$$

says that each $v_1 \in a$ appears as a first component exactly once,

$$\varphi_3(\underline{a},\underline{b},\underline{c}) = (\varphi_1(\underline{a},\underline{b},\underline{c}) \wedge \varphi_2(\underline{a},\underline{b},\underline{c},))$$

does the job. The transferred statement $\varphi_3(*\underline{a},*\underline{b},*\underline{c})$ clearly says that $*c$ is a function from $*a$ to $*b$.

If I only want to say that c is a function without specifying the domain and the range, I can use

$$\varphi_4(c) = \exists v_4 \in V_0^{(1)}(\underline{c})\ \exists v_5 \in V_0^{(1)}(\underline{c})\ \varphi_3(v_4,v_5,\underline{c})$$

(where, of course, $\varphi_3(v_4,v_5,\underline{c})$ is the formula obtained from $\varphi_3(\underline{a},\underline{b},\underline{c})$ by replacing all occurrences of $\underline{a}$ and $\underline{b}$ by v_4 and v_5, respectively). Since any function from a to b is an element of $V_3^{(2)}$ (a,b), the formula

$$\varphi_5(\underline{a},\underline{b}) = \forall v_4 \in \underline{V}_3^{(2)}(\underline{a},\underline{b})\ (\varphi_3(\underline{a},\underline{b},v_4) \Rightarrow \psi(v_4))$$

claims that all functions from a to b have the property expressed by ψ. The last two examples explain why I have singled out the functions $V_m^{(n)}$; they are just convenient tools for producing the right sets to quantify over.

The $*$-transform of $\varphi_5(\underline{a},\underline{b})$ reads

$$*\varphi_5(*\underline{a},*\underline{b}) = \forall v_4 \in *\underline{V}_3^{(2)}(*\underline{a},*\underline{b})\ (*\varphi_3(*\underline{a},*\underline{b},v_4) \Rightarrow *\psi(v_4))$$

and here something important and interesting has happened: the set $*v_3^{(2)}(*a,*b)$ is internal and contains only internal elements. Hence $*\varphi_5(*\underline{a},*\underline{b})$ doesn't say that all functions from *a to *b satisfy $*\psi$; it just say that all *internal* functions from *a to *b have this property! The phenomenon is general; the *-transformation always turns quantification over arbitrary sets in V(S) into quantification over internal sets in V(*S), and thus the transfer principle doesn't really say that V(S) and V(*S) are alike, but that V(S) and the collection of all internal subsets of V(*S) are. This is in accordance with our experience from the previous chapters, and it will be a major theme in the next two sections.

IV.2 ŁOŠ' THEOREM AND THE TRANSFER PRINCIPLE

In the last section I interpreted formulas in L(V(S)) and L*(V(S)) as statements about V(S) and V(*S) in an informal manner. To prove the transfer principle, I have to define these interpretations more formally. But first of all I have to explain the two ways in which a variable can occur in formula.

If you take a look at

$$\exists v_1 \in \underline{a}\ v_1 \in v_2,$$

you will see that the variables v_1 and v_2 play quite different roles in this formula; the truth of the statement depends on the value of the variable v_2 but not on the value of v_1. I shall say that v_2 is *free*, but that v_1 is *bound* by the quantifier $\exists v_1 \in \underline{a}$. There is nothing new or subtle about this distinction; it's similar to the difference between the "free" variable t and the "bound" or "dummy" variable s in the integral

$$\int_0^t f(s)\, ds$$

Intuitively, an occurence of a variable x in a formula φ is bound if it is within the scope of a quantifier of the form $\exists x \in t$ or $\forall x \in t$; formally, I will define an occurence to be *free* or *bound* by induction on the complexity of the formula:

If φ is of the form $t_1 = t_2$ or $t_1 \in t_2$, then all occurrences in φ are free.

If φ is of the form $\neg\varphi_1$, then an occurrence in φ is free if and only if the corresponding occurrence in φ_1 is free.

If φ is $(\varphi_1 \wedge \varphi_2)$, $(\varphi_1 \vee \varphi_2)$, $(\varphi_1 \Rightarrow \varphi_2)$, or $(\varphi_1 \Leftrightarrow \varphi_2)$, then an occurrence in φ is free if and only if the corresponding occurrence in the relevant formulae φ_1, φ_2 is free.

If φ is the form $\exists x{\in}t\ \varphi_1$ or $\forall x{\in}t\ \varphi_1$, then all occurrences of x are bound. If y is a variable different from x, then all occurrences of y in t are free, while an occurrence in the φ_1-part of φ is free if and only if it is free as an occurrence in φ_1.

Note that a variable can have both free and bound occurrences in the same formula; in

$$(v_1 = v_2 \wedge \exists v_1{\in}\underline{a}\ v_1{\in}v_2)$$

the first occurrence of v_1 is free while the last two are bound.

If φ has no other free variables than $x_1,\ldots,x_n$, it's often convenient to indicate this by writing $\varphi(x_1,\ldots,x_n)$ for φ. Similarly, I shall write $t(x_1,\ldots,x_n)$ for t when t has no other variables than $x_1,\ldots,x_n$.

Given a term $t(x_1,\ldots,x_n)$ or a formula $\varphi(x_1,\ldots,x_n)$ of L(V(S)) and n elements $a_1,\ldots,a_n$ in V(S), I shall now define the value $t(a_1,\ldots,a_n)$, $\varphi(a_1,\ldots,a_n)$ of t or φ at $a_1,\ldots,a_n$. The value of a term will be an element in V(S), while the value of a formula will be either T (true) or ⊥ (false).

IV.2.1 **Definition**

The *value* $t(a_1,\ldots,a_n)$ of the term $t = t(x_1,\ldots,x_n)$ is defined inductively by:

(i) If t consists of the single variable x_i, then $t(a_1,\ldots,a_n) = a_i$, and if t consists of a single constant symbol $\underline{a}$, then $t(a_1,\ldots,a_n) = a$.

(ii) If t is of the form $\underline{F}(t_1 \ldots t_k)$, then $t(a_1,\ldots,a_n) = F(t_1(a_1,\ldots,a_n),\ldots,t_k(a_1,\ldots,a_n))$.

The *value* $\varphi(a_1,\ldots,a_n)$ of the formula $\varphi = \varphi(x_1,\ldots,x_n)$ is defined inductively by:

(iii) If φ is of the form $t_1 = t_2$, then $\varphi(a_1,\ldots,a_n) = T$ iff $t_1(a_1,\ldots,a_n) = t_2(a_1,\ldots,a_n)$. If φ is of the form $t_1 \in t_2$, then $\varphi(a_1,\ldots,a_n) = T$ iff $t_1(a_1,\ldots,a_n) \in t_2(a_1,\ldots,a_n)$.

(iv) If φ is $\neg\varphi_1$, then $\varphi(a_1,\ldots,a_n) = T$ iff $\varphi_1(a_1,\ldots,a_n) = \bot$.

(v1) If φ is $(\varphi_1 \wedge \varphi_2)$, then $\varphi(a_1,\ldots,a_n) = T$ iff $\varphi_1(a_1,\ldots,a_n) = \varphi_2(a_1,\ldots,a_n) = T$.

(v2) If φ is $(\varphi_1 \vee \varphi_2)$, then $\varphi(a_1,\ldots,a_n) = T$ unless $\varphi_1(a_1,\ldots,a_n) = \varphi_2(a_1,\ldots,a_n) = \bot$.

(v3) If φ is $(\varphi_1 \Rightarrow \varphi_2)$, then $\varphi(a_1,\ldots,a_n) = T$ unless $\varphi_1(a_1,\ldots,a_n) = T$ and $\varphi_2(a_1,\ldots,a_n) = \bot$.

(v4) If φ is $(\varphi_1 \Leftrightarrow \varphi_2)$, then $\varphi(a_1,\ldots,a_n) = T$ iff $\varphi_1(a_1,\ldots,a_n) = \varphi_2(a_1,\ldots,a_n)$.

(vi) If φ is of the form $\exists y \in t(x_1,\ldots,x_n)\varphi_1(y,x_1,\ldots,x_n)$, then $\varphi(a_1,\ldots,a_n) = T$ iff there is an element $b \in t(a_1,\ldots,a_n)$ such that $\varphi_1(b,a_1,\ldots,a_n) = T$. If φ is of the form $\forall y \in t(x_1,\ldots,x_n)\ \varphi_1(y,x_1,\ldots,x_n)$, then $\varphi(a_1,\ldots,a_n) = T$ iff for all elements $b \in t(a_1,\ldots,a_n)$, $\varphi_1(b,a_1,\ldots,a_n) = T$.

There, is of course, a parallel definition of the values $t(a_1,\ldots,a_n)$, $\varphi(a_1,\ldots,a_n)$ of terms t and formulas φ in the language L*(V(S)) given elements $a_1,\ldots,a_n \in V(^*S)$. Note that since the *-version $^*F$ of a tame function F always maps internal elements to internal elements, the value $t(a_1,\ldots,a_n)$ is always internal when $a_1,\ldots,a_n$ are.

We are ready to take the first step toward the transfer principle.

IV.2.2 **Lemma**

Let $t = t(x_1,\dots,x_n)$ *be a term in* $L(V(S))$, *and let* $*t = *t(x_1,\dots,x_n)$ *be its *-transform. If* $a^{(1)}=\langle a_i^{(1)}\rangle,\dots, a^{(n)}=\langle a_i^{(n)}\rangle$ *are internal sets in* $V(*S)$, *then*

$$*t(a^{(1)},\dots,a^{(n)}) = \langle t(a_i^{(1)},\dots\dots,a_i^{(n)})\rangle$$

Proof. If t is a variable or a constant symbol, the lemma is obviously true. Assume that it holds for $t_1,\dots,t_k$, and that $t = F(t_1,\dots,t_k)$. Then $*t = *F(*t_1,\dots,*t_k)$ and

$$*t(a^{(1)},\dots,a^{(n)}) = *F(*t_1(a^{(1)},\dots,a^{(n)}),\dots,*t_k(a^{(1)},\dots,a^{(n)}))$$

$$*F(\langle t_1(a_i^{(1)},\dots,a_i^{(n)})\rangle,\dots,\langle t_k(a_i^{(1)},\dots,a_i^{(n)})\rangle) = \langle t(a_i^{(1)},\dots,a_i^{(n)})\rangle. ◂$$

This lemma has a natural generalization to formulas:

IV.2.3 **Łoś' Theorem**

Let $\varphi = \varphi(x_1,\dots,x_n)$ *be a formula in* $L(V(S))$ *with *-transform* $*\varphi = *\varphi(x_1,\dots,x_n)$. *If* $a^{(1)} = \langle a_i^{(1)}\rangle,\dots, \langle a^{(n)}\rangle = \langle a_i^{(n)}\rangle$ *are internal sets, then* $*\varphi(a^{(1)},\dots,a^{(n)}) = \top$ *if and only if* $\varphi(a_i^{(1)},\dots,a_i^{(n)}) = \top$ *for almost all* i.

Proof. The proof is by induction on complexity of formulas. I shall not go through all the various induction steps, but just show you a few typical cases.

If φ is $t_1 = t_2$ or $t_1 \in t_2$ for terms t_1 and t_2, the result follows immediately from the lemma.

Assume that φ is $\neg\varphi_1$, and that the theorem holds for φ_1. Then

$$\begin{aligned} *\varphi(a^{(1)},\dots,a^{(n)}) = \top &\iff *\varphi_1(a^{(1)},\dots,a^{(n)}) = \bot \\ &\iff m\{i: \varphi_1(a_i^{(1)},\dots,a_i^{(n)}) = \bot\} = 1 \\ &\iff m\{i: \varphi(a_i^{(1)},\dots,a_i^{(n)}) = \top\} = 1 \end{aligned}$$

Assume that φ is of the form $(\varphi_1 \wedge \varphi_2)$ where the result holds for φ_1 and φ_2. By using that the intersection of two sets of measure one has measure one, we get:

$$*\varphi(a^{(1)},\ldots,a^{(n)}) = T$$

$$\Leftrightarrow [*\varphi_1(a^{(1)},\ldots,a^{(n)}) = T \text{ and } *\varphi_2(a^{(1)},\ldots,a^{(n)}) = T]$$

$$\Leftrightarrow [m\{i: \varphi_1(a_i^{(1)},\ldots,a_i^{(n)} = T\} = 1 \text{ and } m\{i: \varphi_2(a_i^{(1)},\ldots,a_i^{(n)}) = T\} = 1]$$

$$\Leftrightarrow m\{i: \varphi(a_i^{(1)},\ldots,a_i^{(n)}) = T\} = 1.$$

Finally, assume that φ is $\exists y \in t\ \varphi_1$, where the theorem holds for φ_1. If $*\varphi(a^{(1)},\ldots,a^{(n)}) = T$, then there is an element $b \in *t(a^{(1)},\ldots,a^{(n)})$ such that $*\varphi_1(b,a^{(1)},\ldots,a^{(n)})$. Since $*t(a^{(1)},\ldots,a^{(n)}) = \langle t(a_i^{(1)},\ldots,a_i^{(n)})\rangle$, b is an internal element of the form $b = \langle b_i\rangle$, and hence for almost all i, $b_i \in t(a_i^{(1)},\ldots,a_i^{(n)})$ and $\varphi_1(b_i,a_i^{(1)},\ldots,a_i^{(n)} = T$. It follows that $\varphi(a_i^{(1)},\ldots,a_i^{(n)}) = T$ for almost all i.

Assume for the converse that $\varphi(a_i^{(1)},\ldots,a_i^{(n)}) = T$ for almost all i. For each such i, pick an element $b_i \in t(a_i^{(1)},\ldots,a_i^{(n)})$ such that $\varphi_1(b_i,a_i^{(1)},\ldots,a_i^{(n)}) = T$. It is easy to check that the sequence $\{t(a_i^{(1)},\ldots,a_i^{(n)})\}$ is bounded in V(S) (in the sense of Section II.1), and hence $\{b_i\}$ defines an internal element $b = \langle b_i\rangle$ in V(*S). By construction, $b \in *t(a^{(1)},\ldots,a^{(n)})$ and $*\varphi_1(b,a^{(1)},\ldots,a^{(n)}) = T$, from which it follows that $\varphi(a^{(1)},\ldots,a^{(n)}) = T$.◂

A *sentence* is a formula with no free variables. The Transfer Principle is just Łoś' Theorem for sentences.

IV.2.4 Transfer Principle

A sentence φ in L(V(S)) *is true if and only if its *-transform $*\varphi$ is true.*

An important consequence of the Transfer Principle is the following precise formulation of the Internal Definition Principle (recall the discussion of internal sets in Section II.1).

IV.2.5 **Internal Definition Principle**

If $b, a_1, \ldots, a_n$ *are internal elements in* $V({}^*S)$ *and* ψ *is a formula in* $L^*(V(S))$, *then*

$$d = \{c \in b\colon \psi(c, a_1, \ldots, a_n) = T\}$$

is an internal set.

Proof. Let φ be the $L(V(S))$-formula having ψ as its *-transform, and choose $m \in \mathbb{N}$ so large that $b, a_1, \ldots, a_n \in V_m({}^*S)$. The following sentence is obviously true in $V(S)$

$$\forall x_1 \in \underline{V_m(S)} \ldots \forall x_n \in \underline{V_m(S)}\ \forall y \in \underline{V_m(S)}\ \exists u \in \underline{V_m(S)}\ \forall z \in \underline{V_m(S)}$$
$$(z \in u \Leftrightarrow (z \in y \wedge \varphi(z, x_1, \ldots, x_n)))$$

and by the Transfer Principle its *-transform

$$\forall x_1 \in \underline{{}^*V_m(S)} \ldots \forall x_n \in \underline{{}^*V_m(S)}\ \forall y \in \underline{{}^*V_m(S)}\ \exists u \in \underline{{}^*V_m(S)}\ \forall z \in \underline{{}^*V_m(S)}$$
$$(z \in u \Leftrightarrow (z \in y \wedge \psi(z, x_1, \ldots, x_n)))$$

holds in $V({}^*S)$. Choosing $a_1, \ldots, a_n$ for $x_1, \ldots, x_n$ and b for y, we get the existence of a set $u \in {}^*V_m(S)$ such that

$$z \in u \Leftrightarrow z \in b \text{ and } \psi(z, a_1, \ldots, a_n) = T.$$

Clearly, $u = d$, and since $u \in {}^*V_m(S)$, d is internal.◂

There is an alternative proof of the Internal Definition Principle which is perhaps more illustrative from our point of view (although it will be less helpful in the next section). If $b = \langle b^{(i)} \rangle_{i \in I}$, $a_1 = \langle a_1^{(i)} \rangle_{i \in I}, \ldots, a_n = \langle a_n^{(i)} \rangle_{i \in I}$, one simply shows by induction on the complexity of ψ that

$$\{c \in b\colon \psi(c, a_1, \ldots, a_n)\} = \langle \{c \in b^{(i)}\colon \psi(c^{(i)}, a_1^{(i)}, \ldots, a_n^{(i)})\} \rangle_{i \in I}.$$

The importance of the Transfer Principle and the Internal Definition Principle is that the kinds of argument that have so far been carried out by means of equivalence classes of sequences can now be handled by systematic use of these two theorems. I'll use the remainder of the present section to illustrate how this is done in practice. The next section contains a brief look at the theoretical consequences.

I shall assume that $\mathbb{R} \subset S$, and that $K: V(S)^2 \to V(S)$, $L: V(S)^2 \to V(S)$, $M: V(S)^2 \to V(S)$, and $N: V(S) \to V(S)$ are tame functions such that for all $x,y \in \mathbb{R}$

$$K(x,y) = x+y,\ L(x,y) = xy,\ M(x,y) = x-y,\ N(x) = |x|.$$

Moreover, let

$$D = \{\langle x,y\rangle : x,y \in \mathbb{R},\ x \le y\}$$

$$E = \{\langle x,y\rangle : x,y \in \mathbb{R},\ x < y\}.$$

To increase readability, I shall systematically abuse notation and write $L(V(S))$-formulas in terms of $+$, $-$, $\cdot$, $|\ |$, $\le$, and $<$ instead of the correct $\underline{K}$, $\underline{L}$, $\underline{M}$, $\underline{N}$, $\underline{D}$, and $\underline{E}$. As an example,

$$\forall x \in \underline{\mathbb{R}}\ \forall y \in \underline{\mathbb{R}}\ \forall z \in \underline{\mathbb{R}}\ ((\underline{0} < x \wedge y < z) \Rightarrow (xy < xz))$$

ought to read

$$\forall x \in \underline{\mathbb{R}}\ \forall y \in \underline{\mathbb{R}}\ \forall z \in \underline{\mathbb{R}}\ ((\underline{Q}(\underline{0},x) \in \underline{E} \wedge \underline{Q}(y,z) \in \underline{E}) \Rightarrow \underline{Q}(\underline{L}(x,y),\underline{L}(x,z)) \in \underline{E}), \quad (1)$$

where $Q(x,y) = \langle x,y\rangle$ is the ordered pairing function of the last section.

The formula (1) is obviously true in $V(S)$, and thus the $*$-transform

$$\forall x \in {}^*\underline{\mathbb{R}}\ \forall y \in {}^*\underline{\mathbb{R}}\ \forall z \in {}^*\underline{\mathbb{R}}(({}^*\underline{0} < x \wedge y < z) \Rightarrow (xy < xz))$$

holds in $V({}^*S)$. This argument should be compared to Example I.1.4, where the same result is obtained by means of equivalence classes of sequences. All the other ordered field axioms can be treated in the same way, and hence ${}^*\mathbb{R}$ is an ordered field. But when we turn to the least upper bound principle, we shall find the situation more subtle.

Since

$$\varphi(x,X) \equiv \forall y \in X\ (y \le x)$$

says that x is an upper bound for X, the least upper bound principle can be formulated as

$$\forall X \in \underline{\mathcal{P}(\mathbb{R})}\ ((\exists v \in \underline{\mathbb{R}}\ v \in X \wedge \exists x \in \underline{\mathbb{R}}\ \varphi(x,X)) \Rightarrow$$
$$\Rightarrow \exists y \in \underline{\mathbb{R}}\ (\varphi(y,X) \wedge \forall z \in \underline{\mathbb{R}}\ (z < y \Rightarrow \neg\varphi(z,X)))).$$

The $*$-transform of this is

$$\forall X \in {}^*\underline{\mathcal{P}(\mathbb{R})}\ ((\exists v \in {}^*\underline{\mathbb{R}}\ v \in X \wedge \exists x \in {}^*\underline{\mathbb{R}}\ {}^*\varphi(x,X)) \Rightarrow$$
$$\Rightarrow \exists y \in {}^*\underline{\mathbb{R}}\ ({}^*\varphi(y,X) \wedge \forall z \in {}^*\underline{\mathbb{R}}\ (z < y \Rightarrow \neg{}^*\varphi(z,X)))),$$

and since ${}^*\mathcal{P}(\mathbb{R})$ is the set of all *internal* subsets of ${}^*\mathbb{R}$, we have obtained the least upper bound principle I.2.3 for internal sets. This is a typical example of the phenomenon I mentioned at the end of the last section - that the *-transform turns quantification over arbitrary sets in V(S) into quantification over internal sets in V(*S).

As a slightly more advanced example, let me show you how to use the Transfer Principle to prove the characterization of continuity given in Proposition I.3.1. Assume first that f is continuous and that $x \approx a$. I must show that $*f(x) \approx f(a)$. Given an $\varepsilon \in \mathbb{R}_+$, there is a $\delta \in \mathbb{R}_+$ such that

$$\forall y \in \underline{\mathbb{R}} \; (|y-\underline{a}| < \underline{\delta} \;\Rightarrow\; |\underline{f}(y)-\underline{f(a)}| < \underline{\varepsilon})$$

holds. By the Transfer Principle, the *-transform

$$\forall y \in *\underline{\mathbb{R}} \; (|y-*\underline{a}| < *\underline{\delta} \;\Rightarrow\; |*\underline{f}(y)-*\underline{f(a)}| < *\underline{\varepsilon})$$

must also hold, and since $|x-a| < \delta$, I get $|*f(x)-f(a)| < \varepsilon$. Since $\varepsilon \in \mathbb{R}_+$ is arbitrary, $*f(x) \approx f(a)$.

Assume for the converse that $*f(x) \approx f(a)$ for all $x \approx a$. Then for each $\varepsilon \in \mathbb{R}_+$, the sentence

$$\exists \delta \in *\underline{\mathbb{R}} (\delta > *\underline{0} \wedge \forall y \in *\underline{\mathbb{R}} \; (|y-*a| < \delta \;\Rightarrow\; |*\underline{f}(y)-*\underline{f(a)}| < *\underline{\varepsilon}))$$

is true in V(*S) (just choose δ to be any positive infinitesimal), and hence its inverse *-transform

$$\exists \delta \in \underline{\mathbb{R}} \; (\delta > \underline{0} \wedge \forall y \in \underline{\mathbb{R}} \; (|y-\underline{a}| < \delta \;\Rightarrow\; |\underline{f}(y) - \underline{f(a)}| < \underline{\varepsilon}))$$

holds in V(S). Since ε is an arbitrary element of $\mathbb{R}_+$, this means that f is continuous at a.

Finally, as a realistic example of how the internal definition principle is used, I would like to show that the set E appearing in the proof of Lemma II.2.3 really is internal. Recall the situation; Ω is an internal set, $\mathcal{A}$ is an internal family of subsets of Ω, and $C: *\mathbb{N} \to *\mathcal{P}(\Omega)$, $\mu: \mathcal{A} \to *\mathbb{R}$ are internal functions. The set E is defined by

$$E = \left\{n \in *\mathbb{N}: \bigcup_{k=1}^{n} C(k) \in \mathcal{A} \text{ and } \mu\left(\bigcup_{k=1}^{n} C(k)\right) < \alpha + \frac{\varepsilon}{2}\right\};$$

i.e.

$$E = \{n \in *\mathbb{N}: \varphi(\underline{n},\underline{\Omega},\underline{\mathcal{A}},\underline{C},\underline{\mu})\}$$

where

$$\varphi(\underline{n},\underline{\Omega},\underline{\mathcal{A}},\underline{C},\underline{\mu}) \equiv \exists D\in\underline{\mathcal{A}}(\forall x\in\underline{\Omega}\ (x\in D \Leftrightarrow \exists k\in{}^*\underline{\mathbb{N}}\ (k \le \underline{n} \wedge x \in \underline{C}(k))) \wedge \underline{\mu}(D) < {}^*\underline{\alpha} + \frac{{}^*\varepsilon}{\underline{2}})$$

This isn't genuine L(V(*S)), but if we rewrite

$$x \in \underline{C}(k) \text{ as } \exists y \in {}^*\underline{\mathcal{P}(\Omega)}\ (\langle k,y\rangle \in \underline{C} \wedge x \in y)$$

and

$$\underline{\mu}(D) < {}^*\underline{\alpha} + \frac{{}^*\varepsilon}{\underline{2}} \text{ as } \exists z\in{}^*\mathbb{R}\ (\langle D,z\rangle \in \underline{\mu} \wedge z < {}^*\underline{\alpha} + \frac{{}^*\varepsilon}{\underline{2}})$$

φ becomes an L(V(*S))-formula (modulo the usual abbreviations and conventions). By the Internal Definition Principle, E is internal.

Examples of this kind may seem technical to the beginner, but with a little practice they become trivial routine exercises. As the saying goes, "you just put stars on everything" and you have the information you need about internal sets in V(*S).

IV.3 AXIOMATIC NONSTANDARD ANALYSIS

The approach to nonstandard analysis which I have taken in this paper may be called constructive in the sense that I have constructed a specific class of nonstandard models and restricted myself to a study of their properties. With the results of the last section in mind, I can now turn the situation around and give an axiomatic description of nonstandard models as superstructure extensions satisfying the Transfer Principle. This change in point of view leads to a shift in emphasis; the specific models become of secondary interest (they just guarantee the consistency of the axioms), while the language L(V(S)) gains primary importance as the main technical tool of the theory. I shall just sketch the fundamental ideas of the axiomatic approach; as almost all other introductions to nonstandard analysis are presented from this point of view, there should be no problem finding assistance in filling in the details.

Let S be an infinite set and consider the superstructure V(S) over S. A *nonstandard model of* V(S) consists of a superstructure V(*S) and an embedding $*: V(S) \to V({}^*S)$ satisfying the following two axioms:

Extension Principle. S is a proper subset of *S, and $^*s = s$ for all $s \in S$.

Transfer Principle. A sentence φ in L(V(S)) is true in V(S) if and only if its *-transform $^*\varphi$ is true in V(*S).

It may come as a surprise to learn that the fundamentals of nonstandard analysis can be developed on the basis of these two simple principles alone. After all, so many of the most important concepts above have been introduced by means of equivalence classes of sequences - a procedure which can not in any obvious way be expressed in terms of the Extension and Transfer Principles. How, for instance, does one define the internal sets? Well, if at first we are a little less ambitious and only ask for a natural definition of standard sets, there are no problems; a set in V(*S) is *standard* if it is of the form *A for some $A \in V(S)$. Recalling the simple observation II.1.3, we can then define a set to be *internal* if and only if it is an element of some standard set.

As a next step, a number of interesting notions can be defined in terms of internal sets and functions; for example, a set A is *hyperfinite* if and only if there is an $N \in {}^*\mathbb{N}$ and an internal bijection $f: A \to \{1,2,\ldots,N\}$ (recall Proposition I.2.12). To define sums over hyperfinite sets another common trick is useful; let $\mathcal{F}$ be the set of all functions $f: \mathbb{R} \to \mathbb{R}$ and $\mathcal{A}$ the set of all finite subsets of $\mathbb{R}$, and think of summation as the function

$$\Sigma : \mathcal{F} \times \mathcal{A} \to \mathbb{R}$$

given by $\Sigma(f,a) = \sum_{x \in A} f(x)$. Summation of internal functions f over hyperfinite sets A is defined by the *-version of the function Σ:

$$\sum_{x \in A} f(x) = {}^*\Sigma(f,A).$$

By using the Transfer Principle one can check that this makes sense; i.e., that $^*\Sigma$ is a function from $^*\mathcal{F} \times {}^*\mathcal{A}$ to $^*\mathbb{R}$, that $^*\mathcal{F}$ is the set of all internal functions from $^*\mathbb{R}$ to $^*\mathbb{R}$, and that $^*\mathcal{A}$ is the set of all hyperfinite subsets of $^*\mathbb{R}$.

Continuing in this manner, and using the Transfer Principle the way I indicated in the last section, it is possible to develop a nonstandard treatment of large parts of classical mathematics (for example, nonstandard calculus as presented in Section I.3) on the basis of the Extension and Transfer Principles alone. However, to obtain the more abstract theory in Chapters II and III, we need to add saturation:

Saturation Principle. Let κ be an infinite cardinal. The nonstandard model V(*S) is *κ-saturated* if whenever card $\Gamma < \kappa$ and $\{A_\gamma\}_{\gamma\in\Gamma}$ is a set of internal sets with the finite intersection property, then

$$\bigcap_{\gamma\in\Gamma} A_\gamma \neq \emptyset.$$

V(*S) is *polysaturated* if it is card(V(S))-saturated.

In polysaturated models all the results of the first three chapters hold and - once the elementary stages are passed - with the same proofs. Thus whether one prefers a "constructive" or an "axiomatic" approach is of little or no consequence and mainly a matter of personal background and taste. The axiomatic version is perhaps conceptually cleaner and more elegant, while the constructive one is, at least in my opinion, more direct and closer to intuition. A potential advantage of the axiomatic approach is that it keeps open the possibility of using models with special properties that ultrapowers (i.e., the kinds of model I have used in this paper) can't have.

I'll stop here. My intention has been to provide you with a sufficient background in nonstandard analysis to be able to read research papers in the field without difficulty and with only an occasional need to look things up in the literature. Needless to say, I have failed if you feel no urge to continue your study of the modern theory of infinitesimals either by reading other people's papers and books, or by applying the techniques presented here to your own problems. There are references for further study and a few suggestions of things waiting to be done in the Notes.

APPENDIX. ULTRAFILTERS

The purpose of this appendix is to prove the existence of the finitely additive measure m in Definition I.1.2 and its countably incomplete, κ^+-good relative in Theorem III.1.2. The first task is easy, the second requires considerable combinatorial ingenuity.

A *filter* over a set I is a collection $\mathcal{F}$ of subsets of I such that $\emptyset \notin \mathcal{F}$, and

(i) if $A, B \in \mathcal{F}$, then $A \cap B \in \mathcal{F}$,

(ii) if $A \in \mathcal{F}$ and $A \subset B$, then $B \in \mathcal{F}$.

If I is infinite, the set of all $A \subset I$ such that $I \backslash A$ is finite forms a filter. Filters of this kind which don't contain any finite sets are called *free*. An *ultrafilter* is a filter $\mathcal{U}$ such that for all $A \subset I$, either $A \in \mathcal{U}$ or $I \backslash A \in \mathcal{U}$.

Note that if m is the measure in Definition I.2.1, the set

$$\mathcal{M} = \{A \subset \mathbb{N}: m(A) = 1\} \tag{1}$$

is a free ultrafilter over $\mathbb{N}$, and that, conversely, any free ultrafilter $\mathcal{M}$ over $\mathbb{N}$ defines a measure m satisfying I.1.2 by

$$m(A) = \begin{cases} 1 & \text{if } A \in \mathcal{M} \\ 0 & \text{if } A \notin \mathcal{M}. \end{cases} \tag{2}$$

A.1 **Proposition**

Any free filter over I *can be extended to a (free) ultrafilter.*

Proof. Let $\mathcal{S}$ be the set of all filters over I extending $\mathcal{F}$, and order $\mathcal{S}$ by inclusion. By Zorn's lemma, $\mathcal{S}$ has a maximal element $\mathcal{U}$, and I shall show that $\mathcal{U}$ is an ultrafilter.

If $E \subset I$, note that there can't be sets $F_1, F_2 \in \mathcal{U}$ such that both $F_1 \cap E$ and $F_2 \cap (I\setminus E)$ are finite since this would imply that $F_1 \cap F_2 \in \mathcal{U}$ is finite. Hence either $F \cap E$ is infinite for all $F \in \mathcal{U}$ or $F \cap (I\setminus E)$ is. But if $F \cap E$ is infinite for all $F \in \mathcal{U}$, then $E \in \mathcal{U}$, since otherwise $\{D: \exists F\in\mathcal{U}\ (F\cap E \subset D)\}$ is a proper extension of the maximal filter $\mathcal{U}$. Similarly, if $F \cap (I\setminus E)$ is infinite for all $F \in \mathcal{U}$, then $I\setminus E \in \mathcal{U}$. Consequently, either $E \in \mathcal{U}$ or $I\setminus E \in \mathcal{U}$, and thus $\mathcal{U}$ is an ultrafilter.◂

It should now be clear how to obtain the measure m in Definition I.1.2; begin with the filter $\mathcal{F}$ of cofinite sets, extend to an ultrafilter $\mathcal{M}$, and define m by (2).

An ultrafilter is *countably incomplete* and *κ-good* if the associated finitely additive measure is (recall the definition in section III.1). To prove that such things exist, I shall need two lemmas.

A.2 Lemma

Let X be a set of infinite cardinality κ. Given a family $\{Y_x\}_{x\in X}$ of sets Y_x of cardinality κ, there exists another family $\{Z_x\}_{x\in X}$ such that:

(i) *For each $x \in X$, Z_x is a subset of Y_x of cardinaltiy κ.*

(ii) $Z_x \cap Z_y = \emptyset$ *whenever* $x \neq y$.

Proof. I can clearly assume that $X = \kappa$. For each ordinal $\alpha \le \kappa$, let

$$L_\alpha = \{\langle\gamma,\beta\rangle : \gamma \le \beta \text{ and } \beta < \alpha\}.$$

The idea is to define an injective function f with domain L_κ such that

$$\text{whenever } \gamma \le \beta < \kappa,\ f(\gamma,\beta) \in Y_\gamma, \qquad (*)$$

and then put $Z_\gamma = \{f(\gamma,\beta) : \gamma \le \beta < \kappa\}$.

To construct f, I'll define a sequence $\{f_\alpha\}_{\alpha<\kappa}$ of injective functions such that f_α has domain L_α and satisfies (*) with κ replaced by α. Moreover, f_α will be an extension of f_β when $\alpha > \beta$.

To carry out the construction, assume that f_β has been defined for all $\beta < \alpha$. If α is a limit ordinal, put $f_\alpha = \bigcup_{\beta<\alpha} f_\beta$. If $\alpha = \gamma + 1$, define f_α to agree with f_γ on L_γ, and choose $f_\alpha(\eta,\gamma)$, $0 \le \eta \le \gamma$, to be distinct elements of $Y_\eta \backslash \{f_\gamma(\xi,\lambda)\colon (\xi,\lambda) \in L_\gamma\}$ (since Y_η has cardinality κ and L_γ and γ have cardinality less than κ, there are more than enough elements too choose among). To complete the proof, let $f = \bigcup_{\alpha<\kappa} f_\alpha$.◂

If $\mathcal{F}$ is a filter over κ and $\mathcal{E}$ is a family of subsets of κ such that $\mathcal{F} \cup \mathcal{E}$ has the finite intersection property, then

$$(\mathcal{F},\mathcal{E}) = \{D \subset \kappa\colon D \supset F \cap E_1 \cap \ldots \cap E_n \text{ for some } F \in \mathcal{F},\ E_1,\ldots,E_n \in \mathcal{E}\}$$

is the smallest filter containing both $\mathcal{F}$ and $\mathcal{E}$ and is called the *filter generated by $\mathcal{F}$ and $\mathcal{E}$*. When $\mathcal{F}$ is the trivial filter $\{\kappa\}$, $(\mathcal{F},\mathcal{E})$ is called the *filter generated by $\mathcal{E}$*.

Assume that κ is an infinite cardinal and that Π is a family of partitions of κ such that each $P \in \Pi$ has exactly κ equivalence classes. If $\mathcal{F}$ is a filter, we say that the pair $(\Pi,\mathcal{F})$ is *consistent* if whenever we pick an $X \in \mathcal{F}$ and partition classes $X_1,\ldots,X_n$ from distinct partitions $P_1,\ldots,P_n \in \Pi$ then $X \cap X_1 \cap \ldots \cap X_n \neq \emptyset$.

The following technical lemma contains exactly the information needed to prove the existence of countably incomplete, κ^+-good ultrafilters over κ.

A.3 **Lemma**

(i) *Let $\mathcal{F}$ be a filter over κ generated by a set $\mathcal{E}$ of cardinality at most κ, and assume that each element of $\mathcal{F}$ has cardinality κ. There exists a family Π of partitions of κ such that card$(\Pi) = 2^\kappa$ and $(\Pi,\mathcal{F})$ is consistent.*

(ii) *Assume that $(\Pi,\mathcal{F})$ is consistent and that J is a subset of κ. Then either $(\Pi,(\mathcal{F},\{J\}))$ is consistent, or there is a subset Π' of Π such that $(\Pi',(\mathcal{F},\{\kappa\backslash J\}))$ is consistent and $\Pi\backslash\Pi'$ is finite.*

(iii) *Suppose that $(\Pi,\mathcal{F})$ is consistent, and let $p\colon \mathcal{P}_\omega(\kappa) \to \mathcal{F}$ be a reversal (in the sense of Section III.1). For every $P \in \Pi$, there exists a filter $\mathcal{F}'$ extending $\mathcal{F}$ and a strict reversal $q\colon \mathcal{P}_\omega(\kappa) \to \mathcal{F}'$ such that $q \le p$ and $(\Pi\backslash\{P\}, \mathcal{F}')$ is consistent.*

The best way to understand the significance of this lemma is to see how its three parts combine to prove the main theorem.

A.4 **Theorem**

If I *is a set of infinite cardinality* κ, *then there is a countably incomplete,* κ^+*-good ultrafilter over* I,

Proof. I may obviously let $I = \kappa$. Choose $\{I_n\}_{n\in\mathbb{N}}$ to be a decreasing family of subsets of κ such that each I_n has cardinality κ, but $\bigcap_{n\in\mathbb{N}} I_n = \emptyset$, and let $\mathcal{F}_0$ be the filter generated by the family $\{I_n\}_{n\in\mathbb{N}}$. According to the first part of Lemma A.3, there is a family Π_0 of partitions of κ such that $(\Pi_0,\mathcal{F}_0)$ is consistent and $\mathrm{card}(\Pi_0) = 2^\kappa$. Using transfinite induction I shall now define a sequence $(\Pi_\xi,\mathcal{F}_\xi)_{\xi<2^\kappa}$ such that

(i) $\Pi_\xi \subset \Pi_\eta$ and $\mathcal{F}_\xi \supset \mathcal{F}_\eta$ when $\eta \le \xi < 2^\kappa$,

(ii) $\mathrm{card}(\Pi_\xi) = 2^\kappa$, $\Pi_\xi \backslash \Pi_{\xi+1}$ is finite, and $\Pi_\lambda = \bigcap_{\eta<\lambda} \Pi_\eta$ when λ is a limit ordinal,

(iii) $(\Pi_\xi,\mathcal{F}_\xi)$ is consistent,

and such that $\mathcal{F} = \bigcup_{\xi\in 2^\kappa} \mathcal{F}_\xi$ satisfies the theorem.

Fix an enumeration $(p_\xi)_{\xi<2^\kappa}$ of all reversals from $\mathcal{P}_\omega(\kappa)$ to $\mathcal{P}(\kappa)$ and an enumeration $(J_\xi)_{\xi<2^\kappa}$ of $\mathcal{P}(\kappa)$. Assume that ξ is an ordinal less than 2^κ and that for all $\eta < \xi$, a pair $(\Pi_\eta,\mathcal{F}_\eta)$ satisfying (i)-(iii) has been defined.

If ξ is a limit ordinal, just let

$$\Pi_\xi = \bigcap_{\eta<\xi} \Pi_\eta, \quad \mathcal{F}_\xi = \bigcup_{\eta<\xi} \mathcal{F}_\eta.$$

If $\xi = \lambda+2n+1$ for some limit ordinal λ and some natural number n, let J be the first element in $\mathcal{P}(\kappa)$ such that neither J nor $\kappa\backslash J$ belong to $\mathcal{F}_{\xi-1}$. By the second part of Lemma A.3, there is a consistent pair $(\Pi_\xi,\mathcal{F}_\xi)$ such that $\Pi_{\xi-1}\backslash\Pi_\xi$ is finite, card $\Pi_\xi = 2^\kappa$, and either J or $\kappa\backslash J$ belongs to $\mathcal{F}_\xi$.

If $\xi = \lambda+2n+2$, let $p: \mathcal{P}_\omega(\kappa) \to \mathcal{F}_{\xi-1}$ be the first function in the list $(p_\eta)_{\eta<w^\kappa}$ that hasn't yet been encountered in the construction. By the third part of Lemma A.3, there exist a consistent pair $(\Pi_\xi,\mathcal{F}_\xi)$, a partition $P \in \Pi_{\xi-1}$, and a strict reversal $q: P_\omega(\kappa) \to \mathcal{F}_\xi$ such that $\Pi_{\xi-1}\backslash\Pi_\xi = \{P\}$, $F_\xi \supset F_{\xi-1}$, and $q \le p$.

It only remains to show that $F = \bigcup_{\xi<2^\kappa} \mathcal{F}_\xi$ is a countably incomplete, κ^+-good ultrafilter. The countable incompleteness is obvious since $\{I_n: n \in \mathbb{N}\} \subset \mathcal{F}_0 \subset \mathcal{F}$, and $\mathcal{F}$ is an ultrafilter since for each $J \subset \kappa$, there is a step in the construction where either J or $\kappa\backslash J$ is added to $\mathcal{F}$. Finally, let $p: \mathcal{P}_\omega(\kappa) \to \mathcal{F}$ be a reversal. Since the cofinality of 2^κ is greater than κ, there is an ordinal $\xi < 2^\kappa$ such that $p: \mathcal{P}_\omega(\kappa) \to \mathcal{F}_\xi$. By the construction there is a strict reversal $q: \mathcal{P}_\omega(\kappa) \to \mathcal{F}$ such that $q \le p$.◂

Proof of Lemma A.3. (i) Let $(J)_{\alpha<\kappa}$ be an enumeration of the elements in ξ. By Lemma A.2 there is a family $(I_\alpha)_{\alpha<\kappa}$ such that $I_\alpha \subset J_\alpha$, card $I_\alpha = \kappa$, and $I_\alpha \cap I_\beta = \emptyset$ when $\alpha \neq \beta$. The set

$$B = \{\langle s,r\rangle: s \in \mathcal{P}_\omega(\kappa) \text{ and } r: \mathcal{P}(s) \to \kappa\}$$

obviously has cardinality κ, and we fix a listing $(\langle s_\xi,r_\xi\rangle)_{\xi<\kappa}$ such that for each $\alpha < \kappa$

$$B = \{\langle s_\xi,r_\xi\rangle: \xi \in I_\alpha\}. \tag{3}$$

For each subset J of κ, define $f_J: \kappa \to \kappa$ by

$$f_J(\xi) = \begin{cases} r_\xi(J \cap s_\xi) & \text{if } \xi \in \bigcup_{\alpha<\kappa} I_\alpha \\ \emptyset & \text{otherwise,} \end{cases}$$

and let Π consist of the partitions $P_J = \{f_J^{-1}(\gamma): \gamma < \kappa\}$.

To prove that Π has cardinality 2^κ, it suffices to show that if $J_1 \neq J_2$, then $f_{J_1} \neq f_{J_2}$. If $x \in J_1\backslash J_2$, let $s = \{x\}$ and define $r: \mathcal{P}(s) \to \kappa$ by $r(\{x\}) = 0$, $r(\emptyset) = 1$. If ξ is an ordinal such that $\langle s_\xi,r_\xi\rangle = \langle s,r\rangle$, then $f_{J_1}(\xi) = 0$, $f_{J_2}(\xi) = 1$, and thus $f_{J_1} \neq f_{J_2}$.

Assume that ordinals $\alpha, \beta_1, \dots, \beta_n \in \kappa$ and sets $J_1, \dots, J_n \subset \kappa$ are given. To prove that $(\Pi, \mathcal{F})$ is consistent, we must prove that

$$I_\alpha \cap f_{J_1}^{-1}(\beta_1) \cap \dots \cap f_{J_n}^{-1}(\beta_n) \neq \emptyset;$$

i.e., we must find a $\xi \in I_\alpha$ such that $f_{J_i}(\xi) = \beta_i$ for all $i \le n$. But this is easy; just let s be a finite subset of κ such that $s \cap J_i \neq s \cap J_j$ when $i \neq j$, and choose $r\colon \mathcal{P}(s) \to \kappa$ such that $r(s \cap J_i) = \beta_i$. By (3), there is a $\xi \in I_\alpha$ such that $\langle s_\xi, r_\xi \rangle = \langle s, r \rangle$.

(ii) Suppose that $(\Pi, (\mathcal{F}, \{J\}))$ isn't consistent, and pick $X \in \mathcal{F}$ and elements $X_1, \dots, X_n$ from distinct partitions $P_1, \dots, P_n \in \Pi$ such that

$$J \cap X \cap X_1 \cap \dots \cap X_n = \emptyset. \tag{4}$$

Let $\Pi' = \Pi \setminus \{P_1, \dots, P_n\}$. If $\hat{P}_1, \dots, \hat{P}_m$ are distinct elements of Π' and $\hat{X}_i \in \hat{P}_i$ for each i, then by assumption

$$X \cap X_1 \cap \dots \cap X_n \cap \hat{X}_1 \cap \dots \cap \hat{X}_m \neq \emptyset \tag{5}$$

By combining (4) and (5), we get

$$(\kappa \setminus J) \cap X \cap \hat{X}_1 \cap \dots \cap \hat{X}_m \neq \emptyset$$

which guarantees the consistency of $(\Pi', (F, \{\kappa \setminus J\}))$.

(iii) Choose an enumeration $(X_\delta)_{\delta<\kappa}$ of the partition P without repetitions, and an enumeration $(t_\delta)_{\delta<\kappa}$ of $\mathcal{P}_\omega(\kappa)$. For each $\delta<\kappa$, define $q_\delta\colon \mathcal{P}_\omega(\kappa) \to \mathcal{P}(\kappa)$ by

$$q_\delta(s) = \begin{cases} p(t_\delta) \cap X_\delta & \text{if } s \subset t_\delta \\ \emptyset & \text{if } s \not\subset t_\delta. \end{cases}$$

Obviously, $q_\delta(s_1 \cup s_2) = q_\delta(s_1) \cap q_\delta(s_2)$ since $s_1 \cup s_2 \subset t_\delta$ if and only if both $s_1 \subset t_\delta$ and $s_2 \subset t_\delta$. Define $q\colon \mathcal{P}_\omega(\kappa) \to \mathcal{P}(\kappa)$ by

$$q(s) = \bigcup_{\delta<\kappa} q_\delta(s);$$

clearly, q is a strict reversal and $q \le p$.

Put $\mathcal{F}' = (\mathcal{F}, \text{Range}(q))$. It remains to show that $(\Pi \setminus \{P\}, \mathcal{F}')$ is consistent. Let $X \in \mathcal{F}$, $s \in \mathcal{P}_\omega(\kappa)$, $X_i \in P_i \in \Pi$, where $P_i \neq P$ for all i. Then $s = t_\delta$ for some $\delta < \kappa$, and since $q(s) \supset q_\delta(s) = p(t_\delta) \cap X_\delta$,

$$X \cap q(s) \cap X_1 \cap \dots \cap X_n \supset X \cap p(t_\delta) \cap X_\delta \cap X_1 \cap \dots \cap X_n \neq \emptyset$$

by the consistency of $(\Pi, \mathcal{F})$.

NOTES

These notes have two purposes; to provide suggestions for further study and to give an impression of the areas in which nonstandard methods have been particularly useful. I have made no attempt to write a balanced history of nonstandard analysis - recent results and developments are given special attention, and where good surveys with extensive bibliographies exist, I have usually referred to these and not to the original papers.

Section I.1. Nonstandard Analysis was discovered - or perhaps "invented" is a better word in this case - by Abraham Robinson (1961). He was not the first who tried to provide infinitesimal calculus with a logical foundation; among his more successful forerunners were Hahn (1907), who gave a consistent interpretation of infinitesimals in nonarchimedean fields, and Laugwitz & Schmieden (1958) (see also Laugwitz (1959, 1961a,b)) who studied infinitesimal analysis in rings with zero divisors. But although it was intended for quite different purposes, it was probably Skolem's (1934) construction of a nonstandard model of the natural numbers which had the strongest direct influence on Robinson's work. For more information about the history of infinitesimals, see the final chapter of Robinson's book (1966) and the many examples and comments in Laugwitz (1978, 1986).

Robinson originally obtained a set $^*\mathbb{R}$ of hyperreals by referring to the compactness theorem of mathematical logic. The so-called "ultrapower construction" given here was first used in a nonstandard context by Luxemburg (1962), but it was at that time already a well-known technique in model theory and topology (in fact, Hewitt (1948) contains exactly the construction in this section, but he and his school never used the hyperreals for analytic purposes). The "philosophy" about the asymptotic behaviour of sequences remains to a large extent just philosophy; the book by Lightstone & Robinson (1975) lays the foundations for nonstandard asymptotic analysis, but until recently (see van den Berg (1987)) this work has not been followed up.

Let me finally mention a few of the important contributions of nonstandard methods to algebra and algebraic number theory: Robinson and Roquette (1975), Roquette (1975), van den Dries & Schmidt (1984), van den Dries & Wilkie (1984); see also Richter's paper in this volume.

Section I.2. Most of the results of this section are due to Robinson, but the important theorem on $\aleph_1$-saturation is from Keisler (1960). I shall have more to say about saturation in connection with Section III.1.

One of the applications of hyperfinite sets which I unfortunately haven't found room for in this paper is to mathematical economics and the theory of exchange economies. By using a hyperfinite set of consumers, it is possible to combine both the necessary combinatorial properties and the relevant asymptotic information in the same structure; see e.g. Brown & Robinson (1974a,b)), Brown (1976) and Anderson (1981, 1985) for new results obtained by this approach.

Section I.3. Nonstandard calculus is an integral part of Robinson's original conception of nonstandard analysis, and without having checked it in detail, I would guess that all the results of this section (and many more of the same nature) are due to him. Most of the books on nonstandard analysis listed in the references have a chapter on nonstandard calculus, but almost all of them from an advanced point of view. An exception is Keisler (1976a) which is an elementary text written for a typical American first-year calculus course. The book has been tried in class at a number of universities and reports are definitely positive; the students find infinitesimals conceptually simpler than limits and are able to follow more complicated arguments. Also, the nonstandard approach gives the subject a pleasant algebraic and geometric flavour. The instructor's manual (Keisler (1976b)) provides the necessary theoretical background (and is also a good introduction to nonstandard analysis in its own right). For slightly more advanced students, the nice little book by Henle and Kleinberg (1979) is excellent, while Hurd and Loeb (1985) is written as a first text in real analysis.

This may be a natural place to mention the interesting work on ordinary differential equations and dynamical systems which has been carried out over the last ten years mainly in France and Algeria; see for example, Benoit, Callot, Diener & Diener (1980), Lutz & Goze (1981), the surveys by Cartier (1982) and Zvonkin & Shubin (1984), and Francine Diener's paper in this volume.

I said above that freshmen calculus students find infinitesimals conceptually simpler than limits, and a similar phenomenon is noticeable in a number of more complex situations; the nonstandard objects seem to have a lower complexity than the corresponding standard ones. For a theoretical formulation and confirmation of this phenomenon, see Henson, Kaufmann & Keisler (1985) and Henson & Keisler (1986).

Section II.1. As far as I can tell, the superstructure formulation of nonstandard analysis originates with Robinson and Zakon (1969). Earlier treatments (such as Robinson (1966)) were phrased in terms of the finite types over S, which is a much more complicated setting.

Section II.2. Loeb measures were introduced by Loeb (1975) using the Caratheodory Extension Theorem. The uniqueness of infinite Loeb measures (which doesn't follow from Caratheodory's result) was proved later by Henson (1979a). I don't know who first came up with the elementary approach presented here; I must have seen it for the first time either in Hrbacek (1979) or in an early draft of Stroyan and Bayod (1986). My exposition follows Stroyan and Bayod closely. The integration theorem II.2.8 for finite functions is from Loeb's original paper, while the notion of S-integrability is due to Anderson (1976). Loeb (1975) also introduced liftings; the proof of Theorem II.2.11 that I have given here generalizes immediately to functions taking values in separable spaces and is due to Anderson (1977).

The Loeb measure is a relatively new invention, and the only books covering it are the most recent ones: Albeverio et al (1986), Hurd & Loeb (1985), and Stroyan & Bayod (1986) (Keisler (1984) contains a brief sketch but no proofs). In addition, there is an excellent survey paper by Cutland (1983a) with an almost complete bibliography up to the summer of 1983. Hurd & Loeb (1985) use a functional (Daniell-type) approach to Loeb measures and integrals (see also Loeb (1984 a,b) and Loeb's paper in the present volume) while the other expositions follow the more traditional method presented here.

Loeb measures have been applied in a variety of contexts, and I'll just mention a few here to whet your appetite (applications to the theory of stochastic processes will be treated separately in the notes to the next section). Loeb (1979), Anderson & Rashid (1978), Lindstrøm (1982) and others have studied different kinds of limit measures and measure extensions. Closely related is work on statistical mechanics by Hurd (1981), Helms & Loeb (1979, 1982), Helms (1983), Kessler (1984, 1986). Particularly noteworthy is a long series of papers by Arkeryd (1981-1986) on the Boltzmann equation which contain both the best existence results and the best asymptotic estimates to date (see Arkeryd's paper in this volume). In a very recent paper, Hurd (1987) has taken an important step towards the mathematical deduction of the Boltzmann equation from first principles. Most of this material is reviewed in Albeverio, Fenstad, Høegh-Krohn, & Lindstrøm (1986) which also brings new results on quantum field theory. Henson (1979b) studied the relationship between Loeb measurable sets and analytic sets, and his results and techniques are important for Stroyan & Bayod's (1986) approach to the foundations of infinitesimal stochastic analysis (see also Stroyan (1986)). The first application of Loeb measures to analysis was probably Loeb's paper (1976) on potential theory (see also Loeb (1978, 1980, 1982)). Recently Arkeryd's results on the Boltzmann equation have led to an interest in abstract Loeb-Sobolev spaces Arkeryd (1984b), Arkeryd & Bergh (1986). Other fields which deserve more attention are ergodic theory and measurable transformations; the interesting papers by Henson (1972), Kamae (1982), and Ross (1983) haven't really been followed up. Behind many of these applications is a method for representing measures on topological spaces as inverse standard parts of Loeb-measures; see Landers & Rogge (1987) for the latest developments. Finally, I can't refrain from mentioning Keisler's (1987) suprising use of Loeb-measures in classical model theory.

Section II.3. Although Loeb's original paper (1975) contained a nonstandard construction of Poisson processes, it was Anderson's study (1976) of hyperfinite random walks and Itô integration that got nonstandard stochastic analysis off the ground. Keisler (1984) combined

Anderson's construction with the old hyperfinite difference approach to differential equations (remember the proof of Peano's existence theorem in Section I.3) to obtain new, strong existence results for stochastic differential equations of the Itô type (see also Kosciuk (1982, 1983) and Cutland (1982, 1985) for extensions and refinements.) In a series of papers (Cutland 1983b,c, 1984, 1986, 1987. where (1986) is probably the best place to start), Cutland mixed Keisler's ideas with innovations of his own to study optimal controls in both deterministic and stochastic settings. Hoover & Perkins (1983), Lindstrøm (1980, 1983, 198?a) and Stoll (1985) extended the nonstandard theory of stochastic integration to include integration with respect to semi-martingales, infinite dimensional Brownian motion, and white noise. These papers also contain applications to stochastic differential equations and Levy Brownian motion. For a unified presentation of all this material, see chapter 4 of Albeverio et al. (1986).

But Keisler's seminal paper (1984) contains much more than existence results for stochastic differential equations; for example, a study of intrinsic properties of Loeb spaces such as *universality* (whatever happens on some probability space, happens on Loeb space) and *homogeneity* ("similar" processes on Loeb space are connected through a measure preserving transformation); see Keisler's paper in this volume for more information. These ideas have been refined and extended by Hoover & Keisler (1984). There are close connections with the rapidly developing field of probability logic (see the survey by Keisler (1986)), and to the theory of exchangeability (Hoover 1982). Stroyan & Bayod (1986) (see also Stroyan (1986)) have combined Keisler's theory with Henson's results on nonstandard descriptive set theory and created a "Strasbourg-school" setting for nonstandard stochastic analysis. Their book also contains a detailed study of liftings of stochastic processes and an exposition of stochastic integration with respect to semi-martingales.

One of the deepest and most impressive contributions to nonstandard probability theory is Perkins' work on Brownian local time with applications; see for example Perkins (1981, 1982a,b,c, 1983a). Lawler (1980, 1986) and Stoll (1985, 1987) have studied different versions of self-avoiding random walks. Both local time and self-avoiding random walks are important in quantum field theory; these connections are explained and exploited in Albeverio, Fenstad, Høegh-Krohn & Lindstrøm (1986).

Keisler's memoir (1984) also contains a study of Markov solutions to stochastic differential equations. A different approach to Markov processes is taken in Albeverio, Fenstad, Høegh-Krohn & Lindstrøm (1986) where a hyperfinite theory of Dirichlet forms and stationary Markov processes is developed with applications to quantum mechanics and (in Lindstrøm (1986)) to diffusions on manifolds and fractals. Diffusions on manifolds is a topic which deserves more attention from nonstandard probabilists; see Stroyan (1977) for an elementary account of how naturally concepts of differential geometry can be formulated in nonstandard terms.

For more references to the nonstandard literature on stochastic processes, see Albeverio, Fenstad, Høegh-Krohn & Lindstrøm (1986), Cutland (1983), and Stroyan & Bayod (1986). Two short surveys of the field are Perkins (1983b) and Lindstrøm (198? b).

Section III.1. Saturation is a concept of model theory (see Chang & Keisler (1973)). It was introduced in nonstandard analysis by Luxemburg (1969b) and has gradually replaced older notions such as enlargements and countably comprehensive models (see Stroyan & Luxemburg (1976) for definitions). There are other κ^+-saturated models (in the axiomatic sense of Section IV.3) that are technically much simpler to construct than the ultrapowers I have chosen to work with here (see, for example, the ultralimit construction in Stroyan & Luxemburg (1976) and Keisler (1976b)). I feel, however, that the conceptual clarity of the ultrapower models more than repays the extra technicalities - especially as no knowledge of the construction is necessary (or even useful) in applications.

The term *polysaturation* was introduced by Stroyan & Luxemburg (1976); it is a typical example of Stroyanesque humour - another example is *Loebotomy* for the Loeb construction.

Section III.2. Most of the results of this section are due to Robinson (see, for example, (1966)), who worked with enlargements and not κ-saturated models. In enlargements the saturation principle holds for families of standard sets, and this is sufficient for almost all the arguments given here. An exception is Proposition III.2.2 which is due to Luxemburg (1969). Nonstandard topology is well described in Stroyan & Luxemburg (1976), Luxemburg (1969), and Machover & Hirschfeld (1969).

For interesting new developments see Richter (1982), Benninghofen (1982a,b), Benninghofen & Richter (198?), and Benninghofen, Stroyan & Richter (1988).

That nonstandard methods are successful in topology isn't very surprising; there is a close connection between ultrapowers and filter convergence; see, for example, Fenstad (1967), Fenstad & Nyberg (1970). Some very interesting consequences of the nonstandard characterization of compactness are the "almost implies near" theorems of Luxemburg & Taylor (1970) and Anderson (1986). For nonstandard approaches to homology and cohomology, see McCord (1972), Reveilles (1984), and Zivaljevic (1987).

Section III.3. There are nonstandard papers on completions and compactifications by Luxemburg (1969), Machover & Hirschfeld (1969), Wattenberg (1971, 1973), Gonshor (1974), Dyre (1982), and others; the book by Stroyan & Luxemburg (1976) gives a unified presentation.

The theory of Banch spaces is one of the most active areas of nonstandard research. Henons's contribution to this volume and the earlier survey paper by Henson & Moore (1983) (containing an extensive bibliography and a list of open problems) presents the field much better than I can do. Let me just mention that nonstandard hulls were introduced by Luxemburg (1969),and that there are tight connections with ultraproduct techniques as presented in Heinrich (1980) for example. Interesting applications to calculus in locally convex spaces can be found in Stroyan (1978, 1983).

The proof of the spectral theorem given here is due to Moore (1976) (although I have chosen a slightly different formulation of the result to give an example of how Loeb measures turn up in analysis). To approximate infinite dimensional spaces from the outside by hyperfinite dimensional ones is a technique which goes back to Robinson. It has a special place in the history of nonstandard anlaysis as the main tool in Bernstein & Robinson's (1966) solution of the invariant subspace problem for polynomially compact operators - the first open standard problem that was ever solved by nonstandard techniques. Davis' book (1976) has hyperfinite approximations of infinite spaces as its main theme.

Albeverio, Fenstad, Høegh-Krohn & Lindstrøm (1986) have developed a theory of hyperfinite quadratic forms with applications to diffusion processes (see also Lindstrøm (1986b)) and Schrödinger operators with singular potentials (see also Albeverio (1984), Albeverio et al (1984), Fenstad (1986), and Lindstrøm (1986a) for shorter accounts). There are interesting open problems concerning the spectral properties of these forms and also concerning their relationship to the Loeb-Sobolev spaces in Arkeryd (1984b) and Arkeryd & Bergh (1986).

Section IV.1. Constant symbols for tame functions are not usually included in L(V(S)). Although they don't increase the expressive power of the language noticeably, they often reduce the complexity of the formulas substantially.

Section IV.2. Los' theorem (1955) is one the fundamental results of model theory. The Transfer Principle is as old as nonstandard analysis, and can best be described as part of Robinson's original conception of what a nonstandard model is. Although implicit in much of Robinson's work, the Internal Definition Principle seems first to have been formulated explicitly by Stroyan & Luxemburg (1976).

Section IV.3. The three axioms presented here are the result of a long evolution; the Transfer Principle (for some suitable language) has always been an essential part of nonstandard analysis; the superstructure setting was first introduced by Robinson & Zakon (1969): and saturation was emphasized by Luxemburg (1969b). Who first realized that these three principles make up the essence of nonstandard theory, I dare not say. Let me add, however, that for special purposes it may occasionally be convenient to require additional properties of the nonstandard model; for example, models satisfying an isomorphism property often referred to as *Henson's Lemma* (see Stroyan & Luxemburg (1976)) are important in Banch space theory.

There is another axiomatic formulation of nonstandard analysis due to Nelson (1977) which is known as Internal Set Theory (see the joint contribution of Diener and Stroyan to this volume). The axioms of this system form a conservative extension of ZFC (Zermelo-Fraenkel set theory with the axiom of choice). There has been a lively discussion (which I do not want to enter) of the relative logical strengths of various formulations of Nelson's theory and the axiomatics presented here. The books by van den Berg (1987), Lutz & Goze (1981), Richter (1982) and Robert (1985, 1988) are written in terms of Internal Set Theory. I would like to recommend Nelson's paper (1977) even to those who are not particularly interested in alternative formulations of nonstandard analysis; it is full of deep insights and interesting suggestions.

Since the first order languages L(V(S)) and L*(V(S)) are the greatest stumbling blocks for non-logicians trying to learn nonstandard analysis, a number of attempts have been made to create an axiomatic approach where the Transfer Principle is phrased in terms of more common mathematical constructions. Although some of these attempts (such as Keisler's *solution principle* (1976a,b)) are useful for pedagogical purposes, I can not see that any of them have been successful on the research level. If one wants to downplay the importance of formal languages, it seems more natural to me to use a constructive approach such as the one I have presented in this paper.

Appendix. The existence of κ^+-good, countably incomplete ultrafilters was first proved by Keisler (1964) assuming the generalized continuum hypothesis. Kunen (1972) later removed this extra condition. The proof here is taken from Chang & Keisler (1973).

REFERENCES

Albeverio, S. (1984). Nonstandard analysis; polymer models, quantum fields, *Acta Phys. Austriaca, Suppl.* **XXVI.**, 233-254.

Albeverio, S., Fenstad, J.E., Høegh-Krohn, R., Karwowski, W., & Lindstrøm, T. (1984). Perturbations of the Laplacian supported by null sets, with applications to polymer measures and quantum fields, *Phys. Lett.* **104**, 396-400.

Albeverio, S., Fenstad, J.E., Høegh-Krohn, R. & Lindstrøm, T. (1986). *Nonstandard Methods in Stochastic Analysis and Mathematical Physics*, Academic Press, New York.

Anderson, R.M. (1976). A nonstandard representation for Brownian motion and Itô integration, *Israel J. Math* **25**, 15-46.

Anderson, R.M. (1977). Star-finite representation of measure spaces, Ph.D.-thesis, Yale University.

Anderson, R.M. (1981). Core theory with strongly convex preferences, *Econometrica* **49**, 1457-1468.

Anderson, R.M. (1982). Star-finite representation of measure spaces, *Trans. Amer. Math. Soc.* **271**, 667-687.

Anderson, R.M. (1985). Strong core theorems with nonconvex preferences, *Econometrica* **53**, 1283-1294.

Anderson, R.M. (1986). "Almost" implies "near", *Trans. Amer. Math. Soc.* **296**, 229-237.

Anderson, R.M. & Rashid, S. (1978). A nonstandard characterization of weak convergence, *Proc. Amer. Math. Soc.* **69**, 327-332.

Arkeryd, L. (1981a). A nonstandard approach to the Boltzmann equation, *Arch. Rational Mech. Anal.* **77**, 1-10.

Arkeryd, L. (1981b). Intermolecular forces of infinite range and the Boltzmann equation, *Arch. Rational Mech. Anal.* **77**, 11-23.

Arkeryd, L. (1981c). A time-wise approximated Boltzmann equation, *IMA J. Appl. Math.* **27**, 373-383.

Arkeryd, L. (1982). Asymptotic behaviour of the Boltzmann equation with infinite range forces, *Comm. Math. Phys.* **86**, 475-484.

Arkeryd, L. (1984a). Loeb solutions of the Boltzmann equation, *Arch. Rational Mech. Anal.* **86**, 85-97.

Arkeryd, L. (1984b). Loeb-Sobolev spaces with applications to variational integrals and differential equations. Preprint, Chalmers Inst. Tech., Gothenburg.

Arkeryd, L. (1986). On the Boltzmann equation in unbounded space far from equilibrium and the limit of zero mean free path, *Commun. Math. Phys.* **105**, 205-219.

Arkeryd, L. & Bergh, J. (1986). Some properties of Loeb-Sobolev spaces, *J. London Math. Soc.* **34**, 317-334.

Benninghofen, B. (1982a). Nichtstandardmethoden und Monaden, Diplomarbeit, RWTH Aachen.

Benninghofen, B. (1982b). Infinitesimalien under Superinfinitesimalien, Dissertation, RWTH Aachen.

Benninghofen, B. & Richter, M.M. (198?). General theory of superinfinitesimals, *Fund. Math*, to appear.

Benninghofen, B., Stroyan, K.D. & Richter, M.M. (1988). Superinfinitesimals in topology and functional analysis, to appear in *Proc. Lond. Math. Soc.*

Benoit, E., Callot, J.-L., Diener, F. & Diener, M. (1980). Chasse au canard, *Collect. Math.* **31**, 3-74.

van den Berg, I. (1987). *Nonstandard Asymptotic Analysis*, Lecture Notes in Mathematics **1249**, Springer-Verlag, Berlin and New York.

Bernstein, A.R. (1974). Nonstandard analysis. In *Studies in Model Theory* (M. Morley, ed.), Math. Ass. of Amer., Providence, Rhode Island, 35-58.

Bernstein, A. & Robinson, A. (1966). Solution of an invariant subspace problem of K.T. Smith and P.R. Halmos, *Pacific J. Math.* **16**, 421-431.

Billingsley, P.(1968). *Convergence of Probability Measures*, Wiley, N.York.

Brown, D.J. (1976). Existence of a competitive equilibrium in a nonstandard exchange economy, *Econometrica* **44**, 537-546.

Brown, D.J. & Robinson, A (1974a). Nonstandard exchange economies, *Econometrica* **43**, 41-55.

Brown, D.J. & Robinson, A. (1974b). The cores of large standard exchange economies, *J. Economic Theory* **9**, 245-254.

Cartier, P. (1982). Perturbations singulières des équations différentielles ordinaire et analyse non-standard, *Asterisque* **92-93**, 21-44.

Chang, C.C. & Keisler, H.J. (1973). *Model Theory*, North-Holland, Amsterdam.

Cutland, N.J. (1982). On the existence of solutions to stochastic differential equations on Loeb spaces, *Z. Wahrsch. Verw. Gebiete* **60**, 335-357.

Cutland, N.J. (1983a). Nonstandard measure theory and its applications, *Bull. London Math. Soc.* **15** (1983), 529-589.

Cutland, N.J. (1983b). Internal controls and relaxed controls, *J. London Math. Soc.* **27**, 130-140.

Cutland, N.J. (1983c). Optimal controls for partially observed stochastic systems: an infinitesimal approach, *Stochastics* **8**, 239-257.

Cutland, N.J. (1984). Partially observed stochastic controls based on cumulative digital readouts of the observations, in *Proc. 4th IFIP Work. Conf. Stochastic Differential Systems, Marseille-Luminy*, Springer-Verlag, 261-269.

Cutland, N.J. (1985). Simplified existence for solutions to stochastic differential equations, *Stochastics* **14**, 319-326.

Cutland, N.J. (1986). Infinitesimal methods in control theory: Deterministic and stochastic, *Acta Appl. Math.* **5**, 105-136.

Cutland, N.J. (1987). Infinitesimals in action, *J. Lond. Math. Soc.* **35**, 202-216.

Davis, M. (1977). *Applied Nonstandard Analysis*, Wiley, New York.

van den Dries L. & Schmidt K. (1984). Bounds in the theory of polynomial rings over fields: A nonstandard approach, *Invent. Math.* **76**, 77-91.

van den Dries L. & Wilkie A.J. (1984). Gromov's theorem on groups of polynomial growth and elementary logic, *J. Algebra* **89**, 349-374.

Dyre J.C. (1982). Non-standard characterizations of ideals in C(X), *Math. Scand.* **50**, 44-54.

Fenstad, J.E. (1967). A note on "standard" versus "nonstandard" topology, *Indag. Math.* **29**, 378-380.

Fenstad, J.E. (1985). Is nonstandard analysis relevant for the philosophy of mathematics, *Synthese* **62**, 289-301.

Fenstad, J.E. (1986). Lectures on stochastic analysis with applications to mathematical physics; in Bertossi et al. (eds.) *Actas del Simposio Chileno de Logica Matematica, Santiago*, 201-282.

Fenstad, J.E. (198?). The discrete and the continuous in mathematics and the natural sciences, to appear.

Fenstad, J.E. & Nyberg A. (1970). Standard versus nonstandard methods in uniform topology, in *Logic Colloquium 1969*, North-Holland, Amsterdam, 353-359.

Gonshor, H. (1979). Enlargements contain various kinds of completions; in Hurd and Loeb (1974), 60-70.

Hahn, H. (1907). Uber die nichtarchimedische Groszensysteme, *S.-B. Wiener Akad. Math.-Natur. Kl.* **116** (Abt. IIa), 601-655.

Heinrich, S. (1980). Ultraproducts in Banach space theory, *J. Reine Angew. Math.* **313**, 72-104.

Helms, L.L. (1983). A nonstandard approach to the martingale problem for spin models; in Hurd (1983), 15-26.

Helms, L.L. & Loeb P.A. (1979). Applications of nonstandard analysis to spin models, *J. Math. Anal. Appl.* **69**, 341-352.

Helms, L.L. & Loeb P.A. (1982). Bounds on the oscillation of spin systems, *J. Math. Anal. Appl.* **86**, 493-502.

Henle J.M. & Kleinberg E.M. (1979). *Infinitesimal Calculus*, MIT Press, Cambridge, Massachusetts.

Henson, C.W. (1972). On the nonstandard representation of measures, *Trans. Amer. Math. Soc.* **172**, 437-446.

Henson, C.W. (1979a). Unbounded Loeb measures, *Proc. Amer. Math. Soc.* **74**, 143-150.

Henson, C.W. (1979b). Analytic sets, Baire sets and the standard part map. *Canad. J. Math.* **XXXI**, 663-672.

Henson, C.W., Kaufmann, M. & Keisler, H.J. (1985). The strength of nonstandard methods in arithmetic, *J. Symbolic Logic* **49**, 1039-1058.

Henson, C.W. & Keisler, H.J. (1986). The strength of nonstandard analysis, *J. Symbolic Logic* **51**, 377-386.

Henson, C.W. & Moore L.C. Nonstandard analysis and the theory of Banach spaces; in Hurd (1983), 27-111.

Hewitt, E. (1948). Rings of real-valued continuous functions I. *Trans. Amer. Math. Soc.* **64**, 45-99.

Hoover D.N. (1982). Row-column exchangeability and a generalized model for probability; in G. Koch & F. Spizzichino, (eds.), *Exchangeability in Probability and Statistics*, North-Holland, Amsterdam, 281-291.

Hoover, D.N. & Keisler, H.J.. (1984). Adapted probability distributions, *Trans. Amer. Math. Soc.* **286**, 159-201.

Hoover, D.N. & Perkins E. (1983). Nonstandard constructions of the stochastic integral and applications to stochastic differential equations I, II, *Trans. Amer. Math. Soc.* **275**, 1-58.

Hrbacek, K. (1979). Nonstandard set theory, *Amer. Math. Monthly* **86**, 659-677.

Hurd, A.E. (1981). Nonstandard analysis and lattice statistical mechanics: A variational principle. *Trans. Amer. Math. Soc.* **263**, 89-110.

Hurd, A.E. (ed) (1983). *Nonstandard Analysis - Recent Developments*, Lecture Notes in Mathematics **983**, Springer-Verlag, Berlin and New York.

Hurd, A.E. (1987). Global existence and validity for the BBGKY hierarchy, *Arch. Rational Mech. Anal.* **98**, 191-210.

Hurd, A.E. & Loeb P.A. (eds.) (1974). *Victoria Symposium on Nonstandard Analysis*, Lecture Notes in Mathematics **369**, Springer-Verlag, Berlin and New York.

Hurd, A.E. & Loeb P.A. (1985). *An Introduction to Nonstandard Real Analysis*, Academic Press, New York.

Kamae, T. (1982). A simple proof of the ergodic theorem using nonstandard analysis, *Israel J. Math.* **42**, 284-290.

Keisler, H.J. (1960). Theory of models with generalized atomic formulas, *J. Symbolic Logic* **25**, 1-26.

Keisler, H.J. (1964). Good ideals in fields of sets, *Ann. Math.* **79**, 338-359.

Keisler, H.J. (1976a). *Elementary Calculus*, Prindle, Weber and Schmidt, Boston.

Keisler, H.J. (1976b). *Foundations of Infinitesimal Calculus*, Prindle, Weber and Schmidt, Boston.

Keisler, H.J. (1984). An infinitesimal approach to stochastic analysis, *Mem. Amer. Math. Soc.* **297**.

Keisler, H.J. (1985). Probability quantifiers; in *Model-Theoretic Logics* (K.J. Barwise and S. Feferman, eds.), Springer-Verlag, Berlin and New York.

Keisler, H.J. (1987). Measures and forking. *Ann. Pure Appl. Logic* **34**, 119-170

Kessler, C. (1984). Nonstandard methods in random fields, Dissertation, Bochum.

Kessler, C. (1986). Nonstandard conditions for the global Markov property for lattice fields, *Acta Applicandae Mathematicae* **7**, 225-256.

Kosciuk, S.A. (1982). Nonstandard stochastic methods in diffusion theory, Ph.D. thesis, Madison.

Kosciuk, S.A. (1983). Stochastic solutions to partial differential equations; in Hurd (1983), 113-119.

Kunen, K. (1972). Ultrafilters and independent sets, *Trans. Amer. Math. Soc.* **172**, 199-206.

Landers D. & Rogge L. (1987). Universal Loeb-measurability of sets and of the standard part map with applications, *Trans. Amer. Math. Soc.* **304**, 229-243.

Laugwitz, D. (1959). Eine Einfuhrung der δ-Funktionen, *S.-B. Bayerische Akad. Wiss.* **4**, 41-59.

Laugwitz, D. (1961a). Anwendungen unendlich kleiner Zahlen I, *J. Reine Angew. Math.* **207**, 53-60.

Laugwitz, D. (1961b). Anwendungen unendlich kleiner Zahlen II, *J. Reine Angew. Math.* **208**, 22-34.

Laugwitz, D. (1978). *Infinitesimalkalkul*, Bibliographisches Inst., Mannheim.

Laugwitz, D. & Schmieden C. (1958). Eine Erweiterung der Infinitesimalrechnung, *Math. Z.* **69**, 1-39.

Laugwitz, D. (1986). *Zahlen und Kontinuum*, Bibliographisches Inst., Mannheim.

Lawler, G.F. (1980). A self-avoiding random walk, *Duke Math. J.* **47**, 655-693.

Lawler, G.F. (1986). Gaussian behavior of loop-erased, self-avoiding random walk in four dimensions, *Duke Math J.* **53**, 249-270.

Lightstone, A. & Robinson A. (1975). *Nonarchimedean Fields and Asymptotic Expansions*, North Holland, Amsterdam.

Lindstrøm, T. (1980). Hyperfinite stochastic integration I, II, III,. *Math. Scand.* **46**, 265-333.

Lindstrøm, T. (1982). A Loeb measure approach to theorems by Prohorov, Sazonov, and Gross, *Trans. Amer. Math. Soc.* **269**, 521-534.

Lindstrøm, T. (1983). Stochastic integration in hyperfinite dimensional linear spaces; in Hurd (1983), 134-161.

Lindstrøm, T. (1986a). Nonstandard analysis and perturbations of the Laplacian along Brownian paths; in *Stochastic Processes - Mathematics and Physics* (Albeverio et al., eds.), Lecture Notes in Mathematics **1158**, Springer, 1986, 180-200.

Lindstrøm, T. (1986b). Nonstandard energy forms and diffusions on manifolds and fractals; in *Stochastic processes in classical and quantum systems* (Albeverio, S. et al., eds.), Springer-Verlag, 363-380.

Lindstrøm, T. (198?a). The structure of hyperfinite stochastic integrals, to appear.

Lindstrøm, T. (198?b). Nonstandard constructions of diffusions and related processes, to appear.

Loeb, P.A. (1975). Conversion from nonstandard to standard measure spaces and applications in probability theory, *Trans. Amer. Math. Soc.* **211**, 113-122.

Loeb, P.A. (1976). Applications of nonstandard analysis to ideal boundaries in potential theory, *Israel J. Math.* **25**, 154-187.

Loeb, P.A. (1978). A generalization of the Riesz-Herglotz theorem on representing measures, *Proc. Amer. Math. Soc.* **71**, 65-68.

Loeb, P.A. (1979a). An introduction to nonstandard analysis and hyperfinite probability theory; in *Probabilistic Analysis and Related Topics II* (A.T. Bharucha-Reid, ed.) Academic Press, New York. 105-142.

Loeb, P.A. (1979b). Weak limits of measures and the standard part map, *Proc. Amer. Math. Soc.* **77**, 128-135.

Loeb, P.A. (1980). A regular metrizable boundary for solutions of elliptic and parabolic differential equations, *Math. Ann.* **251**, 43-50.

Loeb, P.A. (1982). A construction of representing measures for elliptic and parabolic differential equations, *Math. Ann.* **260**, 51-56.

Loeb, P.A. (1984a) A functional approach to nonstandard measure theory; in *Conference on Modern Analysis and Probability* (Beals et al., eds.) Amer. Math. Soc., Providence, Rhode Island, 259-262.

Loeb, P.A. (1984b). Measure spaces in nonstandard models underlying standard stochastic processes; in *Proceedings of the International Congress of Mathematicians 1* (Z. Ciesielski & C. Oleck, eds.) PWN (Warzaw) and North-Holland (Amsterdam), 323-335.

Los, J. (1955). Quelques remarques, theoremes, et problemes sur les classes definissables d'algebras; in *Mathematical Interpretations of Formal Systems* (Skolem et al., eds.), North-Holland, Amsterdam, 98-113.

Lutz, R. & Goze M. (1981). *Nonstandard Analysis*, Lecture Notes in Mathematics **881**, Springer-Verlag, Berlin and New York.

Luxemburg, W.A.J. (1962). *Nonstandard Analysis. Lectures on A. Robinson's Theory of Infinitesimal and Infinitely Large Numbers.* Caltech Bookstore, Pasadena.

Luxemburg, W.A.J. (ed) (1969a). *Applications of Model Theory to Algebra, Analysis, and Probability*, Holt, New York.

Luxemburg, W.A.J. (1969b). A general theory of monads; in Luxemburg (1969a), 18-86.

Luxemburg, W.A.J. (1973). What is nonstandard analysis?, *Amer. Math. Monthly* **80** (supplement), 38-67.

Luxemburg, W.A.J. & Robinson A. (eds.) (1972). *Contributions to Nonstandard Analysis*, North-Holland, Amsterdam.

Luxemburg W.A.J. & Taylor, R.F. (1970). Almost commuting matrices are near commuting matrices. *Proc. Roy. Acad. Amsterdam*, Ser.A73, 96-98.

McCord, M.C (1972). Nonstandard analysis and homology, *Fund. Math* **74**, 21-28.

Machover, M. & Hirschfeld, J. (1969). *Lectures on Nonstandard Analysis*, Lecture Notes in Mathematics **94**, Springer-Verlag, Berlin and New York.

Moore, L.C. (1976). Hyperfinite extensions of bounded operators on a separable Hilbert space, *Trans. Amer. Math. Soc.* **218**, 285-295.

Nelson, E. (1977). Internal set theory. *Bull. Amer. Math. Soc.* **83**, 1165-1193.

Perkins, E. (1981). A global intrinsic characterization of local time, *Ann. Prob.* **9**, 800-817.

Perkins, E. (1982a). Weak invariance principles for local time, *Z. Wahrsch. Verw. Gebiet* **60**, 437-451.

Perkins, E. (1982b). Local time is a semimartingale, *Z. Wahrsch. Verw. Gebiete* **60**, 79-117.

Perkins, E. (1982c). Local time and pathwise uniqueness for stochastic differential equations, *Sem. Prob. XVI*, Lecture Notes in Mathematics **920**, Springer-Verlag, Berlin and New York, 201-208.

Perkins, E. (1983a). On the Hausdorff dimension of the Brownian slow points, *Z. Wahrsch. Verw. Gebiete* **64**, 369-399.

Perkins, E. (1983b). Stochastic processes and nonstandard analysis; in Hurd (1983), 162-185.

Reed, M. & Simon, B. (1972). *Methods of Modern Mathematical Physics I*, Academic Press, New York.

Reveilles, J.P. (1984). *Infinitesimaux et Topologie*, C.R. Acad. Sc. Paris, Serie I, **298**.

Richter, M.M. (1982). *Ideale Punkte, Monaden, und Nichtstandard-Methoden*, Vieweg, Wiesbaden.

Robert, A. (1985). *Analyse Non-Standard*, Presse Polytechnique Romandes, Lausanne.

Robert, A. (1988). *Nonstandard Analysis*, Wiley, New York & Chichester. (English translation of Robert (1985).)

Robinson, A. (1961). Non-standard analysis, *Proc. Roy. Acad. Amsterdam Ser. A.* **64**, 432-440.

Robinson, A. (1966). *Nonstandard Analysis*, North-Holland, Amsterdam (2nd, revised edition 1974).

Robinson, A. (1979). *Selected Papers*, Vol. 2, North-Holland, Amsterdam.

Robinson, A. & Roquette, P. (1975). On the finiteness theorem of Siegel and Mahler concerning Diophantine equations, *J. Number Theory* **7**, 121-176.

Robinson, A. & Zakon, E. (1969). A set-theoretical characterization of englargements; in Luxemburg (1969a), 109-122.

Roquette, P. (1975). Nonstandard aspects of Hilbert's irreducibility theorem; in *Model Theory and Algebra*, Lecture Notes in Mathematics **498**, Springer-Verlag, 231-275.

Ross, D. (1983). Measurable transformations in saturated models of analysis, Ph.D. thesis, Madison.

Skolem, T. (1934). Über die nichtcharakterisierbarkeit der Zahlenreihe mittels endlich oder abzählbar unendlich vieler Aussagen mit ausschliesslich Zahlenvariablen, *Fund. Math.* **33**, 150-161.

Stoll, A. (1985). Self-repellent random walks and polymer measures in two dimensions, Dissertation, Bochum.

Stoll, A. (1986). A nonstandard construction of Lévy Brownian motion, *Prob. Th. and Related Fields* **71**, 321-334.

Stoll, A. (1987). Self-repellent random walks and polymer measures in two dimensions, in *Stochastic Processes - Mathematics and Physics II* (eds. Albeverio et al.), Lecture Notes in Mathematics **1250**, Springer-Verlag, 298-318.

Stroyan, K.D. (1977). Infinitesimal analysis of curves and surfaces; in *Handbook of Mathematical Logic* (K.J. Barwise, ed.), North-Holland, Amsterdam,197-231.

Stroyan, K.D. (1978). Infinitesimal calculus on locally convex spaces. 1. Fundamentals, *Trans. Amer. Math. Soc.* **240**, 363-383.

Stroyan, K.D. (1983). Locally convex infinitesimal calculus 2. Computations on Mackey (ℓ^{∞}), *J. Funct. Anal.* **53**, 1-15.

Stroyan, K.D. (1986). Previsible sets for hyperfinite filtrations, *Prob. Th. and Related Fields* **73**, 183-196.

Stroyan, K.D. & Bayod J.M. (1986). *Foundations of Infinitesimal Stochastic Analysis*, North-Holland, Amsterdam.

Stroyan, K.D. & W.A.J. Luxemburg. (1976). *Introduction to the Theory of Infinitesimals*, Academic Press, New York.

Wattenberg, F. (1971). Nonstandard topology and extensions of monad systems to infinite points, *J. Symbolic Logic* **36**, 463-576.

Wattenberg, F. (1973). Monads of infinite points and finite product spaces, *Trans. Amer. Math. Soc.* **176**, 351-368.

Živaljevic, R. (1987). On a cohomology theory based on hyperfinite sums of microsimplexes, *Pacific J. Math.* **126**.

Zvonkin, A.K. & Shubin, M.A. (1984). Nonstandard analysis and singular perturbations of ordinary differential equations, *Russian Math. Surveys* **39**, 77-127.

INFINITESIMALS IN PROBABILITY THEORY

H. JEROME KEISLER

The aim of this article is to explain how and why the Loeb measure construction can be applied to problems which arise outside of nonstandard analysis. The Loeb measure and the necessary background from nonstandard analysis are presented in the paper of Lindstrøm in this volume. Most of the applications of Loeb measures in probability theory fit the following pattern. First, lift the original classical problem to a hyperfinite setting. Second, make some hyperfinite computations. Third, take standard parts of everything in sight to obtain the desired classical result. In keeping with this pattern, we shall concentrate on hyperfinite Loeb spaces, that is, Loeb spaces on hyperfinite sets. As we go along, we shall present several examples of applications of hyperfinite Loeb spaces to probability theory. However, our main emphasis will be on the general theory which underlies the applications.

We shall begin with a study of random variables on probability spaces, and then pass to a deeper parallel theory of stochastic processes on adapted spaces. The hyperfinite adapted spaces are counterparts of the hyperfinite Loeb spaces. Corresponding to the distribution of a random variable, we introduce the adapted distribution of a stochastic process. We shall focus on three features of hyperfinite adapted spaces: liftings, universality, and homogeneity. These features help to explain why the applications work. The adapted lifting theorem shows that standard stochastic processes can be lifted to hyperfinite processes with almost the same adapted distribution. The universality theorem shows that hyperfinite adapted spaces are so rich that anything that happens on some space happens on the hyperfinite adapted space. The homogeneity theorem shows that hyperfinite adapted spaces are so well behaved that any two stochastic processes which have the same adapted distribution are related by an automorphism of the hyperfinite adapted space.

This research was supported in part by the National Science Foundation and the Vilas Trust Fund.

1. THE HYPERFINITE TIME LINE

Our main object of study will be Loeb spaces over hyperfinite sets. We refer to the article of Lindstrøm (this volume) for the definition and basic properties of Loeb measures. We shall work within a countably saturated nonstandard universe. We shall deal exclusively with probability measures, that is, measures in which the full space has measure one. Often we work with a pair of Loeb spaces, one over a sample set Ω and another over a hyperfinite set of times T. A Loeb space $L(\Omega) = (\Omega, L(\Omega), L(\mu))$ such that Ω is *-finite and every internal subset is measurable is called a *hyperfinite Loeb space*. In the special case that Ω is infinite and every point $\omega \in \Omega$ has the same weight, the Loeb space is called a *hyperfinite counting space*. A *hyperfinite time line* is a hyperfinite counting space $L(T) = (T, L(T), L(\tau))$ where T has the form

$$T = \{n/H! \ : \ n \in {}^*\mathbb{N} \text{ and } n < H!\}$$

for some infinite hyperinteger H. Notice that every standard rational number in [0,1] belongs to T. We shall use bold letters **s**, **t**,... to denote elements of T. The standard part map restricted to T maps T onto [0,1], st: T → [0,1]. For each set $A \subset [0,1]$, let

$$st^{-1}(A) = \{\mathbf{t} \in T: {}^\circ\mathbf{t} \in A\}.$$

The following result, whose proof is given in Example II.2.6 of Lindstrøm (this volume) shows that the hyperfinite time line represents the Lebesgue measure on the real unit interval.

1.1 **Proposition** (Anderson (1976), Henson (1979), E. Fisher)

A set $A \subseteq [0,1]$ *is Lebesgue measurable if and only if* $st^{-1}(A)$ *is* L(T)*-measurable. Moreover, if* A *is Lebesgue measurable, the Lebesgue measure of* A *is equal to the Loeb measure of* $st^{-1}(A)$.

The integration theory for Loeb measures, which is presented in Lindstrøm (this volume), leads to a lifting theorem for the Lebesgue integral. This lifting theorem is often useful in the study of stochastic processes.

Recall from Lindstrøm (this volume) that a *lifting* of a function $f: T \to \mathbb{R}$ is an internal function $F: T \to {}^*\mathbb{R}$ such that ${}^\circ F(t) = f(t)$ almost surely. Here is a corresponding notion of lifting for functions with domain $[0,1]$.

Definition

Let f be a function from $[0,1]$ into $\mathbb{R}$. By a *lifting* of f we mean an internal function F from T into $^*[0,1]$ such that ${}^\circ F(t) = f({}^\circ t)$ for $L(\mu)$-almost all $t \in T$.

1.2 **Proposition**

A function $f: [0,1] \to \mathbb{R}$ *is Lebesgue measurable if and only if it has a lifting.*

Proof. Let $g: T \to \mathbb{R}$ be defined by $g(t) = f({}^\circ t)$. Then a function F is a lifting of f if and only if it is a lifting of g in the earlier sense. f is Lebesgue measurable if and only if g is Loeb measurable, and by Theorem II.2.7 of Lindstrøm (this volume), g is Loeb measurable if and only if it has a lifting.◂

By using the results on integration and lifting from Lindstrøm (this volume), we obtain the following characterization of Lebesgue integrability.

1.3 **Corollary**

A function $f: [0,1] \to \mathbb{R}$ *is Lebesgue integrable if and only if it has an S-integrable lifting. If* F *is an S-integrable lifting of* f, *then*

$$\int_0^1 f(s)\, ds = {}^\circ\Sigma_{t \in T} F(t)/H! \ .$$

These results extend in the obvious way to the Lebesgue measure on the unit n-cube $[0,1]^n$ and the hyperfinite counting measure $L(T^n)$. Proposition 1.1. can be extended much further. The following result of Anderson (1982) shows that any Radon space can be represented by a hyperfinite counting space.

Definition

A *Radon* probability space is a probability space $\Pi = (\Pi, \mathcal{A}, \nu)$ such that $\mathcal{A}$ is the set of Borel sets in a Hausdorff space on Π and for each $A \in \mathcal{A}$, $\nu(A)$ is the supremum of $\nu(B)$ for compact subsets $B \subset A$.

1.4 **Theorem** (Anderson (1982))

Let $\Pi = (\Pi, \mathcal{A}, \nu)$ *be the completion of a Radon probability space. Suppose the nonstandard universe is* κ*-saturated where* Π *has an open basis of size less than* κ*. Let* U *be a hyperfinite partition of* $^*\Pi$ *such that* $U \subset {}^*\mathcal{A}$ *and for each near-standard point* $p \in {}^*\Pi$*, all points in the partition class of* p *have the same standard part. Let* L(U) *be the hyperfinite Loeb space on* U *in which each partition set* $u \in U$ *has weight* $^*\nu\{u\}$*. Then a set* $A \subset \Pi$ *is* ν*-measurable if and only if* $st^{-1}(A)$ *is Loeb measurable. Moreover, if* A *is* ν*-measurable, then* $\nu(A)$ *is equal to the Loeb measure of* $st^{-1}(A)$.

Proof. The proof of Theorem 1.4 is a generalization of the proof of Theorem 1.1. κ-saturation is needed only in the direction that the Loeb measurability of $st^{-1}(A)$ implies the ν-measurability of A.◂

An *atom* of a probability space $\Pi = (\Pi, \mathcal{A}, \nu)$ is a set $A \in \mathcal{A}$ such that $\nu(A) > 0$ and any measurable subset of A has measure $\nu(A)$ or 0, and an *atomless* space is a space which has no atoms. In Theorem 1.4, if Π is an atomless Radon space, then L(U) may be taken to be a hyperfinite counting space.

2. UNIVERSAL AND HOMOGENEOUS PROBABILITY SPACES

In this section we show that every hyperfinite counting space Ω has two properties which we call universality and homogeneity. We begin with some terminology.

A *Polish space* is a complete separable metric space. The most important example of a Polish space is the real line $\mathbb{R}$. Throughout this paper the letter M will denote an arbitrary Polish space. By a *random variable* on a probability space Π with values in a Polish space M we mean a Π-measurable function x from Π into M. Two random variables x and y with values in M are said to have the same *distribution* if for every Borel subset $B \subseteq M$, the probability that $x(\cdot) \in B$ is equal to the probability that $y(\cdot) \in B$, or equivalently, for any bounded continuous function $\varphi: M \to \mathbb{R}$, $E[\varphi(x(\cdot))] = E[\varphi(y(\cdot))]$

The universality property for hyperfinite counting spaces is easy.

2.1 Proposition

Every hyperfinite counting space is atomless.

***Proof*.** Every infinite internal subset can be cut in half except for at most one extra element.◂

By matching inverse images of subsets of M, one can show that every atomless probability space has the following *universality* property.

2.2 Proposition

Let $\mathbf{\Pi} = (\Pi, \mathcal{A}, \nu)$ *be an atomless probability space. Then for every probability space* $\mathbf{\Lambda} = (\Lambda, \mathcal{B}, \lambda)$ *and every random variable* x *from* $\mathbf{\Lambda}$ *into* M *there is a random variable* y *from* $\mathbf{\Pi}$ *into* M *which has the distribution as* x.

The homogeneity property for hyperfinite counting spaces is more difficult and depends on a hyperfinite computation.

Definition

By an *automorphism* of a probability space $\mathbf{\Pi} = (\Pi, \mathcal{A}, \nu)$ we mean a bijection h mapping Π onto Π such that for each set $A \subseteq \Pi$,

$$A \in \mathcal{A} \text{ if and only if } h(A) \in \mathcal{A}$$

and

$$\text{if } A \in \mathcal{A} \text{ then } \nu(h(A)) = \nu(A).$$

We say that an automorphism h of $\mathbf{\Pi}$ *sends* a random variable x to y if

$$y\,(h(\pi)) = x(\pi) \text{ for } \nu\text{-almost all } \pi \in \Pi.$$

***Remark*.** If there is an automorphism that sends x to y, then x and y have the same distribution. On a hyperfinite counting space $L(\mathbf{\Omega})$, every internal bijection of Ω is an automorphism because it preserves *-cardinality.

Definition

A probability space $\Pi = (\Pi, \mathcal{A}, \nu)$ is said to be ***homogeneous*** if for every Polish space M and every pair of random variables x, y from Π into M which have the same distribution, there is an automorphism h of Π which sends x to y.

The real unit interval with Lebesgue measure is an example of a probability space which is atomless but not homogenous (the real functions $x(t) = t$ and $y(t) = 2 \cdot t - [2 \cdot t]$ have the same distribution but are not automorphic).

It follows from von Neumann (1932) that there exist probability spaces which are atomless and homogenous. We shall show that any hyperfinite counting space is homogenous. Our proof will not involve the results from von Neumann (1932). Other results concerning automorphisms of hyperfinite counting spaces are obtained by Ross (1988).

2.3 **Theorem** (Keisler (1984))

Every hyperfinite counting space is homogenous. In fact, two random variables x *and* y *on a hyperfinite counting space have the same distribution if and only if there is an internal automorphism that sends* x *to* y.

Proof. Let Ω be a hyperfinite counting space and let x and y be two random variables from Ω into M which have the same distribution. We shall find an internal automorphism sending x to y. Let S_n, $n \in \mathbb{N}$, be a countable open basis for the Polish space M. Recall from Lindstrøm (this volume), Theorem II.2.5, that for every Loeb measurable set S there is an internal measurable set C such that the symmetric difference $C \Delta S$ has Loeb measure zero; we call such a set C an *internal approximation* of S. By induction on n we can construct two sequences of internal sets C_n, D_n, $n \in \mathbb{N}$, such that

(1) C_n is an internal approximation of $x^{-1}(S_n)$,

(2) D_n is an internal approximation of $y^{-1}(S_n)$,

and

(3) each Boolean combination of $C_1,\ldots,C_n$ has the same internal cardinality as the corresponding Boolean combination of $D_1,\ldots,D_n$.

By (3), for each $n \in \mathbb{N}$ there is an internal bijection h_n of Ω that maps C_m onto D_m for each $m \le n$. By ω_1-saturation there is an internal bijection h of Ω that maps C_n onto D_n for each $n \in \mathbb{N}$. This h is an internal automorphism sending x to y.◂

Theorem 2.3 suggests that two random variables which have the same distribution are alike.

One of the advantages in studying random variables with values in an arbitrary Polish space M rather than with values in $\mathbb{R}$ is that we can prove a theorem about a given Polish space by using a previous result about another Polish space. Any closed subspace of a Polish space is also Polish. The following three constructions of new Polish spaces from a given space M are especially useful. For the proofs see the book of Billingsley (1968).

(a) The space $M^{\mathbb{N}}$ of all countable sequences in M with the product topology has a metric such that $M^{\mathbb{N}}$ is a Polish space.

(b) The space C(M) of all continuous functions from [0,1] into M with the sup metric is a Polish space.

(c) The space D(M) of all right continuous functions with left limits (rcll functions) from [0,1] into M has a metric such that D(M) is a Polish space and the function $f(x,t) = x(t)$ is a Borel function from $D(M) \times [0,1]$ into M.

Our principal aim in the remainder of this paper is to upgrade the universality and homogeneity results of this section to stochastic processes on Loeb spaces with an adapted structure. We shall first deal with the easier case of stochastic processes on Loeb probability spaces.

3. STOCHASTIC PROCESSES

In this section we shall apply the results of Section 2 to stochastic processes.

Let β be the Borel measure on the unit real interval [0,1]. By a *stochastic process* on a probability space $\Pi = (\Pi,\mathcal{A},\nu)$ we shall mean a $\nu \times \beta$ measurable function from $\Pi \times [0,1]$ into a Polish space M. Thus a stochastic process is a random variable which changes with time. We shall say that two stochastic processes x and y have the same *finite dimensional distribution*, in symbols $x \equiv_{fdd} y$, if for each n-tuple $t_1,\ldots,t_n$ of elements of $[0,1]$, the random variables $(x(\cdot,t_1),\ldots,x(\cdot,t_n))$ and $(y(\cdot,t_1),\ldots,y(\cdot,t_n))$ have the same distribution.

We state without proof a lemma from Hoover & Keisler (1984), Proposition 2.24, which is often helpful in extending results concerning random variables to results concerning stochastic processes. The proof is by a purely standard measure-theoretic argument.

3.1 **Lemma**

Let $x(\lambda,t)$ *be a stochastic process on a probability space* Λ *with values in* M. *There is a sequence* t_n, $n \in \mathbb{N}$ *of elements in* $[0,1]$ *and two Borel functions*

$$\psi\colon [0,1] \times M^{\mathbb{N}} \to 2^{\mathbb{N}}, \qquad \varphi\colon 2^{\mathbb{N}} \to M$$

such that for all $t \in [0,1]$,

$$x(\lambda,t) = \varphi(\psi(t,\langle x(\lambda,t_n)\colon n \in \mathbb{N}\rangle)) \text{ almost surely.}$$

We can now extend the results from Section 2 to stochastic processes.

3.2 **Proposition**

Let Π *be an atomless probability space. Then for every stochastic process* x *on a probability space* Λ *there is a stochastic process* y *on* Π *such that* x *and* y *have the same finite dimensional distribution.*

Proof. Let x have values in M. Let t_n, ψ, and φ be as in Lemma 3.1. $\langle x(\lambda,t_n)\colon n \in \mathbb{N}\rangle$ is a random variable with values in $M^{\mathbb{N}}$. Since Π is atomless, there is a random variable z on Π with the same distribution as $\langle x(\lambda,t_n)\colon n \in \mathbb{N}\rangle$. Therefore the stochastic process $y(\pi,t) = \varphi(\psi(t,z(\pi)))$ on Π has the same finite dimensional distribution as x.◂

Remark. In Proposition 3.2, if almost every path of x is continuous, or rcll (*right-continuous with left limits*), then y can be taken with the same property. This follows from the fact that the set of continuous, or rcll, functions from [0,1] into M forms a Polish space.

An automorphism h of a probability space Π is said to *send* a stochastic process x to y if for all times $t \in [0,1]$, $y(h\pi,t) = x(\pi,t)$ for almost all π. The next result, due to S. Fajardo, is an analogue of Proposition 3.2. for homogeneity.

3.3 Proposition

Let Π *be a homogeneous probability space. Then two stochastic processes* x *and* y *on* Π *have the same finite dimensional distribution if and only if there is an automorphism* h *of* Π *which sends* x *to* y.

Proof. We prove the nontrivial direction. Assume that x and y have the same finite dimensional distribution. Let t_n, ψ, and φ be as in Lemma 3.1, so that for each $t \in [0,1]$,

$$x(\pi,t) = \varphi(\psi(t,\langle x(\pi,t_n): n \in \mathbb{N}\rangle)) \text{ almost surely.}$$

Since y has the same finite dimensional distribution as x, and the functions ψ, φ are Borel, we also have

$$y(\pi,t) = \varphi(\psi(t,\langle y(\pi,t_n): n \in \mathbb{N}\rangle)) \text{ almost surely}$$

for all $t \in [0,1]$. Since Π is homogeneous, there is an automorphism h of Π which sends the random variable

$$\langle x(\pi,t_n): n \in \mathbb{N}\rangle$$

to

$$\langle y(\pi,t_n): n \in \mathbb{N}\rangle.$$

It follows that h sends the stochastic process x to y, as required.◄

4. PRODUCTS OF LOEB SPACES

There are two ways to form a product space from two Loeb spaces. We can either form the Loeb spaces first and then take the product, or take the product of the internal spaces first and then form the Loeb space. Anderson (1976) observed that if Ω_1 and Ω_2 are internal probability spaces, then the product of the Loeb spaces $L(\Omega_1) \times L(\Omega_2)$ is a subspace of the Loeb space of the product, $L(\Omega_1 \times \Omega_2)$. That is, every set that is $L(\Omega_1) \times L(\Omega_2)$ measurable is $L(\Omega_1 \times \Omega_2)$ measurable, and the measures are the same. The following example of Hoover shows that the inclusion is strict.

4.1 Example

Let T be a hyperfinite time line with the counting measure τ, and let Ω be the set of all internal subsets of T with the counting measure μ. Then the relation

$$E = \{(t,\omega): t \in \omega\}$$

is internal but is not measurable in the completion of the product $L(T) \times L(\Omega)$.

Proof. Let ν be the counting measure on $T \times \Omega$. Since each point $t \in T$ belongs to exactly half of the elements of Ω, we have

$$\nu(E) = L(\nu)(E) = ½.$$

Let A and B be internal subsets of T and Ω respectively with positive Loeb measure. By the weak law of large numbers, the set of internal $C \subseteq T$ such that $\tau(A \cap C) \approx \tau(A)/2$ is a subset of Ω of $L(\mu)$ measure one. Regarding C as an element of Ω, we have

$$A \cap C = (A \times \{C\}) \cap E.$$

Therefore

$$\nu[(A \times B) \cap E] \approx \nu(A \times B)/2.$$

Since this holds for all internal rectangles $A \times B$, we have

$$L(\nu)[(A \times B) \cap E] = L(\nu)(A \times B)/2$$

for all Loeb measurable sets A and B. It can then be shown by means of the monotone class theorem that for every $L(T) \times L(\Omega)$ measurable set D,

$$L(\nu)[D \cap E] = L(\nu)(D)/2.$$

Putting D = E, we see that E is not measurable in the completion of the product.◄

Here is an analogue of the Fubini theorem for $L(\Omega_1) \times L(\Omega_2)$. Even though $L(\Omega_1 \times \Omega_2)$ may be larger than the completion of the product of the Loeb measures, the double integral of a $L(\Omega_1 \times \Omega_2)$ measurable function is still equal to the iterated integral. The proof will use theorem II.2.11 in Lindstrøm (this volume) which states that a function is Loeb measurable if and only if it has a lifting.

4.2 Fubini Theorem for Loeb Measures (Keisler(1984))

Let Ω_1 *and* Ω_2 *be hyperfinite probability spaces and let* $\Omega = \Omega_1 \times \Omega_2$ *be the internal product space. Let* f *be a real valued Loeb measurable function on* Ω.

(a) *For* $L(\mu_2)$*-almost all* ω_2, $f(\cdot,\omega_2)$ *is* $L(\mu_1)$*-measurable.*

(b) *If* f *is* $L(\mu)$*-integrable, then*:

(i) *For* $L(\mu_2)$*-almost all* ω_2, $f(\cdot,\omega_2)$ *is* $L(\mu_1)$*-integrable;*

(ii) $\int f(\omega_1,\omega_2)dL(\mu_1)$ *is* $L(\mu_2)$*-integrable;*

(iii) $\int\int f(\omega_1,\omega_2)dL(\mu_1)dL(\mu_2) = \int f dL(\mu)$.

Proof. We shall prove (b) for the case that f is bounded. Let S be the set of μ-measurable sets. If A is a subset of $\Omega_1 \times \Omega_2$ and $\omega_2 \in \Omega_2$, let

$$A(\cdot,\omega_2) = \{\omega_1 \in \Omega_1 : (\omega_1,\omega_2) \in A\}.$$

For each A, the following are equivalent:

(1) $L(\mu)(A) = 0$.

(2) For each $n \in \mathbb{N}$ there is a set $A_n \in S$ such that $A \subseteq A_n$ and

$$\mu(A_n) < 1/n.$$

(3) For each $n \in \mathbb{N}$ there is a set $A_n \in S$ such that $A \subseteq A_n$ and

$$\mu_2\{\omega_2\colon \mu_1(A_n(\cdot,\omega_2)) < 1/n\} \geq 1 - 1/n.$$

(4) For $L(\mu_2)$-almost all ω_2, $L(\mu_1)(A(\cdot,\omega_2) = 0$.

Since f is bounded and Loeb measurable, f has a finitely bounded lifting F. Let A be the set of $\omega \in \Omega$ such that ${}^\circ F(\omega) \neq f(\omega)$. Then (1) holds, and thus (4) holds. By (4), $F(\cdot,\omega_2)$ is a finitely bounded lifting of $f(\cdot,\omega_2)$ for $L(\mu_2)$-almost all ω_2. For each such ω_2, $f(\cdot,\omega_2)$ is Loeb integrable and

$$\int f(\omega_1,\omega_2)dL(\mu_1) = {}^\circ\!\int F(\omega_1,\omega_2)d\mu_1.$$

This proves (i). It follows that $\int F(\omega_1,\cdot)d\mu_1$ is a finitely bounded lifting of $\int f(\omega_1,\cdot)dL(\mu_1)$, so (ii) holds. By taking hyperfinite sums we have

$$\int\int F(\omega_1,\omega_2)d\mu_1 d\mu_2 = \int F(\omega_1,\omega_2)d\mu,$$

and (iii) follows by taking standard parts.◂

Let $L(\Omega)$ be a hyperfinite Loeb space and let $L(T)$ a hyperfinite time line. Our next result relates the Loeb measure on $\Omega \times T$ to the product of the Loeb measure on Ω and the Borel measure β on $[0,1]$. As an analogue of the standard part map we may define the function $st: \Omega \times T \to \Omega \times [0,1]$ defined by

$$st(\omega,t) = (\omega,{}^{\circ}t).$$

4.3 **Theorem** (Keisler (1984))

For each set $A \subset \Omega \times [0,1]$, *the following are equivalent*:

(i) A *is* λ*-measurable where* λ *is the completion of the product of the Loeb measure on* Ω *and* β.

(ii) $st^{-1}(A)$ *is* α*-measurable where* α *is the completion of the product of the Loeb measures on* Ω *and on* T.

(iii) $st^{-1}(A)$ *is* ν*-measurable where* ν *is the Loeb measure on the space* $\Omega \times T$.

Moreover, if (i) *holds then* $\lambda(A) = \alpha(st^{-1}(A)) = \nu(st^{-1}(A))$.

Proof. This result is proved in Keisler (1984), p.99. The key step, needed in showing that (iii) implies (i), is to prove that if $C \subseteq \Omega \times T$ is internal then $st(C)$ is λ-measurable. To see this let $I_1, I_2, \ldots$ be a list of all open intervals in $[0,1]$ with rational endpoints, and let

$$A(m) = \{\omega: (\exists t \in {}^*I_m)\ (\omega,t) \in C\}.$$

Each $A(m)$ is internal. Then the set

$$B = \bigcap_m\ [(A(m) \times [0,1]) \cup ((\Omega \setminus A(m)) \times ([0,1] \setminus I_m))]$$

is λ-measurable. A point (ω,s) belongs B if and only if for every open interval I containing s there is a $t \in {}^*I$ with $(\omega,t) \in C$. By saturation, it follows that $st(C) = B$.◂

5. LIFTINGS OF STOCHASTIC PROCESSES

We now introduce a mixed lifting notion for functions with domain $\Omega \times [0,1]$. We shall call an internal function X from $\Omega \times T$ into ${}^*M$ an *internal stochastic process*.

Definition

Let x be a function from $\Omega \times [0,1]$ into M. By a *lifting* of x

we mean an internal stochastic process X such that

$$x(\omega, {}^{o}t) = {}^{o}X(\omega, t)$$

almost surely in the space $L(\Omega \times T)$.

5.1 Proposition

Let $L(\Omega)$ *be a hyperfinite Loeb space and let* β *be the Borel measure on* $[0,1]$. *A function* $x\colon \Omega \times [0,1] \to M$ *has a lifting if and only if* x *is measurable with respect to the completion of* $L(\Omega) \times \beta$.

Proof. Define $y\colon \Omega \times T \to \mathbb{R}$ by $y(\omega,t) = x(\omega,{}^{o}t)$. For each set $U \subset M$, we have $y^{-1}(U) = st^{-1}(x^{-1}(U))$. Thus in the notation of Theorem 4.3, x is λ-measurable iff y is ν-measurable iff y has a lifting iff x has a lifting.◄

In the above result, the fact that every stochastic process has a lifting is due to Anderson (1976), and the converse is in Keisler (1984). We remark that a function x is measurable with respect to the completion of $L(\Omega) \times \beta$ if and only if there is a stochastic process $\hat{x}$ such that $x(\omega,t) = \hat{x}(\omega,t)$ everywhere on a set of measure one in $L(\Omega) \times \beta$.

There are analogous lifting theorems for various classes of stochastic processes. We shall need one such theorem here, for processes which are right continuous in probability.

Definition

An internal stochastic process is said to be a *right lifting* of x if for each real $t \in [0,1]$ there is an element $\mathbf{t} \approx t$ such that whenever $\mathbf{s} \approx t$ and $\mathbf{s} \geq \mathbf{t}$, $X(\cdot,\mathbf{s})$ is a lifting of $x(\cdot,t)$. We shall call $\mathbf{t}$ a *right lifting point* for $x(\cdot,t)$. Notice that a right lifting places no restriction on the value of x at time 1.

5.2 Lemma

Any right lifting of stochastic process x *is a lifting.*

Proof. Let X be a right lifting of x. Let U be the set of (ω,t) such that $X(\omega,t) \approx x(\omega,{}^{o}t)$ and let V be the set of t such that $X(\cdot,t)$ lifts $x(\cdot,{}^{o}t)$. The U is Loeb measurable. We must show that U has Loeb measure one. By the Fubini Theorem 4.2, it suffices to prove that V has Loeb measure one.

By the Fubini Theorem, V is Loeb measurable. Let A be an internal set containing V. By overspill, for each $t \in [0,1)$ there is a real $r(t) \in (t,1]$ such that $st^{-1}(t,r(t)) \subseteq A$. It follows that for all but countably many $t \in [0,1]$, $st^{-1}(t) \subseteq A$. Therefore A has Loeb measure one, and hence V has Loeb measure one.◂

Definition

A stochastic process x is said to be *right continuous in probability* if for every $t \in [0,1)$ and every real $\varepsilon > 0$, the limit as $u \downarrow t$ of the probability that $x(u,\cdot)$ is within ε of $x(t,\cdot)$ is one.

5.3 **Proposition**

A stochastic process is right continuous in probability if and only if it has a right lifting.

Proof. First suppose X is a right lifting of a stochastic process x. Let $t \in [0,1)$ and let t be a right lifting point for $x(\cdot,t)$. By overspill, for each real $\varepsilon > 0$ there is a real $u > t$ such that whenever $t \le s < u$, the probability that $X(\omega,s)$ is within ε of $X(\omega,t)$ is at least $1 - \varepsilon$. Thus whenever $t < s < u$, the probability that $x(\omega,s)$ is within 2ε of $x(\omega,t)$ is at least $1 - 2\varepsilon$. Therefore x is right continuous in probability.

Now assume x is right continuous in probability. For each natural number n, choose an internal stochastic process Y_n such that for each multiple t of 1/n, $Y_n(\cdot,t)$ is a lifting of $x(\cdot,t)$, and each path $Y_n(\omega,\cdot)$ is a step function which steps at multiples of 1/n. Extend this to an internal sequence $\{Y_K: K \in {}^*\mathbb{N}\}$. By saturation there is an infinite K such that for each rational interval (q,r) and finite n, if the probability that $x(\omega,\cdot)$ varies by 1/n over (q,r) is less than 1/n, then the probability that $Y_K(\omega,\cdot)$ varies by 2/n over (q,r) is less than 2/n. It follows from right continuity in probability of x that Y_K is a right lifting of x.◂

Hoover & Perkins (1983) introduced a stronger "two-sided" lifting called an SDJ lifting, and proved that a stochastic process has an

SDJ lifting if and only if almost every path is rcll. SDJ liftings have the advantage that they correspond to the standard part mapping in the Polish space D(M) of rcll functions. On the other hand, the right lifting notion is simpler and is the notion which is needed for the Adapted Lifting Theorem in Section 7.

All the notions and results in this section have analogues for n-*fold stochastic processes*, that is, $\nu \times \beta^n$-measurable functions on $\Pi \times [0,1]^n$.

A *right lifting* of an n-fold stochastic process x is an internal n-fold stochastic process X such that for all $t_1,\dots,t_n$ in [0,1) there exist $\mathbf{t}_1 \approx t_1,\dots,\mathbf{t}_n \approx t_n$ such that whenever $\mathbf{s}_1,\dots,\mathbf{s}_n$ are infinitely close to but greater than $\mathbf{t}_1,\dots,\mathbf{t}_n$, the random variable $X(\cdot,\mathbf{s}_1,\dots,\mathbf{s}_n)$ is a lifting of $x(\cdot,t_1,\dots,t_n)$.

Right continuity in probability for an n-fold process is defined exactly as before, with the convergence $\vec{u}\downarrow\vec{t}$ meaning that for each coordinate j, $u_j \downarrow t_j$.

In several important classical cases, a stochastic process with a certain finite dimensional distribution can be constructed in a simple canonical way on a hyperfinite counting space $L(\Omega)$. Since $L(\Omega)$ is atomless and homogeneous, it follows that every stochastic process on $L(\Omega)$ with that finite dimensional distribution is automorphic to the simple one. This often allows one to prove a result about an arbitrary process with a certain finite dimensional distribution by making a hyperfinite computation involving the simple process.

Here are three examples. The last two examples were discussed in the article Lindstrøm (this volume). For each of these examples, we let T be a hyperfinite time line, and consider a hyperfinite counting space on a set of the form Ω^T where Ω is internal.

5.4 Example

A *Poisson process* on a probability space Π is a stochastic process $x(\pi,t)$ such that $x(\pi,0) = 0$, each path $x(\pi,\cdot)$ is a right continuous step function with values in $\mathbb{N}$, $x(\pi,t)-x(\pi,s) = n$ with probability $(t-s)^n/(n!e^{t-s})$ whenever $s < t$, and changes in x over disjoint intervals in [0,1] are independent. It is known that all Poisson processes have the same finite dimensional distribution. It follows that on a homogeneous probability space, there is an automorphism between any two Poisson processes. *Loeb's coin tossing process* from Loeb (1975) is a particular Poisson process defined as follows. Let H be a infinite hyperinteger and let T be the hyperfinite time line of size H. Form the hyperfinite counting space $L(\Omega)$ where $\Omega = T^T$. Let $X(\omega,t)$ be equal to the number of $s < t$ in T such that $\omega(s) = 0$. This corresponds to a coin which comes up 0 (heads) with probability 1/H and nonzero (tails) with probability 1 - 1/H. Then X is a right lifting of a Poisson process x on $L(\Omega)$. It follows that for every Poisson process y on $L(\Omega)$ there is an internal automorphism from y to the Loeb coin tossing process x.

5.5 **Example**

A (standard one dimensional) *Brownian motion* on a probability space Π is continuous stochastic process x with values in $\mathbb{R}$ such that $x(\pi,0) = 0$, $x(\cdot,t) - x(\cdot,s)$ has the normal distribution with mean zero and variance $t - s$ whenever $s < t$, and the changes in x over disjoint intervals in [0,1] are independent. All Brownian motions have the same finite dimensional distribution, so on a homogeneous probability there is an automorphism between any two Brownian motions. *Anderson's Brownian motion* from Anderson (1976) is a particular Brownian motion on a hyperfinite probability space of the form Ω^T, where Ω has at least 2 elements, constructed as follows. For simplicity let $\Omega = \{-1,1\}$. The *Anderson random walk* $B(\omega,t)$ is defined by

$$B(\omega,t) = \Sigma_{s<t}\ \omega(s)/\sqrt{H}.$$

Anderson showed that B is a uniform lifting of a Brownian motion b on $L(\Omega^T)$. We shall call b *Anderson's Brownian motion*. It follows that for every Brownian motion W on $L(\Omega^T)$, there is an internal automorphism from W to the Anderson Brownian motion b. A similar result holds for n dimensional Brownian motions (with values in $\mathbb{R}^n$).

5.6 **Example**

Two classical generalizations of Brownian motion to n-fold stochastic processes (with n time variables) are the Brownian sheet and Lévy Brownian motion. Each of these notions determines a unique finite dimensional distribution. Stoll (1982, 1986) has given simple hyperfinite constructions of a Brownian sheet and a Lévy Brownian motion in the style of Anderson's Brownian motion construction, on a hyperfinite probability space of the form $L(\Omega^U)$ where $U = T^n$ (see also Manevitz & Merzbach (1986)). It follows that every Brownian sheet on $L(\Omega^U)$ is automorphic to the one constructed by Stoll by an internal bijection, and similarly for Lévy Brownian motion.

6. ADAPTED PROBABILITY SPACES

Many important properties of stochastic processes depend not only on the probability space but on a richer structure called an *adapted (probability) space*. Intuitively, an adapted space is a probability space in which one can keep track of the time at which an event occurs. Formally, an *adapted space* is a structure

$$\Pi = (\Pi, \mathcal{B}_t, \nu)_{t\in[0,1]}$$

where $(\Pi, \mathcal{B}_1, \nu)$ is a probability space and the $\mathcal{B}_t$ are σ-algebras such that

$$\mathcal{B}_t = \cap\{\mathcal{B}_s : s > t\} \text{ for each } t \in [0,1).$$

Thus the $\mathcal{B}_t$ are increasing and right continuous in t. Intuitively, $\mathcal{B}_t$ is the set of events which occur at or before time t.

Given an integrable real valued random variable y on Π, the *conditional expectation* of y with respect to $\mathcal{B}_t$ is denoted by $E[y(\cdot) \mid \mathcal{B}_t](\pi)$. A stochastic process x is said to be *adapted* if for each $t \in [0,1]$, $x(\cdot,t)$ is $\mathcal{B}_t$-measurable. By definition, $E[y(\cdot) \mid \mathcal{B}_t](\pi)$ is the almost surely unique adapted stochastic process $x(\pi,t)$ on Π such that for each t and $A \in \mathcal{B}_t$, the integrals of $x(\cdot,t)$ and $y(\cdot)$ over A are the same.

Many of the central notions about stochastic processes, such as martingales and Markov processes, can be defined using equations involving the conditional expectation. A stochastic process x is adapted iff for each t and bounded continuous function $\Phi: M \to \mathbb{R}$,

$$E[\Phi(x(\cdot,t)) \mid \mathcal{B}_s](\pi) = \Phi(x(\pi,t)) \text{ almost surely.}$$

A *martingale* on Π is a real valued adapted process x such that for each t, $x(\cdot,t)$ is integrable and for all $s \le t$,

$$E[x(\cdot,t) \mid \mathcal{B}_s](\pi) = x(\pi,s) \text{ almost surely.}$$

A stochastic process x is said to be a *Markov process* if x is adapted and for each bounded continuous function Φ: $M \to \mathbb{R}$ there is a Borel function

$$\Psi: M \times [0,1] \times [0,1] \to \mathbb{R}$$

called the *transition function* such that for all $s < t$,

$$E[\Phi(x(\cdot,t)) \mid \mathcal{B}_s](\pi) = \Psi(x(\pi,s),s,t).$$

We shall need the following fact about conditional expectation (see, for example, the book Doob (1953)).

6.1 **Proposition**

For any integrable real valued random variable $y(\pi)$ *on an adapted space* Π *there is a process* $x(\pi,t)$ *on* Π *which is right continuous in probability such that for each* $t \in [0,1]$,

$$x(\pi,t) = E[y(\cdot) \mid \mathcal{B}_t](\pi) \text{ almost surely.}$$

In fact, x is a martingale and almost every path of x is rcll.

Hereafter we shall let $E[y(\cdot)|\mathcal{B}_t](\pi)$ denote the process $x(\pi,t)$ in the above proposition. For each time t, $x(\cdot,t)$ is unique up to a set of measure sero.

Definition

Let Ω be a *-finite set with at least two elements and let T be a hyperfinite time line. By the *hyperfinite adapated space* on Ω^T we mean the adapted space

$$L(\Omega^T) = (\Omega^T, \mathcal{A}_t, L(\mu))_{t\in[0,1]}$$

where $(\Omega^T, \mathcal{A}_1, L(\mu))$ is the hyperfinite counting space and for each t, $\mathcal{A}_t$ is the σ-algebra of all Loeb measurable sets A such that if $\omega \in A$ and $\omega'(s) = \omega(s)$ for all $s \in T$ with ${}^\circ s \le t$, then $\omega' \in A$.

It is useful to have an internal counterpart of the conditional expectation for a hyperfinite adapted space.

For $\omega \in \Omega^T$ and $t \in T$, let $(\omega|t)$ be the set of all $\omega' \in \Omega^T$ such that $\omega'(s) = \omega(s)$ for all $s \le t$.

Definition

Let Y: $\Omega^T \to {}^*\mathbb{R}$ be an internal random variable and let $t \in T$. The *internal conditional expectation* of Y at time t is defined as the average value of Y over the set $(\omega|t)$,

$$E[Y(\cdot) \mid (\omega|t)] = \Sigma_{\omega'\in(\omega|t)} Y(\omega')/\#(\omega|t).$$

Here is a lifting theorem for conditional expectations on hyperfinite adapted spaces, from Keisler (1984), p.115.

6.2 Theorem

Let y *be a bounded random variable on* Ω^T *with values in* $\mathbb{R}$, *let* Y *be a finitely bounded lifting of* y, *and let* x *be the stochastic process* $x(\omega,t) = E[y(\cdot) \mid \mathcal{A}_t](\omega)$. *Then* $E[Y(\cdot) \mid (\omega|t)]$ *is a right lifting of* x.

Proof. For each $t \in T$ let $\mathcal{C}_t$ be the σ-algebra of subsets of Ω^T generated by the sets $(\omega|t)$. Then $\mathcal{C}_t \subseteq \mathcal{A}_s$ where $s = {}^\circ t$. Moreover, $\mathcal{C}_t$ increases with t and $\mathcal{A}_s$ is the intersection of $\mathcal{C}_u$ where $s < {}^\circ u$. We show first that $E[Y(\cdot)|(\omega|t)]$ is a lifting of $E[y(\cdot)|\mathcal{C}_t](\omega)$. Let $S \in \mathcal{C}_t$. Applying the internal approximation property (Theorem II.2.5 part (iii) in Lindstrøm (this volume)), there is an internal set $A \in \mathcal{C}_t$ such that the symmetric difference $A \vartriangle S$ has Loeb measure zero. Then

$$\begin{aligned}\int_S {}^\circ E[Y(\cdot)|(\omega|t)]dL(\mu) &= {}^\circ\Sigma_{\omega\in A}\, E[Y(\cdot)\ |(\omega|t)]\mu\{\omega\} \\ &= {}^\circ\Sigma_{\omega\in A}\, Y(\omega)\mu\{\omega\} \quad = \quad \int_A {}^\circ Y(\omega)dL(\mu) \\ &= \int_S E[y(\cdot) \mid \mathcal{C}_t](\omega)dL(\mu).\end{aligned}$$

Therefore

$${}^\circ E[Y(\cdot) \mid (\omega|t)] = E[y(\cdot) \mid \mathcal{C}_t](\omega)$$

almost surely, so $E[Y(\cdot)|(\omega|t)]$ lifts $E[y(\cdot)|\mathcal{C}_t]$.

We next show that for any set $U \in \mathcal{A}_s$ there is a $t \approx s$ and a set $V \in \mathcal{C}_t$ such that $U \vartriangle V$ has Loeb measure zero. Let $\mathbf{s} \approx s$. For each $n \in \mathbb{N}$, we have $U \in \mathcal{C}_{\mathbf{s}+1/n}$ and hence there is an internal set $V_n \in \mathcal{C}_{\mathbf{s}+1/n}$ such that $U \vartriangle V_n$ has Loeb measure zero. Then each $V_m \vartriangle V_n$ has internal measure less than $1/(m+n)$. By saturation there is an infinite K such that, putting $t = \mathbf{s}+1/K$, $V_K \in \mathcal{C}_t$ and $U \vartriangle V_K$ has Loeb measure zero.

The function $z(\omega) = E[y(\cdot)|\mathcal{A}_s](\omega)$ is $\mathcal{A}_s$-measurable. Thus for each basic open subset $S \subseteq M$ there is a $t \approx s$ such that $z^{-1}(S)$ belongs to the completion of $\mathcal{C}_t$. By saturation there is a $u \approx s$ such that z is measurable with respect to the completion of $\mathcal{C}_u$. Then whenever $u \le t \approx s$,

$$z(\omega) = E[y(\cdot) \mid \mathcal{A}_s](\omega) = E[y(\cdot) \mid \mathcal{C}_t](\omega)$$

almost surely, and the result follows.◂

7. ADAPTED DISTRIBUTIONS

In this section we shall prepare the groundwork for showing that a hyperfinite adapted space is universal and homogeneous in an appropriate sense for adapted spaces. In order to formalize this idea we need adapted analogues of the notions of an automorphism from x to y and of the finite dimensional distribution. The analogue of a finite dimensional distribution will be called the *adapted distribution*. At the end of this section we shall prove a lifting theorem which shows that if X is a lifting of x, then the internal adapted distribution of X is a lifting of the adapted distribution of x.

Definition

Let $\mathbf{\Pi} = (\Pi,\mathcal{B}_t,\nu)_{t\in[0,1]}$ be an adapted space. By an *adapted automorphism* of $\mathbf{\Pi}$ we mean an automorphism h of the probability space $(\Pi,\mathcal{B}_1,\nu)$ such that for each t, h maps $\mathcal{B}_t$ onto $\mathcal{B}_t$.

The analogue of the finite dimensional distribution of a stochastic process for an adapted space is defined using expressions formed by repeated use of conditional expectations, and is called the *adapted distribution*.

Definition

The set of *conditional expressions* in a Polish space M is defined inductively as follows, where $u_1,\dots,u_n,v$ are "time" variables which range over [0,1] and need not be distinct.

(a) If Φ is a bounded continuous function from M^n into $\mathbb{R}$, then $\Phi(u_1,\dots,u_n)$ is a conditional expression.

(b) Each time variable v is conditional expression.

(c) If f is a conditional expression, then $E[f|v]$ is a conditional expression.

(d) If $f_1,\dots,f_m$ are conditional expressions and φ is a bounded continuous function from $\mathbb{R}^m$ into $\mathbb{R}$, then $\varphi(f_1,\dots,f_m)$ is a conditional expression.

Properties of conditional expressions are usually proved by induction on the complexity of the expression. To show that every conditional expression has a property P, we first show that all conditional expressions of the types (a) and (b) have the property P, and then show that the property P is preserved under the closure rules (c) and (d). Similarly, functions of conditional expressions are usually defined by induction on complexity. As a first illustration, given a conditional expression f and a stochastic process x, we now define the new stochastic process fx. fx is the interpretation of the conditional expression f applied to the process x.

Definition

For each adapted space Π, stochastic process $x(\pi,t)$ from Π into M, and conditional expression f, fx is an n-fold real valued stochastic process defined by the following rules:

(a) If f is $\Phi(u_1,\dots,u_n)$ then

$$fx(\pi,s_1,\dots,s_n) = \Phi(x(\pi,s_1),\dots,x(\pi,s_n)).$$

(b) If f is a time variable v, then $fx(\pi,t) = t$.

(c) If f is $E[g(u_1,\dots,u_n) \mid v]$ then

$$fx(\pi,s_1,\dots,s_n,t) = E[gx(\cdot,s_1,\dots,s_n) \mid \mathcal{B}_t](\pi).$$

(d) If f is $(\varphi(f_1,\dots,f_m))$ then

$$fx = \varphi(f_1x,\dots,f_mx).$$

In keeping with our convention from the preceding section concerning conditional expectations, rule (c) requires that when v is not among $u_1,\dots,u_n$,

$$fx(\pi,s_1,\dots,s_n,t) = E[gx(\cdot,s_1,\dots,s_n) \mid \mathcal{B}_t](\pi)$$

is right continuous in probability in (π,t) for each fixed $s_1,\dots,s_n$. The proof that there always exists an n-fold stochastic process fx which satisfies rules (a)-(d) in the definition is harder than one might guess, and is given in Keisler (1985), Lemma 4.3.8.

If y is a random variable on Π, then we define fy to be fx where x is the *constant stochastic process* $x(\pi,t) = y(\pi)$.

As the preceding defintion shows, a conditional expression f gives a uniform way to make a new stochastic process fx out of an old process x. The same conditional expression may be applied to different processes on different adapted spaces. For this reason, conditional expressions can be used to find common features shared by two stochastic processes. These common features are captured by the adapted distribution of a process, which we define now.

Definition

Two stochastic processes x and y on adapted spaces with values in M *have the same adapted distribution*, in symbols $x \equiv y$, if $fx \equiv_{fdd} fy$ for each conditional expression f.

As a second example of an inductive definition on complexity of conditional expressions, for each internal stochastic process X and conditional expression f, we define a new internal stochastic process fX.

Definition

Let $L(\Omega^T)$ be a hyperfinite adapted space and let X be an internal stochastic process on Ω^T. Then for each conditional expression f with n time variables, we define an internal n-fold stochastic process fX on Ω^T by rules which are the internal counterparts of the rules for standard processes:

(a') If f is $\Phi(u_1,\dots,u_n)$ then

$$fX(\omega,s_1,\dots,s_n) = {}^*\Phi(X(\omega,s_1),\dots,X(\omega,s_n)).$$

(b') If f is a time variable v, then $fX(\omega,t) = t$.

(c') If f is $E[g(u_1,\dots,u_n) \mid v]$ then

$$fX(\omega,s_1,\dots,s_n,t) = E[gX(\cdot,s_1,\dots,s_n) \mid (\omega|t)].$$

(d') If f is $(\varphi(f_1,\dots,f_m))$ then

$$fX = {}^*\varphi(f_1X,\dots,f_mX).$$

Again if Y is an internal random variable on Ω, fY is defined to be fX where $X(\omega,t) = Y(\omega)$.

To introduce our next theorem, we notice that if X is a lifting of a random variable x on a hyperfinite Loeb space, then $\Phi(X(\cdot))$ is a lifting of $\Phi(x(\cdot))$ for each bounded continuous function $\Phi: M \to \mathbb{R}$, because we always have $\Phi({}^{\circ}m) = {}^{\circ}\Phi(m)$.

We now prove a much harder analogous result for hyperfinite adapted spaces. It shows that properties expressible in terms of conditional processes are preserved by liftings. The result is implicit in Keisler (1986), p. 81. We take the opportunity to state and prove it explicitly here.

7.1 **Adapted Lifting Theorem**

Let Ω^T *be a hyperfinite adapted space,* x *a stochastic process mapping* $\Omega^T \times [0,1]$ *into* M, *and* X *a lifting of* x. *Then for every conditional expression* f *with values in* M, fX *is a lifting of* fx. *Moreover, if* X *is a right lifting of* x *then each* fX *is a right lifting of* fx.

Proof. The proof is by induction on the complexity of f. Every step of the induction is routine except for the conditional expectation step, which we now give. To simplify notation let f(u,v) be a conditional expression with at most the variables u,v, and let $g(u,v) = E[f(u,v)|v]$. Suppose that fX lifts fx. We shall show that gX is a lifting of gx. For each $\mathbf{s}$ let $U_{\mathbf{s}}$ be the set of all $\mathbf{t}$ such that $fX(\cdot,\mathbf{s},\mathbf{t})$ lifts $fx(\cdot,{}^{\circ}\mathbf{s},{}^{\circ}\mathbf{t})$. Let U be the set of $\mathbf{s}$ such that $U_{\mathbf{s}}$ has Loeb measure one. It follows from the Fubini Theorem 4.2 that U has Loeb measure one. Let $s = {}^{\circ}\mathbf{s}$, $t = {}^{\circ}\mathbf{t}$. By Theorem 6.2, whenever $\mathbf{t} \in U_{\mathbf{s}}$, $E[fX(\cdot,s,t)|(\omega|\mathbf{r})]$ is a right lifting of $E[fx(\cdot,s,t)|\mathcal{A}_r]$ as a function of ω and $\mathbf{r}$. We may choose a right lifting point $\mathbf{r}(\mathbf{s},\mathbf{t})$ at $E[fx(\cdot,s,t,)|\mathcal{A}_t]$ so that for $\mathbf{t} \in U_{\mathbf{s}}$, $\mathbf{r}(\mathbf{s},\mathbf{t})$ depends only on s,t, and not on $\mathbf{s},\mathbf{t}$. Let $V_{\mathbf{s}}$ be the set of all $\mathbf{t} \in U_{\mathbf{s}}$ such that $gX(\cdot,\mathbf{s},\mathbf{t}) = E[fX(\cdot,\mathbf{s},\mathbf{t})|(\omega|\mathbf{t})]$ lifts $gx(\cdot,s,t) = E[fx(\cdot,s,t)|\mathcal{A}_t]$. Then $V_{\mathbf{s}}$ contains all $\mathbf{t} \in U_{\mathbf{s}}$ such that $\mathbf{r}(\mathbf{s},\mathbf{t}) \le \mathbf{t}$. Moreover, $V_{\mathbf{s}}$ is Loeb measurable.

We show that whenever $s \in U$, V_s has Loeb measure one. To see this let B be any internal subset of U_s and C be any internal superset of V_s. By overspill, for each $p \in B$ there is a real $q > {}^{o}p$ such that whenever $r(s,p) \le r < q$, if $r \in B$ then $r \in C$. Therefore for any $p \in st(B)$ there is a real $q > p$ such that $st^{-1}(p,q) \subseteq (T \setminus B) \cup C$. Using the fact that st(B) is closed, it follows that for all but countably many $p \in [0,1]$, $st^{-1}\{p\} \subseteq (T \setminus B) \cup C$. Therefore $(T \setminus B) \cup C$ has Loeb measure one. Since U_s has Loeb measure one, B can be taken with Loeb measure arbitrarily close to one. Therefore C has Loeb measure one. Since V_s is Loeb measurable, it follows that V_s has Loeb measure one.

By the Fubini Theorem, the set of pairs (s,t) such that $t \in V_s$ has Loeb measure one, whence gX is a lifting of gx. This completes the proof in the case that X is an ordinary lifting of x.

In the case that X is a right lifting of x, the argument is similar but somewhat easier. At the inductive step $g(u,v) = E[f(u,v)|v]$, we assume that fX is a right lifting of fx and must prove that gX is a right lifting of gx. Let $s, t \in [0,1]$ and let (s,t) be a right lifting point for $fx(\cdot,s,t)$. By 6.2, $E[fX(\cdot,s,t)|(\omega|r)]$ is a right lifting of $E[fx(\cdot,s,t)|\mathcal{A}_r]$ in the variables ω and r. Let r be a right lifting point for $E[fx(\cdot,s,t)|\mathcal{A}_r]$ at $r = t$ such that $r > t$. Then for any $s' \ge s$ and $t' \ge r$ in the monad of (s,t), for Loeb almost all ω we have

$$fX(\omega,s,r) \approx fx(\omega,s,t) \approx fX(\omega,s',t'),$$

so

$$gx(\omega,s,t) = E[fx(\cdot,s,t)|\mathcal{A}_t](\omega) \approx E[fX(\cdot,s,r)|(\omega|t')]$$
$$\approx E[fX(\cdot,s',t')|(\omega|t')] = gX(\omega,s',t').$$

This shows that gX is right lifting of gx.◂

The Adapted Lifting Theorem has various special lifting theorems as consequences. As illustrations we prove lifting theorems for adapted processes and martingales. These results are closely related to lifting theorems for continuous adapted processes in Keisler (1984) and for rcll adapted processes and martingales in Hoover & Perkins (1983).

Let us call an internal stochastic process $Y: \Omega \times T \to {}^{*}M$

nonanticipating if for each $t \in T$ and $\omega, \omega' \in \Omega$, if $(\omega|t) = (\omega'|t)$ then $Y(\omega,t) = Y(\omega',t)$.

7.2 Proposition

Let x *be a stochastic process on a hyperfinite adapted space with values in* M.

(i) *There is an adapted process* y *which equals* x *almost surely in* $\Omega^T \times [0,1]$ *if and only if* x *has a nonanticipating lifting.*

(ii) x *is adapted and right continuous in probability if and only if it has a nonanticipating right lifting.*

Proof. We first prove (i). Let Φ_k, $k \in \mathbb{N}$ be a countable set of bounded continuous functions from M into $\mathbb{R}$ such that if $\Phi_k(m) = \Phi_k(m')$ for all k then $m = m'$. By countable comprehension, extend this sequence to an internal sequence Φ_k, $k \in {}^*\mathbb{N}$. Let X be a lifting of x. For each k let

$$Y_k(\omega,t) = E[\Phi_k(X(\cdot,t)) \mid (\omega|t)].$$

By the Adapted Lifting Theorem, for each $k \in \mathbb{N}$, $\Phi_k(X(\omega,t))$ is a lifting of $\Phi_k(x(\omega,t))$, and $Y_k(\omega,t)$ is a lifting of $E[\Phi_k(x(\cdot,t))|\mathcal{A}_t](\omega)$.

Suppose first that the lifting X is nonanticipating. Then

$$\Phi_k(X(\omega,t)) = Y_k(\omega,t).$$

By the Fubini theorem,

$$\Phi_k(x(\omega,t)) = E[\Phi_k(x(\cdot,t)) \mid \mathcal{A}_t]$$

for all $k \in \mathbb{N}$ and all t in a set U of Lebesgue measure one. Define $y(\omega,t)$ to be $x(\omega,t)$ for $t \in U$ and 0 otherwise. Then y is adapted and x equals y almost surely in the product.

Now let x be an adapted process. Then $Y_k(\omega,t)$ as well as $\Phi_k(X(\omega,t))$ is a lifting of $\Phi_k(x(\omega,t))$ for each $k \in \mathbb{N}$. Let $h(k,\omega,t)$ be the first element ω' of $(\omega|t)$, if there is one, such that

$$(\forall n \le k)\ |Y_n(\omega,t) - \Phi_n(X(\omega',t))| < 1/k. \tag{1}$$

Then h is internal and $h(k,\omega,t)$ depends only on k and $(\omega|t)$. Let Z_k be the nonanticipating internal process

$$Z_k(\omega,t) = X(h(k,\omega,t),t).$$

For each $k \in \mathbb{N}$, and Loeb almost all (ω,t), there is an $\omega' \in (\omega|t)$ which satisfies (1). Therefore each $k \in \mathbb{N}$,

$$(\forall n \le k)\ |Y_n(\omega,t) - \Phi_n(Z_k(\omega,t))| < 1/k$$

with internal probability at least $1 - 1/k$ in the product measure. By overspill there is an infinite K such that with Loeb probability one, for all finite n,

$$Y_n(\omega,t) \approx \Phi_n(S_k(\omega,t)) = \Phi_n(x(\omega,{}^\circ t)).$$

Then Z_k is a lifting of x, as required.

The proof of (ii) is similar, but starting and ending with a right lifting of x.◂

7.3 **Proposition**

A real-valued stochastic process x on a hyperfinite adapted space is a martingale if and only if it has a right lifting X such that:

(i) *each $X(\cdot,t)$ is S-integrable*

and

(ii) *for all $s \ge t$ in T and all $\omega \in \Omega^T$,*

$$E[X(\cdot,s)|(\omega|t)] = X(\omega,s).$$

Proof. First suppose x is a martingale. Since the random variable $x(\cdot,1)$ is integrable, it has an S-integrable lifting Y. Define $X(\omega,t)$ by

$$X(\omega,t) = E[Y(\cdot)\ |\ (\omega|t)].$$

Then (i) follows easily and (ii) holds by the associative law for addition. Since x is a martingale, we have

$$x(\omega,t) = E[x(\cdot,1)\ |\ \mathcal{A}_t](\omega) \text{ almost surely}$$

for all t. By 6.2, X is a right lifting of $E[x(\cdot,1)|\mathcal{A}_t]$ and hence of x.

Now suppose X is a right lifting of x which satisfies (i) and (ii). Then for each t, $x(\cdot,t)$ is Loeb integrable. By the Adapted Lifting Theorem, $E[X(\cdot,s)|(\omega|t)]$ is a right lifting of $E[x(\cdot,s)|\mathcal{A}_t](\omega)$. It follows that x is a martingale.◂

8. UNIVERSAL AND HOMOGENEOUS ADAPTED SPACES

We now begin the study of universal and homogeneous adapted spaces. We shall need two standard lemmas about conditional processes, which are proved in Hoover & Keisler (1984). The first lemma is a Stone Weierstrass theorem for conditional expressions.

Definition

Let us call a set F of conditional expressions *dense* for the Polish space M if for any two stochastic processes x and y with values in M, if fx = fy for every f ∈ F then x ≡ y. We call a set C of bounded continuous real valued functions *dense* for M if the set F of all conditional expressions built from functions in C is dense.

8.1 **Lemma**

Let M *be a Polish space. There is a countable set* C *of bounded continuous real valued functions which is dense for* M.

8.2 **Lemma**

Let x *be a stochastic process which is right continuous in probability. Then for every conditional expression* f, fx *is right continuous in probability.*

Proof. This lemma is proved in Hoover & Keisler (1984), p.165. The proof is by induction on the complexity of conditional expressions, using Proposition 6.1 at the conditional expectation step.◂

We are now ready for the main universality and homogeneity results for hyperfinite adapted spaces. The following universality theorem was proved in Hoover & Keisler (1984). We give a different proof here.

8.3 **Theorem**

Every hyperfinite adapted space $L(\Omega)^T$ *is universal in the sense that for every stochastic process* x *on an adapted space* Λ *there is a stochastic process* y *on* $L(\Omega^T)$ *such that* x ≡ y. *Moreover, if* x *is a random variable then* y *may be taken to be a random variable.*

Proof. We first give the proof for the case that x is a random variable with values in M. Let C be a countable set of functions which is dense for M and let

$$T_n = \{1/n, \ldots, 1-1/n, 1\}.$$

Let F_n be the set of all expressions of the form $f(t_1, \ldots, t_m)$ where

$f(v_1,\ldots,v_m)$ is a conditional expression built from functions in C and $t_1,\ldots,t_m$ are elements of T_n. For each $t \in T_n$, let $F_n|t$ be the set of all $f \in F_n$ of the form $E[g|s]$ where $s \le t$. By induction on $t \in T_n$, we construct a family of internal random variables Y_f, $f \in F_n$, with the following properties for each $t \in T_n$.

(1) If $f \in F_n|t$ then $Y_f(\omega)$ depends only on $(\omega|t)$.

(2) The families of random variables

$$({}^oY_f\colon f \in F_n|t) \text{ and } (fx\colon f \in F_n|t)$$

have the same joint distributions, that is, the same distributions as random variables in $M^{\mathbb{N}}$.

The induction step from s to $t = s + 1/n$ is as follows. We already have chosen Y_f for $f \in F_n|s$. For each $m \in \mathbb{N}$, partition $\mathbb{R}$ into half-open intervals of length $1/m$. For each $\omega \in \Omega^T$, $m \in \mathbb{N}$ and finite subset G of $F_n|s$, let $J(G,m,\omega)$ be the set of all $\lambda \in \Lambda$ such that for each $g \in G$, $gx(\lambda)$ is in the same $1/m$-interval as $Y_g(\omega)$. For each G and m and almost all ω, $J(G,m,\omega)$ has positive measure in Λ. The sets $J(G,m,\omega)$ depend only on G, m, and $(\omega|s)$, and form a partition of Λ. Fix G and m. When $J(G,m,\omega)$ has positive measure, we may choose internal random variables U_g, $g \in G$, on the hyperfinite counting space $(\omega|s)$ such that the family $({}^oU_g\colon g \in G)$ on $(\omega|s)$ and the family $(gx\colon g \in G)$ on $J(G,m,\omega)$ have the same joint distribution, and in addition, $U_g(\omega')$ depends only on $(\omega'|t)$. There are only countably many distinct sets $J(G,m,\omega)$ of positive measure. By countable comprehension we may piece the U_g together to obtain a family of random variables U_g, $g \in G$, over Ω such that the families

$$({}^oU_g\colon g \in G) \quad \text{and} \quad (gx\colon g \in G)$$

have the same distributions over Ω and Λ, $U_g(\omega)$ depends only on $(\omega|t)$, and for almost all ω and all $g \in G \cap F_n|s$, $U_g(\omega)$ is within $1/m$ of $Y_g(\omega)$.

By saturation there is a family of internal random variables Y_f, $f \in F_n|t$, on Ω which agrees with the previous Y_f for $f \in F_n|s$ and satisfies (1) and (2). This completes the induction. We now have a family of internal random variables Y_f, $f \in F_n$ such that (1) holds for all $t \in T_n$ and the families

$$({}^{o}Y_f\colon f \in F_n) \text{ and } (fx\colon f \in F_n)$$

have the same joint distribution.

By saturation that there is an internal random variable Z_n with values in *M such that for each $f \in F_n$, $fZ_n(\omega) \approx Y_f(\omega)$ almost surely. Then for each $f \in F_n$, $E[fZ_n(\cdot)] \approx E[fx(\cdot)]$. By saturation again, there is an infinite K and an internal random variable $Y = Z_K$ such that for each finite n and each $f \in F_n$, $E[fY(\cdot)] \approx E[fx(\cdot)]$. It follows that $Y(\omega)$ is near-standard for almost all ω, and therefore Y is a lifting of a random variable y. We shall show that y has the same adapted distribution as x.

Let g be a conditional process built from the dense set C. By the Adapted Lifting Theorem, gY is a right lifting of gy. By Proposition 7.2, gx and gy are right continuous in probability. For each rational $\vec{t}$, $E[gY(\cdot,\vec{t})] \approx E[gx(\cdot,\vec{t})]$. By approaching a real tuple $\vec{s}$ from the right by rational $\vec{t}$, we obtain arbitrarily large $\vec{u}$ in the monad of $\vec{s}$ such that $E[gY(\cdot,\vec{u})] \approx E[gx(\cdot,\vec{s})]$. It follows that $E[gy(\cdot,\vec{s}) = E[gx(\cdot,\vec{s})]$, so y has the same adapted distribution as x. Notice that right liftings were essential in this paragraph.

The result for the case that x is a stochastic process now follows by the method used to prove Proposition 3.2.◂

Again, if almost every path of x is continuous or rcll, then y may be taken to have the same property.

The next theorem, stating that hyperfinite adapted spaces are homogeneous, is from Keisler (1986).

8.4 **Theorem**

Every hyperfinite adapted space $L(\Omega^T)$ *is homogeneous in the sense that for every pair of stochastic processes* x *and* y *on* $L(\Omega^T)$, $x \equiv y$ *if and only if there is an internal adapted automorphism* h *of* Ω^T *which sends* x *to* y.

Sketch of proof. The flavour of the proof is much like that of Theorem

8.3. The nontrivial direction is to prove that if $x \equiv y$ then there is an internal adapted automorphism sending x to y. Again, the main work is to prove the result for random variables. First let X and Y be liftings of the random variables x and y. Let $n \in \mathbb{N}$. By the Adapted Lifting Theorem, for all conditional expressions f and almost all tuples $\vec{s}$ from T we have $E[fX(\cdot,\vec{s})] \approx E[fY(\cdot,\vec{s})]$. We use this to construct by induction on points $t \in T_n$ an internal bijection h_n of Ω^T such that $X(h_n\omega) \approx Y(\omega)$ almost surely, and for each $t \in T_n$, if $(\omega|t) = (\omega'|t)$ then $(h_n\omega|t) = (h_n\omega'|t)$. h_n is not adapted but "looks ahead" to the next point in T_n. Now countable comprehension is used to get an internal bijection $h = h_K$ for an infinite K. h only "looks ahead" an infinitesimal amount 1/K, and is therefore adapted, that is, it sends each $\mathcal{A}_t$ to itself. Also, $Y(h\omega) \approx X(\omega)$ almost surely. Since X and Y lift x and y, $y(h\omega) = x(\omega)$ almost surely, so h is an adapted automorphism sending x to y.

The general case for stochastic processes x and y is proved from the special case for random variables by the same method as Proposition 3.3.◂

The following result shows that for Markov processes, the adapted distribution reduces to the finite dimensional distribution.

8.5 **Theorem** (Hoover & Keisler (1984))

On an adapted space, any two Markov processes x *and* y *which have the same finite dimensional distribution have the same adapted distribution.*

Proof. The idea of the proof is to show by induction on complexity that for each conditional expression $f(\vec{u})$ and tuple $\vec{t}$, there is a Borel function ψ_f such that

$$fx(\omega,\vec{t}) = \psi_f(x(\omega,t_1),\ldots,x(\omega,t_n)) \text{ almost surely,}$$

and

$$fy(\omega,\vec{t}) = \psi_f(y(\omega,t_1),\ldots,y(\omega,t_n)) \text{ almost surely.}$$

Since x and y have the same fdd, they have the same transition functions. At the conditional expectation step going from a conditional expression f to $g = E[f|v]$, the transition function for ψ_f is used to obtain the new function ψ_g.◂

The examples of Section 3 can be improved to obtain representations of adapted Poisson processes and Brownian motions.

8.6 Example

For every adapted Poisson process x on the hyperfinite adapted space $L(\Omega^T)$ there is an internal adapted automorphism from x to the Loeb coin tossing process. This follows from the previous results and the fact that any adapted Poisson process is a Markov process.

8.7 Example

For every adapted Brownian motion W on the hyperfinite adapted space $L(\Omega^T)$ there is an internal adapted automorphism from W to Anderson's Brownian motion. This follows from our previous results and the fact that any adapted Brownian motion is a Markov process.

In many cases, the above example can be used to reduce a problem about an arbitrary adapted Brownian motion to a hyperfinite computation with the particular Brownian motion constructed by Anderson.

9. APPLICATIONS TO STOCHASTIC ANALYSIS

In this section we give some applications of the methods in the preceding sections to stochastic analysis. We shall give only the simplest examples, for the case of integrals with respect to Brownian motions. In recent years the technique has been used successfully in a wide variety of problems in stochastic analysis.

The classical *Itô integral* is defined as follows. Let Π be an adapted space and let W be an adapted Brownian motion on Π. First let $f(\pi,t)$ be an adapted real valued step function on Π with steps at $t_1,\dots,t_n$. Then the Itô integral is defined as the sum

$$\int_0^1 f(\pi,u)\, dW = \sum_{m=1}^{n-1} f(\pi,t_m)\cdot[W(\pi,t_{m+1}) - W(\pi,t_m)].$$

The Itô integral is then extended to the case where the integrand f is a bounded adapted real valued function on Π by taking limits in mean square.

The Itô integral of f from 0 to t may be defined as the Itô integral of f times the characteristic function of [0,t]. It turns out to be a continuous adapted process.

Anderson (1976) gave the following simple construction of the Itô integral with respect to the Anderson Brownian motion $b(\omega,t)$ on the hyperfinite adapted space $L(\Omega^T)$. Let f be a bounded adapted function. By the proposition 7.2, f has a nonanticipating lifting F with the same finite bound. Form the hyperfinite sum

$$Y(\omega,t) = \sum_{s<t} F(\omega,s)\ \Delta B(\omega,s)$$

where $\Delta B(\omega,s) = \omega(s)/\sqrt{H}$ is the change in $B(\omega,s)$. It can then be shown that Y is a nonanticipating S-continuous process and is a lifting of a continuous adapted process $y(\omega,t)$. This process is the Itô integral,

$$y(\omega,t) = \int_0^t f(\omega,s)\ db(\omega,s).$$

In many cases one can use an easy computation to prove a result for Itô integrals with respect to the Anderson Brownian motion. Then using the universality and homogeneity properties of the hyperfinite adapted space, one case often conclude that the result holds for adapted Brownian motions on arbitrary adapted spaces.

9.1 **Example**

By a direct compuation, Anderson (1976) gave a proof of Itô's formula for his special Brownian motion on a hyperfinite adapted space: If f is bounded and adapted and

$$y(\omega,t) = \int_0^t f(\omega,s)\ db(\omega,s)$$

and φ is C^2 then

$$\varphi(y(\omega,t)) = \varphi(y(\omega,0)) + \int_0^t \varphi'(y(\omega,s))f(\omega,s)db(\omega,s)$$

$$+ ½ \int_0^t \varphi''(y(\omega,s))(f(\omega,s))^2 ds.$$

One can then conclude that Itô's formula holds in general as follows. Suppose Itô's formula has a counterexample on some adapted space. By universality, there is a counterexample to Itô's formula on a hyperfinite adapted space with respect to some adapted Brownian motion.

Then by homogeneity, there is a counterexample to Itô's formula with respect to the Anderson Brownian motion, contradicting the preceding paragraph.

9.2 **Example**

Let $f(\omega,s,x)$ be a bounded function on a hyperfinite adapted space $L(\Omega^T)$ such that $f(\omega,s,\cdot)$ is an adapted stochastic process with values in $C(\mathbb{R})$ and for each ω, $f(\omega,\cdot,\cdot)$ is continuous. Let $b(\omega,t)$ be the Anderson Brownian motion. Consider the stochastic integral equation

$$x(\omega,t) = \int_0^t f(\omega,s,x(\omega,s))\, db(\omega,s). \tag{1}$$

It is shown in Keisler (1984) that (1) has a continuous adapted solution, in the following way. Let B be the Anderson random walk which lifts b. By 7.2, the function f has a finitely bounded nonanticipating right lifting $F(\omega,s,x)$. Let X be the solution of the hyperfinite difference equation

$$X(\omega,t) = \sum_{s<t} F(\omega,s,X(\omega,s))\, \Delta B(\omega,s). \tag{2}$$

It is shown that X is S-continuous and nonanticipating. It follows that X is a lifting of a continuous adapted stochastic process x which is a solution of (1).

It now follows from homogeneity that for any other adapted Brownian motion $c(\omega,t)$ on $L(\Omega^T)$, the stochastic integral equation

$$y(\omega,t) = \int_0^t f(\omega,s,y(\omega,s))\, dc(\omega,s)$$

has a continuous adapted solution. This solution is obtained by taking an adapted automorphism h from b to c, finding a solution x of

$$x(\omega,t) = \int_0^t f(h^{-1}\omega,s,x(\omega,s))\, db(\omega,s),$$

and taking $y(\omega,t) = x(h\omega,t)$.

REFERENCES

Anderson, R.M. (1976). A non-standard representation for Brownian motion and Itô integration, *Israel J. Math* **25**, 15-46.

Anderson, R.M. (1982). Star-finite representations of measure spaces, *Trans. Amer. Math. Soc.* **271**, 667-687.

Billingsley, P. (1968). *Convergence of Probability Measures*, Wiley, New York.

Cutland, N.J. (1983). Nonstandard measure theory and its applications, *Bull. London Math Soc.* **15**, 529-589.

Doob, J.L. (1953). *Stochastic Processes*, Wiley, New York.

Henson, C.W. (1979). Analytic sets, Baire sets, and the standard part map, *Canadian J. Math.* **31**, 663-672.

Hoover, D.N. & Keisler H.J. (1984). Adapted probability distributions. *Trans Amer. Math. Soc.* **286**, 159-201.

Hoover, D.N. & Perkins E. (1983). Nonstandard constructions of the stochastic integral and applications to stochastic differential equations, I, II, *Trans. Amer. Math. Soc.* **275**, 1-58.

Keisler, H.J. (1984). An infinitesimal approach to stochastic analysis, *Memoirs Amer. Math. Soc.* **297**.

Keisler, H.J. (1985). Probability quantifiers; in *Model Theoretic Logics* (J. Barwise and S. Feferman, eds.), Springer-Verlag, Berlin and New York, 506-556.

Keisler, H.J. (1986). Hyperfinite models for adapted probability logic, *Annals of Pure and Applied Logic*, **31**, 71-86.

Lindstrøm, T. (this volume). An invitation to nonstandard analysis.

Loeb, P.A. (1975). Conversion from nonstandard to standard measure spaces and applications to probability theory, *Trans. Amer. Math. Soc.* **211**, 113-122.

Loeb, P.A. (1979). An introduction to nonstandard analysis and hyperfinite probability theory, in *Probabilistic Analysis and Related Topics*, Vol. 2 (ed. A.T. Bharachua-Reid), Academic Press, 105-142.

Manevitz, L. & Merzbach, E. (1986). Multi-parameter stochastic processes via non-standard analysis, preprint.

Ross, D. (1988). Automorphisms of the Loeb algebra, *Fund Math.* **128**, 29-36.

Stoll, A. (1982). A nonstandard construction of Lévy Brownian motion with applications to invariance principles, Diplomarbeit, Univ, Freiburg.

Stoll, A. (1986). A nonstandard construction of Lévy Brownian motion, *Probability Theory and related Fields* **71**, 321-334.

von Neumann, J. (1932). Einige Sätze über messbare abbildungen. *Ann. Math.* **33**, 574-586.

INFINITESIMALS IN FUNCTIONAL ANALYSIS

C. WARD HENSON

The aim of this article is to provide, when combined with the survey paper Henson & Moore (1983), a fairly complete description of the nonstandard hull construction and of the most important ways in which nonstandard methods have been used to solve problems in functional analysis. Most of the material concerning Banach spaces is already covered in that earlier survey; there are a few important recent developments, which we have included in the last two sections of this paper. Here we will, however, concentrate on the general nonstandard hull construction for topological vector spaces and for operators on such spaces. We have also tried to include here some more elementary variations on arguments which appeared there, and it may well be that this paper can in part serve as an introduction to the Banach space survey. (But also the reverse may be true for some readers.)

The nonstandard hull construction applied to topological vector spaces and to continuous operators on them plays very much the same role in functional analysis that the Loeb measure construction does in probability theory. It provides a systematic (functorial) procedure for obtaining a topological vector space or continuous operator (in the usual mathematical sense) from internal spaces and operators. Moreover, in the setting of functional analysis there is an elaborate and important structure of infinitesimals and finite points, which provide an elegant framework for the expression of complicated topological concepts, as well as for the study of the nonstandard hulls themselves.

In Section 1 of this paper we present the principal nonstandard tools for functional analysis, with emphasis on an explanation of how the ideas are motivated and how they fit together, but few proofs. We have included proofs only when they illustrate an important method of argument, or for other pedagogical purposes. We have assumed that the reader is somewhat familiar with the topological vector space point of view toward functional analysis (and so we have not included many examples just to illustrate standard concepts) but this Section is by no means directed only at experts. Also we have assumed familiarity with the basic concepts and methods of nonstandard analysis itself, as described by Tom Lindstrøm in his contribution to this book, for example, or in any other of the basic books which he has given as references.

In Section 2 we summarize what has been accomplished using nonstandard methods in the study of operators on topological vector spaces. This is actually an area in which much remains to be done, and the principal applications have concerned continuous linear operators on Hilbert spaces, with some further applications in Banach space theory.

In Section 3 we present recent applications of nonstandard methods to the study of uniform equivalence of locally convex spaces, due to Stefan Heinrich. He has been able to carry over many important nonstandard tools from the Banach space setting to the study of locally convex spaces, including the important Local Duality Theorem of Kürsten and Stern. His applications include a significant generalization of a theorem of Ribe (which covered the Banach space case.) Namely, Heinrich has shown that uniformly equivalent locally convex spaces must be finitely represented in each other. Also, for locally convex spaces in which every bounded set is relatively compact, he shows that uniformly equivalent spaces must actually be linearly isomorphic.

Section 4 concerns indiscernible sequences in Banach spaces. It not only presents arguments which can provide a more elementary and perhaps helpful introduction to Section 10 of Henson & Moore (1983), but gives a brief discussion of the class of stable Banach spaces. These spaces were introduced by Krivine and Maurey and the ideas involved in studying them have proved to be very important in the theory of Banach spaces and also to be closely connected to model theoretic methods. (Heinrich, Krivine and Maurey phrase their arguments in terms of ultrapowers, while we use the language of nonstandard analysis. As we have emphasized often, the two points of view are essentially equivalent. For a more elaborate expression of this point of view see the Introduction to Henson & Moore (1983).)

Finally, in Section 5 we discuss recent results concerning the isometric and isomorphic classification of nonstandard hulls of Banach spaces. Such results can often be important in applications, where one is combining a nonstandard argument with the use of a standard result applied to an appropriate nonstandard hull. We bring the information given in Henson & Moore (1983) up to date.

The lectures for which this paper and the earlier Banach space survey paper provide background could not, of course, cover any large fraction of the basic material which the papers present. Moreover, these two papers do not make any claim to have exhaustively covered all of the interesting ideas and results which fall into the relevant area. It is satisfying to see how successful has been the application of Abraham Robinson's fundamental ideas within functional analysis itself, in solving problems and introducing important concepts within the standard theory itself. Evidently this success owes very much to the genius of Jean-Louis Krivine and Stefan Heinrich, whose understanding and instinct both for functional analysis and for model theory has enabled them to contribute so much to the development and application of these ideas. One sees their names often in any exposition of this subject.

The author owes a particular debt, both intellectual and personal, to L. C. Moore, Jr. and to Stefan Heinrich, without whose friendship his own contribution and enjoyment of this subject would have been much reduced. The writing of this paper was largely done while the author was visiting the Technische Hochschule, Aachen, W. Germany, to which he is grateful for support. His research has also been supported by a grant from the National Science Foundation, to whom the author is grateful.

1. TOPOLOGICAL VECTOR SPACES

Let E be a vector space over the real or complex numbers (which we will refer to simply as "the scalars") and let θ be a vector topology on E. That is, θ is a Hausdorff topology on E such that the mappings of addition (on E) and scalar multiplication are continuous, as functions of two variables. Essentially every important object in linear functional analysis can be construed as a topological vector space (E,θ) or as a continuous operator between two such spaces In this section we will outline (with only a few proofs) the main tools within nonstandard analysis for studying such objects. This was originally presented in Henson & Moore (1972) (see also Chapter 10 of Stroyan &

Luxemburg (1976) for an exposition) and it has recently been extended in important ways by S. Heinrich (1984). For standard references in this area see Kothe (1969) or Schaeffer (1966).

We adopt the usual notation in both nonstandard analysis and the theory of topological vector spaces. In particular, the dual space of (E,θ), denoted by $(E,\theta)'$ or simply E', is the vector space of all continuous (scalar valued) linear functionals on E.

Because the translation functions $T_a(x) = x + a$ are homeomorphisms from E onto itself, the topology θ is completely determined by its behavior at 0. If we let $\mu_\theta(a)$ denote the monad at a determined by θ,

$$\mu_\theta(a) = \cap\{{}^*U\colon U \text{ is an open neighborhood of } a\}$$

then we see that for each $a \in E$

$$\mu_\theta(a) = a + \mu_\theta(0)$$
$$= \{a + p\colon p \text{ is } \textit{infinitesimal} \text{ in } {}^*E\}.$$

Since θ is a vector topology, the monad $\mu_\theta(0)$ must be closed under $+$ and under multiplication by finite scalars. (This expresses the continuity of addition at $(0,0)$ and the continuity of scalar multiplication at $(c,0)$ for each standard scalar c. Here we use also the important fact that every finite nonstandard scalar is near-standard.)

Let $\kappa = \kappa(\theta)$ be the smallest cardinal number of a base at 0 for the topology θ. In using nonstandard methods to study (E,θ) it is usually necessary to work with a nonstandard model which is at least κ^+-saturated. (If (E,θ) is a normed space or more generally, if it is metrizable, then $\kappa = \omega$ and we are requiring that the nonstandard model be $\aleph_1$-saturated.) We will always make the assumption that our nonstandard model is $\kappa(\theta)^+$-saturated, usually without specific mention.

Let $\gamma = \gamma(\theta)$ be the smallest cardinal number of a family $\{B_i\colon i \in I\}$ of θ-bounded sets which is *cofinal*, in the sense that every θ-bounded subset of E is contained in B_i for some i. Often it is necessary to (possibly) strengthen our saturation hypothesis by assuming that the nonstandard model is at least $\max(\kappa,\gamma)^+$-saturated. (If (E,θ) is a normed space then $\kappa = \gamma = \omega$; however, even when (E,θ) is metrizable it can happen that $\gamma(\theta) = 2^\omega$. In general one can show that $\gamma(\theta) \le 2^{\kappa(\theta)}$).

We let $ns_\theta(*E)$ denote as usual the set of near-standard elements of $*E$, and $pns_\theta(*E)$ the set of pre-near-standard elements:

$$ns_\theta(*E) = \{p \in *E\colon (\exists a \in E)(p \in \mu_\theta(a))\}$$

$$pns_\theta(*E) = \{p \in *E\colon \text{ for every } \theta\text{-neighborhood } U \text{ of } 0, \; (\exists a \in E)(p \in a + *U)\}.$$

Note that $pns_\theta(*E)$ looks like a kind of "closure" of E within $*E$; later we will see that this can be given a precise meaning.

For the nonstandard approach to functional analysis there are two further subsets of $*E$ which are of central importance: the set $fin_\theta(*E)$ of *finite* points and the set $bd_\theta(*E)$ of *bounded* points in $*E$.

1.1 Definition

An element p of $*E$ is *θ-finite* if, for each infinitesimal scalar ε, the element εp always lies in $\mu_\theta(0)$. The set of all such points is written $fin_\theta(*E)$.

It is easily shown that $fin_\theta(*E)$ is closed under + and under multiplication by finite nonstandard scalars. Note that in a normed space, $fin_\theta(*E)$ consists simply of the elements of $*E$ whose norm is a finite element of $*\mathbb{R}$.

Since $fin_\theta(*E)$ is a vector space, and $\mu_\theta(0)$ is evidently a subspace of it, we may form the quotient space, which we denote by $\hat{E}$ (suppressing the dependence on θ) or sometines by $\hat{E}_\theta$. So we take

$$\hat{E} = fin_\theta(*E)/\mu_\theta(0)$$

and let $\pi : fin_\theta(*E) \to \hat{E}$ be the associated quotient map. Often when S is a subset of $*E$ we will write $\hat{S}$ for $\pi(S \cap fin_\theta(*E))$. If necessary we sometimes write π_θ in place of π. If $A \subseteq E$, we usually write $\hat{A}$ in place of $\hat{}(*A) = \pi(*A \cap fin_\theta(*E))$.

It is easy to show that $p \in *E$ is θ-finite iff for every θ-neighborhood U of 0 there is a standard integer $n > 0$ with $\frac{1}{n} p \in *U$. (Proof. If p is θ-finite, and U is given, then for every infinite integer H, $\frac{1}{H} p \in \mu_\theta(0) \subseteq {}^*U$. Hence n exists by "underspill". If p is not θ-finite then there is an infinitesimal scalar ε so that $\varepsilon p \notin \mu_\theta(0)$. Hence there is a θ-neighborhood U of 0, which we may assume is closed under multiplication by scalars of absolute value ≤ 1, such that $\varepsilon p \notin {}^*U$. But for any standard integer n, $|\varepsilon| < \frac{1}{n}$ and hence $\frac{1}{n} p \notin {}^*U$.) From this argument can be seen that $\text{fin}_\theta({}^*E)$ corresponds to the filters J on E which are finite in the sense that for any neighborhood U of 0 there exists an integer $n > 0$ so that $nU \in J$. As is usual in nonstandard analysis, it is much easier to deal directly with the set $\text{fin}_\theta({}^*E)$ and the structures it supports, than to work with mappings or topologies on the lattice of these filters.

A natural topology, which we denote $\hat{\theta}$, can be put on $\hat{E}$. A base for $\hat{\theta}$ at 0 consists of the family of sets $\hat{U}$, where U is a θ-neighborhood of 0 in E. The space $(\hat{E},\hat{\theta})$ is referred to as a *nonstandard hull* of (E,θ). Its basic properties and its relation to (E,θ) can be summarized as follows:

1.2 Theorem

Let (E,θ) *be any topological vector space and* $(\hat{E},\hat{\theta})$ *a nonstandard hull of* (E,θ).

(1) $(\hat{E},\hat{\theta})$ *is also a topological vector space, and it is complete.*

(2) (E,θ) *is topologically and linearly embedded into* $(\hat{E},\hat{\theta})$, *by the map which takes each* $a \in E$ *to* $\pi({}^*a)$; *the closure of* E *under this mapping is* $\pi(\text{pns}_\theta({}^*E))$.

(3) *If* A *is any internal subset of* ${}^*E$, *then* $\hat{A}$ *is closed in* $\hat{E}$.

The completeness of $(\hat{E},\hat{\theta})$ and part (3) of Theorem 1.2 come from the $\kappa(\theta)^+$-saturation of the nonstandard model. From (1) and (2) together it follows that $\pi(\text{pns}_\theta({}^*E))$ with the $\hat{\theta}$-topology is the completion of (E,θ). We also see that (E,θ) is complete exactly when $\text{pns}_\theta({}^*E) = \text{ns}_\theta({}^*E)$.

Note that if $\{U_i : i \in I\}$ is a base of neighborhoods of 0 in E, then $\{\hat{U}_i: i \in I\}$ is a base of neighborhoods of 0 in $\hat{E}$. In particular, $\hat{E}$ is a Banach space when E is one, $\hat{E}$ is metrizable when E is, etc. If E is a Banach space with norm $\| \ \|$, then the norm which makes $\hat{E}$ into a Banach space can be defined by

$$\|x\| = st({}^*\|p\|)$$

whenever $p \in fin({}^*E)$ and $x = \pi(p)$.

If (E,θ) is a given topological vector space and if X is a dense subspace of E, then the inclusion map $I: X \to E$ induces a linear topological isomorphism between $\hat{X}$ and $\hat{E}$. In particular, a topological vector space and its completion have the same nonstandard hull.

There is another relationship between $fin_\theta({}^*E)$ and $\mu_\theta(0)$ which is very important in practice, and which is given in the next result: (This is one of the places where our saturation assumption is used.)

1.3 **Lemma**

(a) *If* $x \in \mu_\theta(0)$, *then there is an infinite integer* H *so that also* $Hx \in \mu_\theta(0)$;

(b) $\mu_\theta(0)$ *is the set of all* εy, *where* ε *is an infinitesimal scalar and* y *is* θ*-finite.*

Proof. (a) Let $\{U_i: i \in I\}$ be a base of θ-neighborhoods of 0, such that our nonstandard model is card $(I)^+$-saturated. (This is possible by our blanket saturation assumption.) For each $i \in I$ and each standard integer n, $nx \in {}^*U_i$.

We are looking for $H \in {}^*\mathbb{N}$ satisfying the conditions $n < H$, for each $n \in \mathbb{N}$, and $Hx \in {}^*U_i$, for each $i \in I$. Since each finite set of these conditions is satisfiable (by a large enough standard integer) and since there are $<$ card $(I)^+$ conditions, our saturation assumption yields the desired H.

(b) If $x \in \mu_\theta(0)$, take H as in part (a) and let $y = Hx$ and $\varepsilon = 1/H$; then $x = \varepsilon y$, ε is infinitesimal and y is θ-finite. For the converse, suppose that x is of the form εy as in (b). If

$x \notin \mu_\theta(0)$, then there is a θ-neighbourhood U of 0 with $x \notin {}^*U$. We may assume that U is closed under multiplication by scalars of absolute value ≤ 1. Since y is θ-finite, there is a standard integer n satisfying $y \in n{}^*U$. But $|\varepsilon n| \leq 1$ and hence

$$x = \varepsilon y \in \varepsilon n{}^*U \subseteq {}^*U$$

which is a contradiction.◂

Part (b) of Lemma 1.3 shows that the monad $\mu_\theta(0)$, and hence the entire topology θ, is completely determined by $\mathrm{fin}_\theta({}^*E)$. Also, if θ_1, θ_2 are two vector topologies on E and if $\mathrm{fin}_{\theta_1}({}^*E) \subseteq \mathrm{fin}_{\theta_2}({}^*E)$, then also $\mu_{\theta_1}(0) \subseteq \mu_{\theta_2}(0)$. Hence it also follows that $\theta_2 \subseteq \theta_1$ as topologies on E.

Suppose that θ_1, θ_2 are two distinct vector topologies on E. We may assume that there exists an element x of ${}^*E$ which is in $\mu_{\theta_1}(0)$ but not in $\mu_{\theta_2}(0)$. (If not, then simply interchange θ_1 and θ_2.) Choosing H as in Lemma 1.3(a) we obtain an element Hx of $\mu_{\theta_2}(0)$ which cannot even be in $\mathrm{fin}_{\theta_2}({}^*E)$. That is, if $\theta_1 \neq \theta_2$ then ${}^*E$ contains an element y which is infinitesimal for one topology and infinite for the other. (If $\theta_2 \subset \theta_1$ then such a y will necessarily be θ_1-infinite and θ_2-infinitesimal.)

Now we turn to the θ-bounded elements of ${}^*E$. Recall that a subset B of E is called θ-bounded if for each neighbourhood U of 0 there exists an integer $n > 0$ so that $\frac{1}{n} B \subseteq U$. It is easy to show that B is θ-bounded iff ${}^*B$ is a subset of $\mathrm{fin}_\theta({}^*E)$.

1.4 Definition

An element p of ${}^*E$ is *θ-bounded* if there exists a θ-bounded set $B \subseteq E$ such that $p \in {}^*B$. The set of all such points is written $\mathrm{bd}_\theta({}^*E)$.

From the paragraph above we see that $\mathrm{bd}_\theta({}^*E) \subseteq \mathrm{fin}_\theta({}^*E)$. If (E,θ) is a normed space then $\mathrm{bd}_\theta({}^*E) = \mathrm{fin}_\theta({}^*E)$, since then the unit ball is bounded in E; in general this equality does not hold. It is easy to see that $\mathrm{bd}_\theta({}^*E)$ is closed under $+$ and under multiplication by finite nonstandard scalars. We define the *bounded nonstandard hull* of (E,θ), denoted $\hat{E}^b$, to be the topological linear subspace of (E,θ) with $\hat{E}^b = \pi(\mathrm{bd}_\theta({}^*E))$. We remark that $\mathrm{bd}_\theta({}^*E)$ does not always contain $\mu_\theta(0)$ as a subset. In general the precise relationship between $\hat{E}$ and $\hat{E}^b$ is a subtle one, and both spaces have proved to be important in practice. It is not always true that $\hat{E}^b$ is closed in $\hat{E}$; it is even an open question whether $\hat{E}^b$ could be a proper dense subset of $\hat{E}$. Such possibilities reflect complications concerning the bounded subsets of E, and are worth studying further.

Before turning our attention to locally convex topologies, let us look quickly at one specific non-locally-convex space, about which many interesting things can be said:

1.5 Example

For any probability measure μ, let $L_0(\mu)$ denote as usual the space of (equivalence classes of) μ-measurable, scalar valued functions. Give $L_0(\mu)$ the topology of convergence in measure. Then one can show that the nonstandard hull of $L_0(\mu)$ is precisely $L_0(\hat{\mu})$, where $\hat{\mu}$ is the Loeb measure obtained from the internal probability measure ${}^*\mu$.

For the proof, recall that a base for the neighbourhoods of 0 in $L_0(\mu)$ is given by the family of sets of the form

$$\{f \in L_0(\mu):\ \mu(|f| > \varepsilon) < \varepsilon\}$$

where ε is a positive real number. It follows that an internal ${}^*\mu$-measurable function φ is finite iff

$$\hat{\mu}(\{\omega:\ \varphi(\omega) \text{ is finite}\}) = 1.$$

Moreover, φ is infinitesimal exactly when

$$\hat{\mu}(\{\omega:\ \varphi(\omega) \text{ is infinitesimal}\}) = 1.$$

The claimed identification of $\hat{L}_0(\mu)$ and $L_0(\hat{\mu})$ (as topological vector spaces) now follows by elementary facts about Loeb measure.

It should be noted that the situation for $\hat{L}_p(\mu)$ is somewhat different when $p > 0$. When $0 < p < \infty$ the nonstandard hull of $L_p(\mu)$ is a space of the form $L_p(\mu')$ for a certain measure μ', but μ' is defined on a much larger measure space than is the Loeb measure $\hat{\mu}$. Indeed, μ' is an unbounded measure which can be seen abstractly as the result of taking uncountably many copies of $\hat{\mu}$ located on mutually disjoint sets. The nonstandard hull of $L_\infty(\mu)$ is *not* an $L_\infty(\mu')$ space at all; it is however a space of continuous functions. An analysis of $\hat{L}_p(\mu)$ for $p \geq 1$ and further examples of nonstandard hulls of Banach spaces can be found in Henson & Moore (1983) and in the papers referenced there. For $0 < p < 1$ see the discussion in Schreiber (1972).

For the rest of this section we will assume that (E,θ) is a *locally convex* topological vector space. This means that there is a base of θ-neighborhoods of 0, $\{U_i: i \in I\}$, in which each U_i is an absolutely convex set. (That is, if $a,b \in U_i$ and α,β are scalars with $|\alpha| + |\beta| \leq 1$, then also $\alpha a + \beta b \in U_i$.) Equivalently it means that there is a family $\{\rho_j: j \in J\}$ of seminorms on E which defines the topology θ. This can also be given a simple nonstandard characterization: a topological vector space (E,θ) is locally convex iff $\mu_\theta(0)$ is *hyperconvex*; this means that if $p_1,\ldots,p_\omega$ is any internal sequence with each p_j in $\mu_\theta(0)$, and if $\alpha_1,\ldots,\alpha_\omega$ is any internal sequence of nonstandard scalars with $\Sigma|\alpha_j| \leq 1$, then $\Sigma\alpha_j p_j \in \mu_\theta(0)$.

S. Heinrich (1984) has made a very interesting recent study of the bounded nonstandard hull $\hat{E}^b$ and has used it to solve several problems concerning the uniform equivalence of locally convex spaces (as we discuss in Section 3.) The following definition gives an important new condition on locally convex spaces which was isolated by Heinrich as he analyzed what it means for $\hat{E}^b$ to be dense in $(\hat{E},\hat{\theta})$. We let $U(E,\theta)$ denote the set of all closed, absolutely convex neighborhoods of 0 in (E,θ).

1.6 Definition

(E,θ) is said to satisfy the *density condition* if for each function $\lambda: U(E,\theta) \to \mathbb{R}^+$ and each $V \in U(E,\theta)$, there exists a finite set $\mathcal{U} \subseteq U(E,\theta)$ and a θ-bounded set $B \subseteq E$ so that

$$\cap\{\lambda(U)\,U \ : \ U \in \mathcal{U}\} \subseteq B + V.$$

The initial importance of this condition is simply that it is a standard condition on (E,θ) which is equivalent (as is proved in Heinrich (1984)) to the statement that $\hat{E}^b$ is $\hat{\theta}$-dense in $\hat{E}$. (Hence this condition does not depend on the nonstandard model used.) However, the density condition has subsequently turned out to play a broader role, not only in the context of nonstandard analysis, but even in connection with problems which are phrased wholly within the standard theory of locally convex spaces. (See Bierstedt & Bonet (198?).)

In case (E,θ) is metrizable, then the relation between $\hat{E}^b$ and $\hat{E}$ is especially nice, as shown by the next results (which are due to Heinrich (1984)):

1.7 Theorem

Let (E,θ) *be a metrizable locally convex space. Then* $\hat{E}^b$ *is closed in* $\hat{E}$. *Moreover, the following conditions are equivalent*:

(a) *the density condition on* (E,θ);

(b) $\hat{E}^b = \hat{E}$;

(c) *each* $\hat{\theta}$*-bounded subset of* $\hat{E}$ *is contained in a set of the form* $\hat{B}$, *where* B *is a* θ*-bounded subset of* E.

Note, that whenever B is a θ-bounded subset of E, then $\hat{B}$ (and hence each of its subsets) is $\hat{\theta}$-bounded in $\hat{E}$. Hence in (c) it is the cofinality of such sets among the bounded subsets of $\hat{E}$, which is at issue.

For general locally convex spaces *no* standard condition on (E,θ) is known which is equivalent to the equality $\hat{E}^b = \hat{E}$. Similar ignorance prevails concerning the assertion that $\hat{E}^b$ is closed in $\hat{E}$ and concerning the cofinality condition on bounded sets which is expressed in condition (c) of Theorem 1.7. (This can be as much a cause for optimism as for pessimism: these conditions may be useful and are expressed very simply in their nonstandard formulation.)

Not all metrizable spaces satisfy the density condition: Heinrich (1984) has in fact constructed a Köthe sequence space which fails to satisfy condition (c) of Theorem 1.7 (and hence does not satisfy the other conditions either.)

Note that the equality $E = \hat{E}^b$ has a simple equivalent formulation: $E = \hat{E}^b$ iff for every bounded set $B \subseteq E$, $^*B \subseteq ns_\theta$ iff every bounded set in E is relatively compact. Similarly, E is dense in $\hat{E}^b$ iff every bounded set in E is precompact. Henson & Moore (1972), (1974) initiated the study of locally convex spaces (E,θ) which have *invariant nonstandard hulls* (called (HM) spaces in Stroyan & Luxemburg (1976)), meaning that E is dense in $\hat{E}$. Several important classes of locally convex spaces, including Schwartz spaces, satisfy this condition. The results of Heinrich allow one to conclude that E is dense in $\hat{E}$ iff E satisfies the density condition and all bounded sets in E are precompact. (Examples show that neither of these conditions can be omitted in general.) This equivalence comes simply from analyzing the condition E is dense in $\hat{E}$ as the conjunction of the conditions: E is dense in $\hat{E}^b$ and $\hat{E}^b$ is dense in $\hat{E}$. Among the natural spaces in functional analysis which have invariant nonstandard hulls are several spaces of distributions, spaces of analytic functions under the topology of uniform convergence on compact sets, the class of Schwartz spaces, and a class of interesting inductive limit spaces which have recently been studied and which have importance in mathematical economics . (See Benninghofen, Richter & Stroyan (198?); Stroyan (1983); and Stroyan & Benninghofen (198?).)

Now let F be the dual space of (E,θ), $F = (E,\theta)'$, and let $\langle\ ,\ \rangle$ denote the pairing between E and F given by evaluation of functions: for $x \in E$ and $\varphi \in F$,

$$\langle x,\varphi\rangle = \varphi(x).$$

This pairing gives rise to a corresponding "pairing" between $^*E$ and $^*F$, which takes nonstandard scalars as values. (This should be written as $^*\langle\ ,\ \rangle$, but we will drop the *.)

It is natural to ask which elements of $^*F$ give rise to continuous linear functionals on $\hat{E}$. If $\varphi \in {}^*F$ and $\langle x,\varphi\rangle$ is finite for each $x \in fin_\theta(^*E)$, then we can at least define a scalar valued function on $fin_\theta(^*E)$, by considering $st\langle x,\varphi\rangle$ for θ-finite x. This is clearly linear in x. In order to obtain a well defined linear function on $\hat{E}$, we need to know that $\langle x,\varphi\rangle$ is infinitesimal whenever $x \in \mu_\theta(0)$. But this follows immediately from what has already been assumed, using Lemma 1.3. (Proof. If $x \in \mu_\theta(0)$, take H as in Lemma 1.3; by the assumption on φ, we see that $H\langle x,\varphi\rangle = \langle Hx,\varphi\rangle$ must be finite, since $Hx \in \mu_\theta(0) \subseteq fin_\theta({}^*E)$. Since H is infinite, $\langle x,\varphi\rangle$ must be infinitesimal.) So we may define a linear functional f on $\hat{E}$ by setting

$$f(\pi(x)) = st\langle x,\varphi\rangle$$

for all θ-finite x. As we will now prove, this linear functional f is actually $\hat{\theta}$ continuous. Let $\delta > 0$ be a standard real number. We want to show that there exists a θ-neighborhood U of 0 in E such that $\hat{U} \subseteq \{u \in \hat{E}: |f(u)| \le \delta\}$. Let $\{U_i: i \in I\}$ be a θ-neighborhood base at 0 (and recall that we are assuming that this can be done in such a way that our nonstandard model is at least card $(I)^+$-saturated.) We proved above that whenever $x \in \mu_\theta(0)$, then $\langle x,\varphi\rangle$ is infinitesimal and hence $|\langle x,\varphi\rangle| \le \delta$. Thus

$$\cap\{{}^*U_i: i \in I\} \subseteq \{x \in {}^*E: |\langle x,\varphi\rangle| \le \delta\}.$$

Since the right side is an internal set, a simple application of our saturation assumption shows that there must be a finite sequence $i_1,\dots,i_k \in I$ so that

$${}^*U_{i_1} \cap \dots \cap {}^*U_{i_k} \subseteq \{x \in {}^*E: |\langle x,\varphi\rangle| \le \delta\}.$$

Setting $U = U_{i_1} \cap \dots \cap U_{i_k}$ and taking standard parts yields $\hat{U} \subseteq \{u \in \hat{E}: |f(u)| \le \delta\}$, as desired. The following statement summarizes what we have proved:

1.8 **Theorem**

Suppose $\varphi \in {}^*F$ *is such that* $\langle x,\varphi\rangle$ *is finite for all* $x \in fin_\theta({}^*E)$. *Then there is a unique* $\hat{\theta}$*-continuous linear functional* f *on* $\hat{E}$ *such that* $f(\pi(x)) = st\langle x,\varphi\rangle$ *for all* $x \in fin_\theta({}^*E)$.

Theorem 1.8 suggests the usefulness of the following notation: let A be any subset of ${}^*E$:

$$A^f = \{\varphi \in {}^*F: \langle a,\varphi\rangle \text{ is finite for all } a \in A\}$$

$$A^i = \{\varphi \in {}^*F: \langle a,\varphi\rangle \text{ is infinitesimal for all } a \in A\}.$$

Similarly if B is a subset of $^*F$, then we may define B^f and B^i as corresponding subsets of $^*E$.

1.9 Definition

An element φ of $^*F$ is *θ-equicontinuous* if it is in $\text{fin}_\theta(^*E)^f$. The set of all such elements will be written $\text{eq}_\theta(^*F)$.

We will also use the notation $m_\theta(^*F)$ for the set $\text{fin}_\theta(^*E)^i$. Note that $m_\theta(^*E)$ is precisely the set of $\varphi \in \text{eq}_\theta(^*F)$ such that the continuous linear functional on $\hat{E}$ obtained from φ (as in Theorem 1.8) is identically 0. Hence $\text{eq}_\theta(^*F)/m_\theta(^*F)$ can be identified with a subset of the dual space of $(\hat{E},\hat{\theta})$.

The reason for the terminology in Definition 1.9 (which was suggested by Heinrich (1984) and is different from the original notation in Henson & Moore (1972)) is given in the following result: (For the proof see Henson & Moore (1972).)

1.10 Theorem

(a) *Let* $\varphi \in {}^*F$. *Then* $\varphi \in \text{eq}_\theta(^*F)$ *iff there is a set* $B \subseteq F$ *such that* B *is θ-equicontinuous and* $\varphi \in {}^*B$.

(b) *The relationship among* $\text{fin}_\theta(^*F)$, $\mu_\theta(0)$, $\text{eq}_\theta(^*F)$ *and* $m_\theta(^*F)$ *under the operations* i *and* f *is given in the following diagram*:

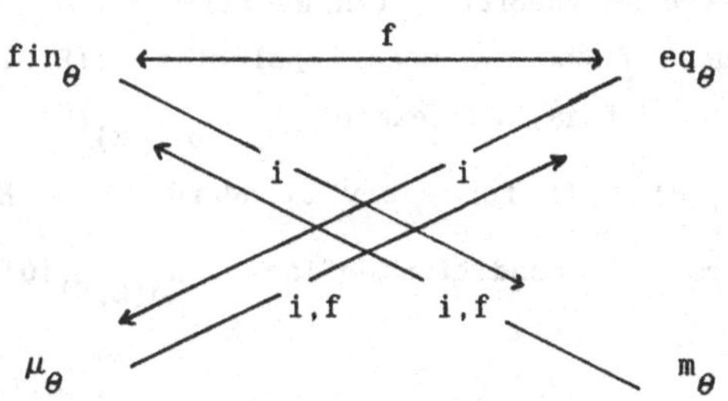

(Thus $(\text{eq}_\theta)^i = \mu_\theta$, $(m_\theta)^f = \text{fin}_\theta$, etc.)

It is natural to ask whether or not every continuous linear functional on $(\hat{E},\hat{\theta})$ arises as in Theorem 1.8. For Banach spaces it is known that this is true iff (E,θ) is super-reflexive, but for general spaces it is not known exactly what this means.

Henson & Moore (1974) showed that eq_θ/m_θ is $\beta(\hat{E}',\hat{E})$-dense in $\hat{E}'$ exactly when $(\hat{E},\hat{\theta})$ is semi-reflexive. Heinrich (1984) observed that $m_\theta(*F)$ is contained in the $\beta(F,E)$ monad of 0 in $*F$, and hence there is a natural linear mapping τ from eq_θ/m_θ into the nonstandard hull $\hat{F}_\beta$ of $(F,\beta(F,E))$. Actually τ maps into the bounded hull $\hat{F}_\beta^{\,b}$. The mapping τ is 1-1 exactly when (E,θ) satisfies the density condition. Moreover, if (E,θ) is metrizable and satisfies the density condition, then τ gives a linear homeomorphism from eq_θ/m_θ (given the restriction of the $\beta(\hat{E}',\hat{E})$ topology) onto $\hat{F}_\beta^{\,b}$.

There is so much symmetry in the relationship between $*E$ and $*F$ that it is tempting to look for a topology on F with respect to which m_θ is the monad of 0, eq_θ is the set of finite points, etc. If (E,θ) is a normed space, then this is possible: just take the usual dual-norm topology on F. However, if (E,θ) is not a normed space, then m_θ is not the monad at 0 of any topology on F, as is shown in Henson & Moore (1972).

Using the subsets of $*E$ and $*F$ introduced here, it is possible to give elegant and compact expression to many important results in the theory of locally convex spaces. For example, the following result (proved in Henson & Moore (1972)) gives the essential content of Grothendieck's Completeness Theorem. (In addition to the given topology θ on E, we make use of the two weak topologies $\sigma(E,F)$ on E and $\sigma(F,E)$ on F. We can define, for example, $\mu_{\sigma(F,E)}(0)$ by saying (for $\varphi \in *F$): $\varphi \in \mu_{\sigma(F,E)}(0)$ iff for every standard $x \in E$, $\langle *x,\varphi\rangle$ is infinitesimal; the symmetric condition defines $\mu_{\sigma(E,F)}(0)$ as a subset of $*E$.)

1.11 **Theorem**

Let (E,θ) *be a Hausdorff locally convex space with dual space* F.

(a) $pns_\theta(^*E) = [eq_\theta(^*F) \cap \mu_{\sigma(F,E)}(0)]^i$

(b) $ns_\theta(^*E) = pns_\theta(^*E) \cap ns_{\sigma(E,F)}(^*E)$.

Although the internal linear functionals in $eq_\theta(^*F)$ do not in general yield *all* continuous linear functionals on $(\hat{E},\hat{\theta})$, there are various senses in which the space of linear functionals given by $eq_\theta(^*F)$ is quite large. (Of course the advantage of determining $f \in (\hat{E},\hat{\theta})'$ by some $\varphi \in eq_\theta(^*F)$, as in Theorem 1.8, is that φ is *internal* and hence can be further analyzed using nonstandard methods!) The next three results show this largeness quite clearly. (In all three results we strengthen our blanket saturation assumption as discussed near the beginning of this section, taking into consideration the size of a cofinal family of θ-bounded subsets of E.)

1.12 **Theorem**

Suppose X *is a subspace of* $\hat{E}$ *whose vector space dimension is* $\le \kappa$, *where the nonstandard model is* κ^+*-saturated. For each* $f \in (E,\theta)'$ *there exists* $\psi \in eq_\theta(^*F)$ *so that the* $\hat{\theta}$*-continuous linear functional* g *on* $\hat{E}$, *defined by*

$$g(\pi(x)) = st\langle x,\psi\rangle \quad \textit{for all } x \in fin_\theta(^*E)$$

agrees with f *on* X.

Proof. See Theorem 1 of Henson & Moore (1974).◄

It follows from Theorem 1.12 that all $\sigma(\hat{E},\hat{E}')$-bounded subsets of E are actually $\hat{\theta}$-bounded.

1.13 **Theorem**

Suppose the nonstandard model is $\max(\kappa(\theta),\gamma(\theta))^+$*-saturated. Let* $\varphi \in eq_\theta(^*F)$ *be such that* $\langle x,\varphi\rangle$ *is infinitesimal for all*

$x \in bd_\theta(^*E) \cap \mu_\theta(0)$, *so that we may define a linear functional on* $\hat{E}^b$ *by* $f(\pi(x)) = st\langle x,\varphi\rangle$ *for all* $x \in bd_\theta(^*E)$. *If the linear functional* f *is* $\hat{\theta}$*-continuous on* $\hat{E}^b$, *then there exists* $\psi \in eq_\theta(^*F)$ *so that the* $\hat{\theta}$*-continuous linear functional* g *on all of* $\hat{E}$, *defined by*

$$g(\pi(x)) = st\langle x,\psi\rangle \quad \textit{for all} \quad x \in fin_\theta(^*E),$$

agrees with f *on* $\hat{E}^b$.

Proof. See Heinrich (1984).◂

The proofs of Theorems 1.12 and 1.13 are relatively easy, involving a combination of a Hahn-Banach argument with a saturation argument. The next result is much deeper. In the Banach space setting this Local Duality Theorem was proved independently by K. D. Kürsten and J. Stern, and it has been a very important tool for obtaining applications in the standard theory of Banach spaces of nonstandard methods. (See Henson and Moore (1983) for a discussion.) Recently Heinrich (1984) extended this Local Duality Theorem to the general locally convex setting and showed what an important tool it continues to be.

1.14 Local Duality Theorem

Suppose that the nonstandard model is κ^+*-saturated, where* $\kappa \geq \max(\kappa(\theta), \gamma(\theta))$.

Let X *be a subspace of* $(\hat{E},\hat{\theta})'$ *and* S *a subset of* $\hat{E}$, *with the cardinality of* S *and the vector space dimension of* X *both being* $\leq \kappa$. *Then there exists a linear mapping* T *from* X *into* $eq_\theta(^*F)/m_\theta(^*F)$ *such that*

(a) $\langle u, Tf\rangle = \langle u,f\rangle$ *for all* $u \in S$ *and* $f \in X$;

(b) $T(f) = f$ *for all* $f \in X \cap (eq_\theta/m_\theta)$;

(c) *for every bounded set* $B \subseteq E$, *the supremum of* $|f|$ *on* $\hat{B}$ *is the same as the supremum of* $T(f)$ *on* $\hat{B}$, *for all* $f \in X$.

When (E,θ) is a normed space then we recover the Local Duality Theorem of Kürsten and Stern by considering in (c) just the closed unit ball of E. In that case the supremum referred to is just the dual norm on $(\hat{E})'$ and eq_θ/m_θ is just $\hat{E}'$. The proof of Theorem 1.14 is quite involved and very interesting; it shows successfully how many nonstandard arguments in Banach space theory can be carried over to the locally convex setting, after being suitably generalized to Banach spaces with a finite number of equivalent norms. See Heinrich (1984) for details and for several interesting applications.

In the Banach space setting there is a well understood and important concept of *finite representability*, which happens to fit very nicely into the nonstandard framework: if X, Y are Banach spaces, then X is finitely represented in Y iff X is isomorphically embedded in a nonstandard hull of Y. Moreover, an easy direct proof can be given that this embeddability condition is independent of the particular nonstandard model we are using, as long as we restrict attention to models which are sufficiently saturated. (They should be at least κ^+-saturated, where κ is the density character of X.)

In the locally convex setting it seems best to take this equivalence as a definition:

1.15 Definition

Let E_1 and E_2 be locally convex spaces. We say E_1 is *finitely represented* in E_2 if E_1 can be linearly isomorphically embedded into some nonstandard hull of E_2.

It is not too hard to work out a standard condition on the family of finite dimensional subspaces of E_1 (a certain uniform embedding condition into E_2) which is equivalent to the concept defined here. In order to carry this analysis out, one must first understand the relationship between a locally convex space (E,θ) and the finite dimensional subspaces of its nonstandard hull $(\hat{E},\hat{\theta})$. This is based on the following simple but useful fact:

1.16 Lemma

Let $x_1,\ldots,x_n \in fin_\theta({}^*E)$ *be such that* $\pi(x_1),\ldots,\pi(x_n)$ *are linearly independent in* $\hat{E}$; *then*:

(a) $x_1,\ldots,x_n$ *are linearly independent over the space of nonstandard scalars;*

(b) *For any nonstandard scalars* $\alpha_1,\ldots,\alpha_n$, *the element* $x = \Sigma\alpha_j x_j$ *is* θ*-finite iff all of* $\alpha_1,\ldots,\alpha_n$ *are finite; when* x *is* θ*-finite,*

$$\pi(x) = \Sigma \mathrm{st}(\alpha_j)\pi(x_j).$$

The proof of this is very elementary; see Theorem 1.8 of Henson & Moore (1972). The argument is valid for all Hausdorff topological vector spaces, in fact.

An immediate consequence of Lemma 1.16 is a trivial proof that on each vector space of finite dimension there is exactly one Hausdorff vector topology. (This proof was first noticed by S. Bellenot (1972).) Indeed, suppose E is a vector space of dimension n and that θ is a Hausdorff vector topology on E. Let $x_1,\ldots,x_n$ be a basis for E, so that $^*x_1,\ldots,{}^*x_n$ is an internal basis for $^*E$, by the Transfer Principle. Since $\pi(^*x_1),\ldots,\pi(^*x_n)$ in $\hat{E}$ are the images of $x_1,\ldots,x_n$ under a linear homeomorphism, $\pi(^*x_1),\ldots,\pi(^*x_n)$ must be linearly independent. Thus Lemma 1.16 applies and part (b) gives a characterization of $\mathrm{fin}_\theta(^*E)$ which is independent of θ. (By the discussion after Lemma 1.3, this shows that θ is uniquely determined.)

Now let (E,θ) be any Hausdorff locally convex space and let $\Sigma = \Sigma(\theta)$ be the family of all continuous seminorms on (E,θ). It is clear that $\{\hat{\rho}: \rho \in \Sigma\}$ will be a family of seminorms on $\hat{E}$ which generates the topology $\hat{\theta}$; here we take

$$\hat{\rho}(\pi(x)) = \mathrm{st}(^*\rho(x))$$

for all $x \in \mathrm{fin}_\theta(^*E)$.

Consider an arbitrary finite dimensional subspace X of $(\hat{E},\hat{\theta})$. By the argument above, the restriction of $\hat{\theta}$ to X is uniquely determined. It must be defined by some seminorm of the form $\hat{\rho}, \rho \in \Sigma(\theta)$; $\hat{\rho}$ must be a norm on X. In order to work successfully with subspaces of this form, we must actually consider at one time a finite number of seminorms $\rho_1,\ldots,\rho_k \in \Sigma(\theta)$ such that each $\hat{\rho}_i$ is a norm on X which defines on X the topology which is just the restriction of $\hat{\theta}$ to X. The basic local-geometric relationship between $(\hat{E},\hat{\theta})$ and (E,θ) is given in the following result:

1.17 Theorem

Let X *be a finite dimensional subspace of* $(\hat{E},\hat{\theta})$ *and let* $\rho_1,\ldots,\rho_k \in \Sigma(\theta)$ *be given, with* $\hat{\rho}_1,\ldots,\hat{\rho}_k$ *each defining on* X *the restriction of* $\hat{\theta}$ *to* X. *Then for each real number* $\lambda > 1$ *there is a linear isomorphism* T *from* X *onto a subspace* Y *of* E *such that* T *is a* λ*-isomorphism from* $(X,\hat{\rho}_i)$ *onto* (Y,ρ_i) *for each* $i = 1,2,\ldots,k$. *Moreover, we can assume that* $T(x) = x$ *for all* $x \in X \cap E$.

Proof. The argument is the same as given in Henson & Moore (1983) for Prop. 3.8, except that k norms must be considered at the same time.◂

Theorem 1.17 yields easily a (somewhat complicated) condition on locally convex spaces (E_1,θ_1) and (E_2,θ_2) that is necessary and sufficient for (E_1,θ_1) to be isomorphically embedded in some nonstandard hull $(\hat{E}_2,\hat{\theta}_2)$. We leave this to the reader. (Compare Henson & Moore (1983), Section 3, for the Banach space case. See Heinrich (1984) for other uses of Banach spaces with finitely many equivalent norms in the study and use of nonstandard hulls of general locally convex spaces.)

2. OPERATORS

Let (X,θ) and (Y,τ) be topological vector spaces and let $T: X \to Y$ be a continuous linear operator. We then obtain a corresponding internal linear operator $^*T: {}^*X \to {}^*Y$. The continuity of T at 0 implies that $^*T$ maps $\mu_\theta(0)$ into $\mu_\tau(0)$. Since T is homogeneous, and so therefore is $^*T$, it follows that $^*T$ maps $\mathrm{fin}_\theta({}^*X)$ into $\mathrm{fin}_\tau({}^*Y)$. (Proof. Suppose $x \in \mathrm{fin}_\theta({}^*X)$ and ε is any infinitesimal scalar. Then $\varepsilon{}^*T(x) = {}^*T(\varepsilon x) \in \mu_\tau(0)$. From this it follows that $^*T(x) \in \mathrm{fin}_\tau({}^*Y)$.)

Hence we may define a mapping $\hat{T}: \hat{X} \to \hat{Y}$ by setting

$$\hat{T}(\pi_\theta(x)) = \pi_\tau(*T(x))$$

for all $x \in fin_\theta(*X)$. It is clear that $\hat{T}$ is linear. Moreover it is easy to show that $\hat{T}$ is continuous from $(\hat{X},\hat{\theta})$ to $(\hat{Y},\hat{\tau})$. In addition, $\hat{T}$ maps $\hat{X}^b$ into $\hat{Y}^b$, because the image under T of any θ-bounded subset of X is a τ-bounded subset of Y. Finally we note that the restriction of $\hat{T}$ to X is exactly the original mapping T. (Here we regard X as a subspace of $\hat{X}$ and Y as a subspace of $\hat{Y}$, via the embeddings $u \to \pi(*u)$.)

If we are given two continuous linear mappings $T: X \to Y$ and $S: Y \to Z$, then of course $*(S \circ T) = *S \circ *T$. Hence also $\hat{}(S \circ T) = \hat{S} \circ \hat{T}$. If $I : X \to X$ is the identity mapping of X, then $\hat{I}$ must be the identity mapping on $\hat{X}$, as is easy to verify. Thus we see that the nonstandard hull construction is a functor on the category of topological vector spaces and continuous linear operators.

The extension $\hat{T}$ can also be defined in many cases even when T is a nonlinear operator. One requires some degree of boundedness and uniform continuity of T in order that $*T$ should map $fin_\theta(*x)$ into $fin_\tau(*Y)$ and should also preserve the infinitesimal equivalence relations involved in the construction of $\hat{X}$ and $\hat{Y}$ as quotients. Also we may sometimes replace the standard operator $*T$ by a more general internal operator S, constructing $\hat{S}$ where circumstances permit.

Surprisingly little has been done to study nonstandard hulls of operators in the general setting of topological vector spaces. (Perhaps some readers can be encouraged to remedy this situation?)

In the Banach space setting this nonstandard hull construction for bounded linear operators has been used in important ways by Pietsch and Heinrich in their study of operator ideals. (See Pietsch (1974) (1980) and especially Heinrich (1980 a).) For Banach spaces it makes sense to consider internal Banach spaces V, W and linear operators $T: V \to W$ which are internal and finite in the sense that the norm $\|T\|$ is finite in $*\mathbb{R}$. Then one can define nonstandard hulls $\hat{V}$, $\hat{W}$ and $\hat{T}$, and $\hat{T}$ will be a bounded linear operator between Banach spaces $\hat{V}$ and $\hat{W}$. (See Henson & Moore (1983).)

Several interesting things can be said about the spectrum of a bounded linear operator T on a complex Banach space X, in relation to the spectrum of its nonstandard extension $\hat{T}$ on $\hat{X}$. We refer to the first ten sections of Dowson (1977) for basic definitions and results concerning spectra of operators.

Recall that λ is in the approximate point spectrum of T iff there are elements x_n with $\|x_n\| = 1$ such that $\|T(x_n) - \lambda x_n\| \to 0$. Since we always assume that our nonstandard model is at least $\aleph_1$-saturated, it is clear that every complex number in the approximate point spectrum of $\hat{T}$ is actually an eigenvalue of $\hat{T}$. The same is true if $\hat{T}$ is replaced by $\hat{Q}$, where Q is any internal bounded linear operator on an internal complex Banach space V, with the norm of Q being finite in $^*\mathbb{R}$. In particular, this implies that every such operator $\hat{T}$ or $\hat{Q}$ has at least one eigenvalue.

It also follows by an easy transfer argument that T and $\hat{T}$ have the same approximate point spectrum (and this includes at least the boundaries of the spectra of T and of $\hat{T}$.) It is also clear that T and $\hat{T}$ have the same spectrum: first note that if S is a bounded inverse on X for $T - \lambda I$, then $\hat{S}$ is an inverse for $\hat{T} - \lambda I$. For the converse, suppose $T - \lambda I$ is not invertible, but $\hat{T} - \lambda I$ is invertible. A simple transfer argument shows that the range of $T - \lambda I$ is dense in X. From this it follows that λ is in the approximate point spectrum of T and hence λ is an eigenvalue of $\hat{T}$, by the above argument. This contradiction completes the proof that T and $\hat{T}$ have the same spectrum.

Moore (1976) has observed that if H is an internal Hilbert space whose internal dimension is a nonstandard integer, and if Q is an internal linear operator on H whose norm is finite in $^*\mathbb{R}$, then every complex number in the spectrum of the nonstandard hull operator $\hat{Q}$ on $\hat{H}$ is an eigenvalue of $\hat{Q}$.

Perhaps the fullest investigation of operators using nonstandard methods has been carried out in the context of Hilbert spaces. In his book *Nonstandard Analysis*, Robinson gives a nonstandard description of the spectral decomposition of a compact self adjoint operator on a separable Hilbert space. (He also gives the proof using these ideas of the well known Robinson-Bernstein invariant subspace theorem for polynomially compact operators.) Moore (1976) gave a simple nonstandard description of the spectral decomposition of a general bounded self adjoint operator on a Hilbert space. In all of these arguments, a common theme is the use of simple ideas from nonstandard analysis to reduce the situation to finite dimensional linear algebra, a setting where everything is somewhat simpler.

In unpublished work, Moore has gone further with the study of bounded linear operators on Hilbert space and their invariants under unitary equivalence (or modifications of this equivalence.) First of all he has given a simple and natural representation of the Calkin algebra. Let H be a complex Hilbert space and let $L(H)$ be the Banach space of all bounded linear operators on H. Also let $K(H)$ be the set of compact linear operators on H; this is a closed linear subspace of $L(H)$ and so we may form the quotient space $A(H) = L(H)/K(H)$. Since $K(H)$ is even an ideal in $L(H)$ under composition of operators, $A(H)$ is an algebra, which is referred to as the Calkin algebra on H.

It is easy to show that the nonstandard hull $\hat{H}$ is also a Hilbert space, with an inner product defined by

$$\langle \pi(x), \pi(y) \rangle = st\langle x, y\rangle$$

for all finite x,y in $^*H$. We let $H^\perp$ denote the orthogonal complement of H inside $\hat{H}$.

If $T \in L(H)$, then T is compact iff $^*T$ maps finite points of $^*H$ to nearstandard points. (This is an easy consequence of the nonstandard characterization of relative compactness.) Hence T is compact iff the range of $\hat{T}$ lies wholly in H. Note further that for any $T \in L(H)$, both H and $H^\perp$ are invariant subspaces for $\hat{T}$. (That is, H is a reducing subspace for $\hat{T}$.)

Given $T \in L(H)$, let $\Phi(T)$ be the restriction of $\hat{T}$ to $H^\perp$. This defines Φ as a continuous algebra homomorphism from $L(H)$ into $L(H^\perp)$. Moreover, the remarks above show that the kernel of Φ is exactly $K(H)$. Hence Φ gives rise (by taking the quotient by $K(H)$) to an algebra embedding of the Calkin algebra $A(H)$ into $L(H^\perp)$. It can be verified routinely that this is in fact an isometric embedding. Moore has shown that this representation of the Calkin algebra can be quite useful.

We also note that in Chadwick & Wickstead (1977) it is shown that $T \in L(H)$ is a Fredholm operator exactly when the restriction of $\hat{T}$ to $H^\perp$ is invertible.

In another unpublished piece of work, Moore has given a partial characterization of the pairs T_1, T_2 in $L(H)$ for which $\hat{T}_1$ and $\hat{T}_2$ are unitarily equivalent (in $L(\hat{H})$.) The result is as follows:

2.1 Theorem

Let H *be a separable Hilbert space. For* $T_1, T_2 \in L(H)$ *with* T_1 *essentially normal, the following conditions are equivalent:*

(a) $\hat{T}_1$ *and* $\hat{T}_2$ *are unitarily equivalent;*

(b) *For each real number* $\varepsilon > 0$ *there is a unitary operator* U *on* H *so that* $T_1 - UT_2U^{-1}$ *has norm* $\leq \varepsilon$.

In the event that (a) *and* (b) *hold, the operator* T_2 *must also be essentially normal.*

It has occasionally been noticed that nonstandard methods can be used to study perturbations of operators. For operators between finite dimensional spaces this is outlined in Lutz and Goze (1981). For operators on infinite dimensional spaces the ideas should be quite effective and deserve to be studied further. For example, if X is a Banach space and $T: X \to Y$ is any bounded linear operator, then a perturbation of T is any internal linear operator of $P: {}^*X \to {}^*X$ such that ${}^*T - P$ is infinitesimal. (There are various concepts worthy of study here; one might require that ${}^*T - P$ have infinitesimal norm, or one might take some weaker notion of infinitesimal for internal operators.) The object of the investigation is then to see how the eigenvalues, eigenvectors, and other objects related to T can be recovered from the internal versions of these objects related to P.

This is very close to the point of view expressed for example by Bellman (1970), in which tools from topology and analysis are used to simplify arguments in linear algebra over $\mathbb{R}$ and $\mathbb{C}$. The nonstandard formulation of these ideas is particularly simple and elegant, however, and suggests some interesting directions for research on infinite dimensional spaces.

In Chapters 5 and 6 of the monograph Albeverio, Fenstad, Høegh-Krohn & Lindstrøm (1986) several nonstandard perturbations of operators are studied in detail; in particular perturbations of certain unbounded operators arising in mathematical physics are analyzed in a detailed way.

There are a number of interesting uses of nonstandard analysis which fit into the subject area of this paper, but which we do not have space to discuss in detail. For example, Wolff (1984) introduces a nonstandard hull construction for strongly continuous semigroups of not necessarily bounded operators on a Banach space, and uses this to investigate the spectra of such semigroups. Stroyan (1978) investigates concepts of differentiability for not necessarily linear mappings between locally convex spaces; this is carried further in Neves (198?), with special emphasis on (HM) spaces.

3. UNIFORM EQUIVALENCE

S. Heinrich (1984) has recently used nonstandard methods (in particular his extension of the Kürsten-Stern Local Duality Theorem, Theorem 1.14 above) to obtain several new applications concerning uniform equivalence of locally convex spaces. This extends his earlier work with Mankiewicz on uniform equivalence of Banach spaces.

The general idea in both situations is relatively simple (although there are, of course, many difficult technical steps needed to carry the program out.) Suppose (X,θ) and (Y,τ) are locally convex spaces, and suppose T is a uniform homeomorphism of X onto Y. (That is, T and T^{-1} are uniformly continuous, 1 - 1 and onto; they need not, of course, be linear.) The first step is to consider the internal mapping $P(x) = \frac{1}{H}\,{}^*T(Hx)$, where H is an infinite integer. It can be shown that P maps $bd_\theta({}^*X)$ exactly onto $bd_\tau({}^*Y)$ and that, for $x,y \in bd_\tau({}^*Y)$, $x - y$ is θ-infinitesimal iff $P(x) - P(y)$ is τ-infinitesimal. Hence P gives rise to a well defined bijection from $\hat{X}^b$ onto $\hat{Y}^b$, which we shall denote by $\hat{P}$. Indeed, it can be shown that both $\hat{P}$ and $\hat{P}^{-1}$ are Lipschitz mappings. (For $\hat{P}$ for example, this means that for each continuous seminorm ρ_1 on $\hat{Y}^b$ there is a continuous seminorm ρ_2 on $\hat{X}^b$ such that

$$\rho_1(\hat{P}u - \hat{P}v) \le \rho_2(u-v)$$

for all $u,v \in \hat{X}^b$. For $\hat{P}^{-1}$ the roles of $\hat{X}^b$ and $\hat{Y}^b$ are reversed.)

This is proved from the uniform continuity of T and T^{-1} much as in the Banach space case, as discussed in Section II of Henson and Moore (1983) for example, by a very elementary calculation. The basic idea goes back to Corson and Klee: uniformly continuous mappings on locally convex spaces are Lipschitz at large distances. By applying $^*T$ to infinite multiples of bounded elements of $^*X$, we are pushing them out to where the Lipschitz estimates are valid, unless the points in question were already infinitely close. The scaling factor of $\frac{1}{H}$ turns out to bring the images under $^*T$ back into the bounded part of $^*Y$.

The use of $\hat{X}^b$ and $\hat{Y}^b$ here, in place of the full nonstandard hulls, is quite important. It was for this reason that Heinrich originally carried out his analysis of the bounded nonstandard hulls.

From this simple idea it follows that if (X,θ) and (Y,τ) are uniformly homeomorphic, then their bounded nonstandard hulls $\hat{X}^b$ and $\hat{Y}^b$ are Lipschitz homeomorphic. (One actually obtains a bit more, concerning isometric aspects of the behavior of these Lipschitz homeomorphisms on bounded sets.)

The nice thing about Lipschitz mappings (which makes them easier to work with than more general classes of uniformly continuous mappings) is that they tend to be differentiable at many points, and these derivatives are themselves linear mappings. Under the right circumstances they already are, or can be used to obtain, linear isomorphisms between the spaces in question or perhaps linear embeddings of each space into the other.

One consequence of these ideas which was obtained by Heinrich (1984) is the following generalization of a Banach space result of Ribe. It shows that even in the very general setting of locally convex spaces, the associated uniform structure carries a surprisingly large amount of information about the linear topological structure itself.

3.1 **Theorem**

Suppose (X,θ) *and* (Y,τ) *are uniformly homeomorphic locally convex spaces. Then* X *is linearly isormophic to a subspace of* $\hat{Y}^b$ *and* Y *to a subspace of* $\hat{X}^b$. *In particular, each of* X, Y *is finitely represented in the other.*

When one of the spaces X, Y in theorem 3.1 has the property that all bounded sets are relatively compact (so that the space equals its bounded nonstandard hull) then the conclusion of Theorem 3.1 implies that the other space must also have this property. Hence each space can be linearly embedded in the other, under these conditions. By extending these ideas (in ways which do not involve any further use of nonstandard methods) Heinrich has shown that far more is true:

3.2 **Theorem**

Suppose (X,θ) *and* (Y,τ) *are uniformly homeomorphic locally convex spaces and that all bounded sets in* X *are relatively compact. Then* (X,θ) *and* (Y,τ) *must be linearly homeomorphic.*

The class of spaces considered in Theorem 3.2 includes many spaces that are important for applications such as spaces of distributions or of analytic functions, as well as duals of Banach spaces given the weak topology.

4. INDISCERNIBLES

Let X be a Banach space, with norm $\|\ \|$, and let (x_n) be a sequence in X. The sequence (x_n) is called *indiscernible* (or sometimes *subsymmetric*) if for every $k > 0$, every $n_1 < n_2 < \ldots < n_k$ and every sequence of scalars $\alpha_1,\ldots,\alpha_k$ we have

$$\Big\|\sum_{j=1}^{k} \alpha_j x_{n_j}\Big\| = \Big\|\sum_{j=1}^{k} \alpha_j x_j\Big\|.$$

In other words, the norm value on the left side is independent of the indices $n_1,\ldots,n_k$ as long as they are listed in increasing order. (The same definition can be applied to any $\{x_n : n \in I\}$ as long as I is an index set on which there is given a linear ordering). To avoid the trivial case where (x_n) is simply a constant sequence, we also require $\|x_1 - x_2\| = \delta > 0$. In recent years indiscernible sequences have become ever more important in Banach space theory, particularly in connection with the "local" theory. The methods of nonstandard analysis have proved to be quite useful in producing, studying and using indiscernibles; see especially in the hands of Krivine (1976), (1984) and Krivine & Maurey (1981).

First we explain how the methods of nonstandard analysis can be used to produce indiscernible sequences which are finitely represented in a given Banach space X. (That is, the sequence will actually occur in the nonstandard hull $\hat{X}$.) We start with a separable subspace Y of $\hat{X}$ and a sequence (y_n) in Y. We suppose that Y is infinite dimensional and that (y_n) satisfies some estimates $0 < a \le \|y_n\| \le b$ for all n and $0 < a \le \|y_n - y_m\|$ for all $n \ne m$. By refining (y_n) to a subsequence if necessary (which we still denote by (y_n) to avoid complicated notation) we may assume that for every $b \in Y$, $\lim_{n\to\infty} \|b + y_n\|$ exists. (This can be done because Y is separable, and hence there are really only a countable number of $b \in Y$ which need to be considered.)

When these limits all exist, we can use an $\aleph_1$-saturation argument to show that there exists an element x in $\hat{X}$ such that for all $b \in Y$ we have

$$\|b + x\| = \lim_{n\to\infty} \|b + y_n\|.$$

(Again we use the separability of Y, so that x needs only to satisfy a countable number of conditions.)

We now iterate this process countably many times, thereby generating a sequence (x_n) in $\hat{X}$ and a sequence of separable subspaces $Y = Y_0 \subseteq Y_1 \subseteq \ldots$ of $\hat{X}$ such that for each $n \ge 1$

(a) $x_1, \ldots, x_n \in Y_n$;

(b) for each $b \in Y_n$,

$$\|b + x_{n+1}\| = \lim_{k\to\infty} \|b + y_k\|.$$

At each stage we add the previously chosen x_n's to Y and choose x_{n+1} satisfying (b) as described above. (Actually we must also refine the given sequence (y_n) to a subsequence at each stage, to be sure that the new limits in (b) exist. At the end we can diagonalize, obtaining a single sequence (y_k) which is a subsequence of the originally given sequence and such that condition (b) holds for every $n \geq 1$.) The sequence (x_n) produced by this process is always indiscernible, as we now show.

4.1 Lemma

For each $r \geq 1$, *each* $n_1 < \dots < n_r$, *each sequence of scalars* $\alpha_1, \dots, \alpha_r$ *and each* $b \in Y$,

$$\Big\| b + \sum_{j=1}^{r} \alpha_j x_{n_j} \Big\| = \lim_{k_r \to \infty} \dots \lim_{k_1 \to \infty} \Big\| b + \sum_{j=1}^{r} \alpha_j y_{k_j} \Big\|.$$

Proof. We consider the case $r = 2$. Note that $\alpha_2 = 0$ or $\frac{1}{\alpha_2}(b + \alpha_1 x_{n_1}) \in Y_{n_1}$, so that

$$\| b + \alpha_1 x_{n_1} + \alpha_2 x_{n_2} \| = \lim_{k_2 \to \infty} \| b + \alpha_1 x_{n_1} + \alpha_2 y_{k_2} \|$$

$$= \lim_{k_2 \to \infty} \lim_{k_1 \to \infty} \| b + \alpha_1 y_{k_1} + \alpha_2 y_{k_2} \|.$$

The last step uses the fact that for each k_2, $\alpha_1 = 0$ or $\frac{1}{\alpha_1}(b + \alpha_2 y_{k_2}) \in Y$.◂

The equation in Lemma 4.1 shows clearly that (x_n) is indiscernible. Note that if $m < n$ then

$$\|x_m - x_n\| = \lim_{k \to \infty} \lim_{l \to \infty} \|y_l - y_k\| \geq a > 0$$

so that (x_n) is a nontrivial sequence.

If the conditions in Lemma 4.1 are satisfied by (x_n), (y_k) and Y, with $(x_n) \subseteq \hat{X}$, then (x_n) is not only indiscernible, but it is *indiscernible over* Y: given any $r > 0$, any scalars $\alpha_1, \dots, \alpha_r$ and any element $b \in Y$, the value of

$$\Big\| b + \sum_{j=1}^{r} \alpha_j x_{n_j} \Big\|$$

is independent of the indices $n_1, \dots, n_r$ as long as $n_1 < n_2 < \dots < n_r$.

If (x_n) is indiscernible over Y, then it is very far from Y in a sense. Namely, we have the following lower estimate:

4.2 Lemma

Let (x_n) *be indiscernible over* Y *and let* $\delta = \|x_1 - x_2\| > 0$. *Then for any* $b \in Y$, *any* $r > 0$ *and any scalars* $\alpha_1,\dots,\alpha_r$,

$$\left\| b + \sum_{j=1}^{r} \alpha_j x_j \right\| \geq \frac{\delta}{2} \max_j |\alpha_j|.$$

Proof. For simplicity take $r = 3$ and, to obtain a contradiction, suppose

$$\| b + \alpha_1 x_1 + \alpha_2 x_2 + \alpha_3 x_3 \| < \frac{\delta}{2}|\alpha_2|.$$

Using the indiscernibility of (x_n) with respect to $b \in Y$ we have also

$$\| b + \alpha_1 x_1 + \alpha_2 x_2 + \alpha_3 x_4 \| < \frac{\delta}{2}|\alpha_2|,$$

and

$$\| b + \alpha_1 x_1 + \alpha_2 x_3 + \alpha_3 x_4 \| < \frac{\delta}{2}|\alpha_2|.$$

So taking differences and using the triangle inequality we obtain

$$\delta|\alpha_2| = |\alpha_2|\,\|x_2 - x_3\| = \|\alpha_2 x_2 - \alpha_2 x_3\| < \delta|\alpha_2|$$

which is impossible. The general argument is similar.◂

Saturation arguments can be used to obtain indiscernible sequences in X which are indexed by richer ordered sets than the integers. Let $\{x_n : n \in \mathbb{N}\}$ be a given indiscernible sequence in $\hat{X}$, and choose $p_n \in {}^*X$ so that $\pi(p_n) = x_n$ for each standard n. Then use $\aleph_1$-saturation to extend to an internal sequence $\{p_n : 1 \leq n \leq H\}$, where H is an infinite integer. By reducing H if necessary (but keeping it infinite) we may obtain that $\{\pi(p_n) : 1 \leq n \leq H\}$ is an indiscernible set in $\hat{X}$. (This really only requires $\aleph_1$-saturation; we must satisfy a countable number of conditions of the following form: for all $n_1,\dots,n_r$ with $1 < n_1 \leq \dots < n_r \leq H$,

$$A - \varepsilon < \left\| \sum_{j=1}^{r} q_j p_{n_j} \right\| < A + \varepsilon;$$

in this condition $r \geq 1$ is a fixed integer, $q_1, \dots, q_r$ are fixed numbers from a countable dense set of scalars, $\varepsilon > 0$ is a fixed rational number, and $A = \left\| \sum_{j=1}^{r} q_j x_j \right\|$. The corresponding condition for the entire field of scalars follows by an approximation argument.)

Note that the ordered set of real numbers can be embedded in the ordered set $\{n: 1 \leq n \leq H\}$ of integers, as long as H is infinite. Just map the real number r to the smallest n satisfying $1 \leq n \leq H$ and

$$r \leq -\sqrt{H} + \frac{n}{\sqrt{H}} .$$

Therefore we get directly a set $\{z_r: r \in \mathbb{R}\}$ of indiscernibles in $\hat{X}$ indexed by the real numbers. Moreover, these indiscernibles have the property that for each $r_1 < \dots < r_k$ in $\mathbb{R}$, the sequences $\{z_{r_j}: j = 1,\dots,k\}$ and $\{x_j: j = 1,\dots,k\}$ are isometrically equivalent. That is, we have greatly enriched the index set.

Note that if the original sequence (x_n) was indiscernible over a separable space Y, then we can carry out the "extension" process just described so that the set $\{\pi(p_n): 1 \leq n \leq H\}$ is also indiscernible over Y. In particular if $(y_k) \subseteq Y$ and if (x_n), (y_k) and Y satisfy the condition in Lemma 4.1, then for each $r \geq 1$, each $1 \leq n_1 < \dots < n_r \leq H$, each sequence of scalars $\alpha_1, \dots, \alpha_r$ and each $b \in Y$, we have

$$\left\| b + \sum_{j=1}^{r} \alpha_j \pi(p_{n_j}) \right\| = \left\| b + \sum_{j=1}^{r} \alpha_j x_j \right\|$$

$$= \lim_{k_r \to \infty} \dots \lim_{k_1 \to \infty} \left\| b + \sum_{j=1}^{r} \alpha_j y_{k_j} \right\|$$

since $\pi(p_j) = x_j$ for standard j and $\{\pi(p_n): 1 \leq n \leq H\}$ is indiscernible over Y. That is, the entire set $\{\pi(p_n): 1 \leq n \leq H\}$ satisfies the condition in Lemma 4.1.

Notice what happens if we consider this set with the opposite ordering on the index set: for convenience of notation, let us write $u_n = \pi(p_{H-n+1})$ for $1 \leq n \leq H$. Then $\{u_n: 1 \leq n \leq H\}$ is indiscernible, or indiscernible over Y, if $\{\pi(p_n)\}$ had the corresponding property. The condition in Lemma 4.1 becomes however:

$$\left\| b + \sum_{j=1}^{r} \alpha_j u_{n_j} \right\| = \lim_{k_1\to\infty} \dots \lim_{k_r\to\infty} \left\| b + \sum_{j=1}^{r} \alpha_j y_{k_j} \right\|$$

with the limits taken in the opposite order. A sequence (u_n) with this property is called a *spreading model* for the original space Y and sequence (y_k). A general treatment of spreading models and their properties may be found in Beauzamy & La Preste (1984). A discussion of the relation between spreading models and the kind of "reverse" spreading models obtained in Lemma 4.1 can be found there and also in Rosenthal (1982). Note that the indiscernible sequences (x_n) and (u_n) are reversed geometrically, in that for any $n_1 < \dots < n_r$ the finite sequences $(x_{n_1},\dots,x_{n_r})$ and $(u_{n_r},\dots,u_{n_1})$ are isometrically equivalent.

Now let (x_n) be an indiscernible sequence in $\hat{X}$. We want to show how the sequence can be modified within $\hat{X}$ (by taking linear combinations) to produce another indiscernible sequence (z_n) in $\hat{X}$ which is also 1-unconditional. (That is, for all scalars $\alpha_1,\dots,\alpha_n$ and all $\varepsilon_j = \pm 1$, $j = 1,\dots,n$

$$\left\| \sum_{j=1}^{n} \varepsilon_j \alpha_j z_j \right\| = \left\| \sum_{j=1}^{n} \alpha_j z_j \right\|.)$$

We let $u_k = x_1 - x_2 + \dots + x_{2k-1} - x_{2k}$ for each $k \geq 1$. The first case we consider is when there is a constant C such that $\|u_k\| \leq C$ for all $k \geq 1$. In that case let $v_k = x_{2k-1} - x_{2k}$ for all k, so that $\|v_1 + v_2 + \dots + v_k\| \leq C$ for all k. Note that the sequence (v_k) is also indiscernible. It follows that for every $\varepsilon_j = \pm 1$, $j = 1,\dots,n$

$$\left\| \sum_{j=1}^{n} \varepsilon_j v_j \right\| \leq 2C.$$

To see this, split the sum into two parts, depending on whether $\varepsilon_j = +1$

or $\varepsilon_j = -1$. Then for some l, m with $l + m = n$

$$\left\|\sum_{j=1}^{n} \varepsilon_j v_j\right\| \le \left\|\sum_{j=1}^{l} v_j\right\| + \left\|\sum_{j=1}^{m} v_j\right\| \le 2C,$$

using the indiscernibility of (v_k). But then we achieve a c_o upper estimate

$$\left\|\sum_{j=1}^{n} \alpha_j v_j\right\| \le 2C \max_j |\alpha_j|$$

for all n and all scalars $\alpha_1,\dots,\alpha_n$. Indeed if $|\alpha_j| \le 1$ for $j = 1,\dots,n$ then the vector $(\alpha_1,\dots,\alpha_n)$ is in the absolutely convex hull of the 2^n vectors $(\pm1,\dots,\pm1)$ in $\mathbb{R}^n$.

This estimate together with Lemma 4.2 shows that (v_j) is equivalent to the c_o basis. A classical argument due to Banach now shows that there is a sequence in $Z = \mathrm{span}\,(v_j)$ which is asymptotically equivalent to the c_o basis; this sequence has a spreading model in $\hat{X}$, which must be isometric to the c_o basis (and is hence 1-unconditional.)

It remains to consider the case where the norms $\|u_k\|$ are not uniformly bounded. It is convenient to use a saturation argument here. Choose $p_k \in {}^*X$ for each $k \ge 1$, so that $\pi(p_k) = x_k$. Using $\aleph_1$-saturation as above we may extend the sequence (p_k) to an internal sequence $\{p_k\colon 1 \le k \le H\}$ in $^*X$, where H is an infinite integer, in such a way that the sequence $\{p_k\colon 1 \le k \le H\}$ is almost indiscernible in $^*X$, even when certain hyperfinite linear combinations are concerned. (By "almost" here we mean that $=$ in the norm equations is replaced by the relation $\approx$ of being separated by only an infinitesimal.) Now choose an infinite integer N, small enough so that N/H is infinitesimal, and so that $A = \|p_1 - p_2 + \dots + p_{2N-1} - p_{2N}\|$ is infinite. For each standard integer n define

$$t_n = \frac{1}{A}\sum_{j=1}^{N} (-1)^{j+1} p_{(n-1)N+j}$$

and take $w_n = \pi(t_n)$ for each such n. Since the indices used in defining t_n never get beyond H, the sequence (w_n) is indiscernible. Moreover this sequence is automatically 1-unconditional. To see this, consider for example the norm of an expression like $-w_1 + w_2$. We calculate

$$\|-w_1 + w_2\| \approx \|-t_1 + t_2\| = \frac{1}{A}\Big\|\sum_{j=1}^{N} (-1)^j p_j + \sum_{j=1}^{N} (-1)^{j+1} p_{N+j}\Big\|$$

$$\approx \frac{1}{A}\Big\|\sum_{j=1}^{N} (-1)^j p_j + \sum_{j=1}^{N} (-1)^{j+1} p_{N+j+1}\Big\|$$

and similarly

$$\|w_1 + w_2\| \approx \|t_1 + t_2\|$$

$$= \frac{1}{A}\Big\|\sum_{j=1}^{N} (-1)^{j+1} p_{j+1} + \sum_{j=1}^{N} (-1)^{j+1} p_{N+j}\Big\|$$

$$\approx \frac{1}{A}\Big\|\sum_{j=1}^{N} (-1)^{j+1} p_{j+1} + \sum_{j=1}^{N} (-1)^{j+1} p_{N+j+1}\Big\|.$$

Taking the difference between these expressions and using the triangle inequality in $*X$ we see

$$\big|\|w_1 + w_2\| - \|-w_1 + w_2\|\big| \lessapprox \frac{1}{A} \| -p_1 + (-1)^N p_{N+1}\| \approx 0$$

and hence

$$\|w_1 + w_2\| = \|-w_1 + w_2\|.$$

The general argument to show that (w_n) is 1-unconditional is similar. This gives an elementary proof of the following basic result:

4.3 Theorem

Let X *be any infinite dimensional Banach space over* $\mathbb{R}$. *Then there is a nontrivial indiscernible sequence in* $\hat{X}$ *which is also 1-unconditional.*

If X is separable and we start with a sequence (x_n) which is indiscernible over X, then the argument can be modified to produce a 1-unconditional sequence which is indiscernible over X.

An alternate approach to Theorem 4.3 is described in Beauzamy & La Preste (1984). It uses Rosenthal's ℓ_1-theorem (which depends on somewhat deeper combinatorial arguments) to show that if (x_n) is a spreading model then the sequence $(x_{2n} - x_{2n+1})$ is a 1-conditional spreading model.

What we have described so far in this section provides an introduction to Section 10 of Henson & Moore (1983), where a proof of the following important result due to Krivine (1976) is presented; here we consider spaces over $\mathbb{R}$ only.

4.4 **Theorem**

Let (x_n) *be a sequence in a Banach space* X, *with the linear span of* (x_n) *being infinite dimensional. Then the usual basis of* c_o *or of* ℓ_p *(for some* $1 \le p < \infty$*) is block-finitely represented (isometrically) in* (x_n).

In particular c_o or some ℓ_p $(1 \le p < \infty)$ is (isometrically) finitely represented in X. Since it is not too difficult to show that ℓ_2 is finitely represented in each of c_o or ℓ_p, Krivine's Theorem provides an alternate proof of Dvoretsky's Theorem: *If* X *is any infinite dimensional Banach space, then* ℓ_2 *is (isometrically) finitely represented in* X.

Model theoretic arguments (in the form of ultrapower proofs) have played a continuing role in the study of properties of sequences in Banach spaces. One of the most important of these concerns the class of *stable* Banach spaces introduced by Krivine & Maurey (1981). A Banach space X is *stable* iff whenever (x_n) and (y_n) are bounded sequences such that the limits

$$L_1 = \lim_{n\to\infty} \lim_{m\to\infty} \|x_n + y_m\| \quad \text{and} \quad L_2 = \lim_{m\to\infty} \lim_{n\to\infty} \|x_n + y_m\|$$

both exist, then one has $L_1 = L_2$. Note that if (x_n) and (y_n) are any two bounded sequences in X, then they can be refined to subsequences for which the limits L_1 and L_2 exist. It is easy to show that X is stable iff for every two bounded sequences (x_n) and (y_n) in X

$$\inf_{m<n} \|x_n + y_m\| \le \sup_{m>n} \|x_n + y_m\|.$$

In Krivine & Maurey (1981) it is proved that if X is stable then either c_o or ℓ_p (for some $1 \le p < \infty$) can be $(1 + \varepsilon)$-isomorphically embedded in X for each $\varepsilon > 0$. (Not just finitely represented in X as in Theorem 4.4). Perhaps the most important stable spaces are the L_p spaces, $1 \le p < \infty$, and thus also their subspaces. (The fact that every subspace of L_p embeds c_o or some ℓ_q, $1 \le q < \infty$, had been proved earlier by Aldous (1981), verifying a conjecture of Rosenthal. The proof by Krivine and Maurey uses only elementary Banach space geometry and is somewhat more general.) We refer the reader to Krivine & Maurey (1981) or to Garling (1982) for the proof of this theorem, and to Krivine (1984) for an interesting summary.

5. ISOMORPHIC NONSTANDARD HULLS

In Henson & Moore (1983) there is a full discussion of the equivalence relation $X \equiv_A Y$ between Banach spaces, which is defined to hold exactly when X and Y have nonstandard hulls which are isometrically isomorphic. This relation is important in many cases where applications of nonstandard methods are concerned, and there are now many classes of Banach spaces for which this relation is quite well understood. Moreover, it is understood now that $\equiv_A$ is in fact the relation of elementary equivalence for a certain formalized logic for Banach spaces, which is interesting in its own right. (See Sections 5 and 8 of Henson & Moore (1983).)

We can summarize the state of knowledge concerning the classical Banach spaces as follows. (Here C(K) denotes the space of all continuous, scalar valued functions on K, with the supremum norm, where K is a compact Hausdorff space; $C_o(K,t)$ is the closed subspace of C(K) consisting of functions which vanish at t; when σ is a homeomorphism of K onto itself which is an involution ($\sigma^2 = id_K$), then $C_\sigma(K)$ is the closed subspace of C(K) consisting of functions f satisfying $f(\sigma(x)) = -f(x)$ for all $x \in K$; B(K) is the Boolean algebra of all open subsets of K.)

5.1. Theorem

The following classes of Banach spaces are closed under the equivalence relation $\equiv_A$:

(1) $L_p(\mu)$-*spaces* (p *fixed*, $1 \le p < \infty$);

(2) $C(K)$-*spaces*;

(3) $C_\sigma(K)$-*spaces*.

Proof. See Section 6 of Henson & Moore (1983) and also Heinrich, Henson and Moore (1983) (1986).◂

A fundamental fact is that $X \equiv_A \hat{X}$ holds, for any nonstandard hull $\hat{X}$ of X. Therefore Theorem 5.1 includes the interesting (and non-trivial) information that if $\hat{X}$ is isometric to an $L_p(\mu)$-space ($1 \le p < \infty$), or a $C(K)$-space, or a $C_\sigma(K)$ space, then the same condition must also hold for X.

If Y is a finite dimensional space, then $X \equiv_A Y$ iff X is isometric to Y. Thus in studying $\equiv_A$ it suffices to consider just the situation where X and Y are both of infinite dimension; we do this below without specific mention. In many cases a quite precise classification is known:

5.2 Theorem

Let Y *be given (as listed below in various cases). Then* $X \equiv_A Y$ *if and only if the given condition on* X *is satisfied*:

(1) Y *is* ℓ_2; X *is a Hilbert space*;

(2) Y *is* $L_p(\mu)$, $1 \le p < \infty$; X *is isometric to* $L_p(\lambda)$ *for some measure* λ *such that* λ *has infinitely many atoms, if* μ *does, and* λ, μ *have the same number of atoms otherwise*;

(3) Y *is* $C(K)$, K *a totally disconnected compact Hausdorff space*; X *is isometric to another space* $C(K')$ *of the same kind, where the Boolean algebras* $B(K)$ *and* $B(K')$ *are elementarily equivalent*;

(4) Y *is* ℓ_∞; X *is isometric to a* $C(K)$ *space where* K *is a totally disconnected, compact Hausdorff space which has a dense set of isolated points*;

(5) Y *is* c_o; X *is isometric to a space* $C_\sigma(K)$ *where* K *is as in* (4), σ *is an involutory homeomorphism of* K *onto itself such that* σ *has a unique fixed point* t *and* t *is not isolated in* K;

(6) Y *is* $C(\Delta)$, *where* Δ *is the Cantor set*; X *is isometric to a* $C(K)$ *space where* K *is a totally disconnected, compact Hausdorff space without any isolated points.*

Proof**.** See Henson & Moore (1983), Section 6, and Heinrich, Henson & Moore (1983) (1986). These papers discuss several other examples of this kind. In the second of these papers it is shown that $X \equiv_A c_o$ need *not* imply that X can be given a compatible Banach lattice structure, even though $\hat{c}_o$, and hence $\hat{X}$, are Banach lattices.◄

In the setting of general Banach space theory, it is also important to consider pairs of spaces with nonstandard hulls which are linearly isomorphic (but not necessarily isometric.) We introduce a relation $X \equiv_A^\lambda Y$ for each $\lambda \geq 1$; it is defined to hold iff there are nonstandard hulls $\hat{X}$, $\hat{Y}$ and a linear isomorphism T form X onto Y which is a λ-isomorphism, in the sense that $\|T\| \leq \lambda$ and $\|T^{-1}\| \leq \lambda$. Evidently $\equiv_A^1$ is just equivalent to $\equiv_A$, the isometric case. Each relation $\equiv_A^\lambda$ is symmetric, and for any spaces X, Y, Z and any real numbers $\alpha,\beta \geq 1$,

$$X \equiv_A^\alpha Y \text{ and } Y \equiv_A^\beta Z \text{ implies } \equiv_A^{\alpha\beta} Z.$$

X and Y have isomorphic nonstandard hulls iff $X \equiv_A^\alpha Y$ holds for some real number $\alpha \geq 1$.

It is important to know that the relation $X \equiv_A^\alpha Y$ has a characterization independent of the nonstandard model, and that when $X \equiv_A^\alpha Y$ holds, then $\hat{X}$ and $\hat{Y}$ will be α-isomorphic whenever the nonstandard hulls are constructed using a nonstandard model which satisfies some reasonable property of "richness" (for example a saturation property or one of the isomorphism properties.). The required information is given in the following result:

5.3 Theorem

For any Banach spaces X, Y *and* $\alpha \geq 1$, *the following are equivalent*:

(a) $X \equiv_A^\alpha Y$;

(b) *For each positive bounded sentence* σ *and each* $\lambda > \alpha$, *if* σ *holds in* X *then the approximation* σ_λ *holds in* Y.

Moreover if $X \equiv_A^\alpha Y$ *and both nonstandard hulls are constructed using a nonstandard model which satisfies the* $\aleph_1$*-isomorphism property, then* $\hat{X}$ *and* $\hat{Y}$ *are* α*-isomorphic*.

Proof. See Heinrich & Henson (1986). For the isometric case ($\alpha = 1$) see Henson & Moore (1983). This paper contains a discussion of positive bounded logical sentences and their approximations, as well as the $\aleph_1$-isomorphism property.◂

In Heinrich & Henson (1986) there is given a game-theoretic characterization of the relation $X \equiv_A^\alpha Y$ which displays another aspect of the geometric equivalence which it expresses. (See Theorem 4 of that paper and also Henson & Moore (1983), Section 8.)

Let us write $X \equiv_A^\infty Y$ iff there exists $\alpha \geq 1$ so that $X \equiv_A^\alpha Y$. In the following we consider only infinite dimensional spaces.

5.4 Theorem

(a) *Fix* $1 < p < \infty$. *If* X *and* Y *are any two* $L_p(\mu)$ *spaces, then* $X \equiv_A^\infty Y$; *indeed, for any Banach space* Z, $Z \equiv_A^\infty L_p$ *iff* Z *is a* $\mathcal{L}_p$*-space*.

(b) *If* X *and* Y *are isomorphic to complemented subspaces of* $L_1(\mu)$*-spaces, then* $X \equiv_A^\infty Y$, *in particular if* Z *is an* $\mathcal{L}_1$*-space which is complemented in its second dual space, then* $Z \equiv_A^\infty L_1$.

(c) *If* X *and* Y *are isomorphic to complemented subspaces of abstract* M*-spaces, then* $X \equiv_A^\infty Y$; *in particular, if* X *is a* $\mathcal{L}_\infty$*-space which is complemented in its second dual space, then* $X \equiv_A^\infty C(K)$. (*Here* C(K) *represents any infinite dimensional* C(K)*-space*).

Proof. See Heinrich & Henson (1986), Section 5.◂

Theorem 5.4 shows that $\equiv_A^\infty$ is a resaonably coarse equivalence relation, but that its equivalence classes are often interesting classes of Banach spaces. These facts have yet to be exploited in applications, and also the corresponding equivalence relation for operators on Banach spaces has yet to be given the attention it deserves. (See Heinrich (1980a) for many interesting ideas and Pietsch (1980) for a thorough presentation of the classes of operators to which these nonstandard methods are likely to have applications.)

The equivalence relation $\equiv_A^\infty$ has some reasonable persistence properties, at least under geometric conditions which are often needed when investigating isomorphism of Banach spaces. For example it is persistent under passing to dual spaces:

5.5 **Theorem**

If $X \equiv_A^\infty Y$ *and* X, Y *are each isomorphic to their squares, then* $X' \equiv_A^\infty Y'$. *If* X, Y *are super-reflexive, then the extra condition may be dropped.*

Proof. In the super-reflexive case, this is trivial: $\hat{X}'$ is $\hat{\ }(X')$ and $\hat{Y}'$ is $\hat{\ }(Y')$; when $\hat{X}$ and $\hat{Y}$ are isomorphic, so must be $\hat{X}'$ and $\hat{Y}'$. For the general case see Heinrich & Henson (1986).◂

Note that no isometric version of Theorem 5.5 is possible: if μ is a measure with infinitely many atoms and also a nonatomic part, then ℓ_1 and $L_1(\mu)$ have isometric nonstandard hulls, but their dual spaces ℓ_∞ and $L_\infty(\mu)$ do not. (This follows from Theorem 5.2 (2) and (3).) By Theorem 5.4 or 5.5, ℓ_∞ and $L_\infty(\mu)$ do have linearly isomorphic nonstandard hulls.

REFERENCES

Albeverio, S., Fenstad, J.E., Høegh-Krohn R., & Lindstrøm T. (1986). *Nonstandard Methods in Stochastic Analysis and Mathematical Physics*, Academic Press, New York.

Aldous, D. (1981). Subspaces of L_1 via random measures, *Trans. Amer. Math. Soc.* **267**, 445-463.

Beauzamy, B. & La Preste J.T., (1984). *Models Etales des Espaces de Banach*, Hermann, Paris.

Bellenot, S. (1972). Nonstandard topological vector spaces; in *Lecture Notes in Mathematics* **369**, 37-39.

Bellman, R. (1970). *Introduction to Matrix Analysis*, McGraw-Hill.

Benninghofen, B., Richter M.M. & Stroyan, K.D. (198?). Superinfinitesimals in topology, to appear.

Bierstedt, K.D. & Bonet, J. (198?). Stefan Heinrich's density condition for Frechet spaces and the characterization of the distinguished Kothe echelon spaces, to appear.

Chadwich, J.J.M. & Wickstead, A.W. (1977). A quotient of ultrapowers of Banach spaces and semi-Fredholm operators, *Bull. London Math. Soc.* **9**, 839-873.

Dowson, H.P. (1977). *Spectral Theory of Linear Operators*, Academic Press, New York.

Garling, D.J.H. (1982). Stable Banach spaces, random measures and Orlicz function spaces; in *Lecture Notes in Mathematics* **928**, 121-175.

Heinrich, S. (1980a). Finite representability and super-ideals of operators, *Dissertiones Math.* **172**.

Heinrich, S. (1980b). Ultraproducts in Banach space theory, *J. Reine Angew. Math.* **313**, 72-104.

Heinrich, S. (1984). Ultrapowers of locally convex spaces and applications, I and II, *Math. Nachrichten* **118**, 285-315, and **121**, 211-229.

Heinrich, S. & Henson, C.W. (1986). Model theory of Banach spaces II: isomorphic equivalence, *Math. Nachrichten* **125**, 301-317.

Heinrich, S., Henson C.W. & Moore, L.C. Jr., (1983). Elementary equivalence of L_1-preduals; in *Lecture Notes in Mathematics.* **991**, 79-90.

Heinrich, S., Henson C.W. & Moore, L.C. Jr., (1986). Elementary equivalence of $C_\sigma(K)$-spaces for totally disconnected, compact Hausdorff K, *J. Symb. Logic* **51**, 135-146.

Henson C.W., (1976). Nonstandard hulls of Banach spaces, *Israel J. Math.* **25**, 108-144.

Henson C.W. & Moore, L.C. Jr., (1972). The nonstandard theory of topological vector spaces, *Trans. Amer. Math. Soc.* **172**, 405-435.

Henson C.W. & Moore, L.C. Jr., (1974). Invariance of the nonstandard hulls of a locally convex space; in *Lecture Notes in Mathematics.* **369**, 71-84.

Henson C.W. & Moore, L.C. Jr., (1983). Nonstandard analysis and the theory of Banach spaces; in *Springer Lecture Notes in Math.* **983**, 27-112.

Köthe, G. (1969). *Topological Vector Spaces*, Springer-Verlag.

Krivine, J.-L. (1976). Sous espaces de dimension finie des espaces de Banach réticulés, *Ann. of Math.* **104**, 1-29.

Krivine, J.-L. (1984). Méthodes de théorie des modèles en géométrie des espaces de Banach, *General Logic Seminar (Univ. Paris VII, 1982-83)*, Publ. Math. Univ. Paris VII.

Krivine, J.-L. & Maurey, B. (1981). Espaces de Banach stables, *Israel J. Math.* **39**, 273-281.

Lindstrøm, T. An invitation to nonstandard analysis, this volume.

Lutz, R. & Goze, M. (1981). *Nonstandard Analysis*, *Lecture Notes in Mathematics* **881** (1981).

Moore, L.C. Jr., (1976). Hyperfinite extensions of bounded operators on a separable Hilbert space, *Trans. Amer. Math. Soc.* **218**, 285-295.

Neves. V. (198?) Infinitesimal calculus in HM spaces, to appear.

Pietsch, A. (1974). Ultraprodukte von Operatoren in Banachräumen, *Math. Nachr.* ***61***, 123-132.

Pietsch, A. (1980). *Operator Ideals*, North-Holland, Amsterdam.

Rosenthal, H.P. (1982). Some remarks concerning unconditional basic sequences, *Longhorn Notes* (Univ. of Texas), 15-47.

Schaefer, H.H. (1966). *Topological Vector Spaces*, Macmillan, New York.

Schrieber, M. (1972). Quelques remarques sur les caractérisations des éspaces L^p, $0 \leq p < 1$, *Ann. Inst. Henri Poincaré Sect.* B *(N.S.)*, **8**, 83-92.

Stroyan, K.D. (1978). Infinitesimal calculus in locally convex spaces I, Fundamentals, *Trans. Amer. Math. Soc.* **240**, 363-383.

Stroyan, K.D. (1983). Myopic utility functions on sequential economies, *J. Math. Econ.* **11**, 267-276.

Stroyan, K.D. & Benninghofen, B. (198?) Bounded weak star continuity, to appear.

Stroyan, K.D. & Luxemburg, W.A.J. (1976). *Introduction to the Theory of Infinitesimals*, Academic Press, New York

Wolff M. (1984). Spectral theory of group representations and their nonstandard hulls, *Israel. J. Math.* **48**, 205-224.

APPLICATIONS OF NONSTANDARD ANALYSIS IN MATHEMATICAL PHYSICS

SERGIO ALBEVERIO

Abstract. The aim of this article is to give a short introduction to applications of nonstandard analysis in mathematical physics. Two basic techniques, the hyperfinite and the hypercontinuous are presented, together with illustrations mainly from quantum mechanics, polymer physics and quantum field theory.

1. INTRODUCTION

Nonstandard analysis is a specific mathematical technique as well as a way of thinking: both aspects are also represented in the interaction between nonstandard analysis and mathematical physics, which is the subject of this paper. As in other domains of application of nonstandard analysis, mathematical physics (or the mathematical study of problems of physics) has particular aspects that make some of the nonstandard methods most natural to use. Often in mathematical physics one has to study systems with many interacting components, idealized as systems with *infinitely many degrees of freedom*. To look upon a fluid or a gas as a composed of infinitely many particles might seem at first sight to be a very rough abstraction, but is a useful one for mathematical purposes, being in some sense easier to handle than the more realistic case of finitely many particles. On the other hand, in quantum field theory, for example, the abstraction itself creates its own problems, like the famous ones connected with divergences, about which we will say a few more words below; sometimes also it is only in a limit, like that of infinitely many degrees of freedom, that one "sees" some specific phenomenon, raising challenging problems, like phase transitions[1] in thermodynamic systems (only perceived in the so called "infinite volume" or "thermodynamic limit"), or exact invariance properties (under a

[1] Singularities in some thermodynamic functions in their dependence on physical parameters such as temperature - or in correlation functions.

continuous group of symmetries), in systems idealized as "continua" (as in field theories). Another particularity of mathematical physics is that one often encounters singularities that arise naturally and are often forced upon us by symmetries of the problems (for example, singular interactions like the Coulomb $1/|x|$ potential in classical and quantum physics, or singularities of Green's functions in quantum field theory, essentially forced upon us by the joint requirements of locality and relativistic invariance). On the other hand, singularities sometimes present formal computational advantages and are exploited as such in heuristic computations (we shall see below an example in connection with the so called δ-potential in quantum mechanics); however the mathematical justification for such computations is hard to find. We shall try to make and illustrate the point that the model of the continuum used by nonstandard analysis and the nonstandard techniques themselves provide a convenient setting for handling problems of mathematical physics such as the ones mentioned above, and, more generally, enlarge the range of "natural models" worthwhile considering and studying.[2] Basically we shall distinguish two types of nonstandard approach to problems in mathematical physics, according to the modelling of the equations involved, one which we call the *hyperfinite* or *fine discrete* approach and the other which we call the *hypercontinuous* approach. The first replaces the field $\mathbb{R}$ of real numbers by a lattice with infinitesimal spacing, and correspondingly, for example, differential equations by difference equations; the other replaces $\mathbb{R}$ by the hyperreal number field ${}^*\mathbb{R}$ with singular terms in equations replaced by smooth nonstandard ones. We shall see below many illustrations of these two basic methods. In some of the other papers in this volume there are applications of nonstandard analysis to physics that are complementary to the ones we present; see, for example in the papers of Arkeryd, Diener & Diener, Keisler and Lindstrøm in this volume. We also supplement our paper with a bibliography of topics not discussed.

Basically all the material discussed in this article is drawn from the book Albeverio, Fenstad, et al. (1986). In Section 2 we shall consider as a case study the subject of Schrödinger operators of the form $-\Delta+\lambda\delta$. In section 3 we shall digress a little on nonstandard methods for

---

[2] For a long time in p.d.e.'s, for example, *natural models* were formulated with smooth coefficients. In connection with certain problems, however, distributional or even worse coefficients are suitable.

handling operators on Hilbert spaces. In Section 4 we discuss hyperfinite Dirichlet forms, as a tool to study singular differential operators, with applications to quantum mechanics. In Section 5 we continue the discussion of Section 4 by concentrating on so called energy forms, and we shall mention applications to diffusions and quantum mechanics on fractal sets.

In Section 6 we discuss applications to the polymer measures of physical chemistry. In Section 7 we discuss quantum fields and we relate them to the polymer measures of Section 6. In Section 8 we briefly mention other applications and draw some conclusions.

2. SINGULAR INTERACTIONS IN SCHRÖDINGER OPERATORS: A CASE STUDY

In (low energy) nuclear physics the basic nuclear forces are of very short range. Typically a deuteron (neutron/proton system) is fairly well described quantum mechanically by two point particles interacting by a real-valued potential function V (i.e. a force $F = -\operatorname{grad} V$) very negatively peaked at $x = 0$ and going rapidly to zero for $x \neq 0$,[3] something like $V_\varepsilon(x) = \lambda_\varepsilon \delta_\varepsilon(x)$, with $\delta_\varepsilon(x)$ an ε-approximation of Dirac's delta function (for example $\varepsilon > 0$, $\delta_\varepsilon(x) = \left[\frac{4}{3}\pi\varepsilon^3\right]^{-1} \chi_\varepsilon(|x|)$, with χ_ε the characteristic function of the ball of radius ε centered at the origin), and λ_ε a suitable negative *coupling constant* (giving the "strength" of the attraction). It is well known that in quantum mechanics the dynamics of 2-particle systems (in $\mathbb{R}^3$) is described by the Schrödinger equation, which in suitable units is the evolution equation $i\frac{\partial}{\partial t}\psi = H\psi$, with i the imaginary unit, t = time, and $H = -\Delta + V$ the Hamiltonian or *energy operator* (where Δ is the Laplacian in $\mathbb{R}^3$ and V is a real-valued function, the potential, describing the interaction). The unknown is ψ, a complex-valued function of t and x, with given initial condition $\psi(0,x) = \varphi(x)$, usually taken in $L^2(\mathbb{R}^3)$, so that one looks for solutions ψ which at all times t are in $L^2(\mathbb{R}^3)$. In this way H is also interpreted as an operator in $L^2(\mathbb{R}^3)$. It is well known that the equation has a unique solution if φ is, for example, in Schwartz test function space $\mathcal{S}(\mathbb{R}^3)$ and H is self-adjoint (which is really a restriction upon V; for example, V

---

[3] A mean distance of the order of a few Fermis (10^{-13} cm).

measurable and bounded outside a set of Lebesgue measure zero is enough). The above potential V_ε satisfies these requirements, of course, and the solution here is given by $\psi(t) = e^{-itH}\varphi$, where e^{-itH} is the unitary group in $L^2(\mathbb{R}^3)$ generated by H (Stone's theorem). We thus see that the quantity determining ψ, hence the time evolution (i.e. the dynamics) is the unitary group e^{-itH}. Since $-\Delta$ and V do not commute, it is of course not immediate to compute e^{-itH} even when V is "simple" (computing e^{-itH} is, by the above, equivalent to solving the Schrödinger equation for all time and arbitrary initial condition in the domain of V, for example in $\mathcal{S}(\mathbb{R}^3)$). By "functional calculus" the computation of e^{-itH} is equivalent with computing the resolvent $(H-z)^{-1}$ for Im $z \neq 0$. In the case of the deuteron, $V = V_\varepsilon$, and we really want ε *very* small. In fact, the physicists (starting with Thomas (1935)), thought of ε as infinitesimal and computed $(H-z)^{-1}$ heuristically by treating $V_\varepsilon(x)$ algebraically as if it were $\lambda\delta(x)$ (with δ = Dirac's delta function at zero), and with some suitably chosen λ (which turns out to be a negative infinitesimal). The arugment is like this: for V bounded and smooth,

$$(H-z)^{-1}(x,y) = G_z(x,y) + \sum_{j=1}^{\infty} (-1)^j[(G_zV)^j\ G_z](x,y), \qquad (2.1)$$

with $G_z(x,y) \equiv (-\Delta-z)^{-1}(x,y) \equiv G_z(x-y)$, the Green's function for $-\Delta-z$, and the series converges for |Im z| sufficiently large (for any operator A we write A(x,y) for its distributional kernel, so that $(Af)(x) = \int A(x,y)f(y)dy$, for, say, $f \in C_0^\infty$ (where C_0^∞ denotes the smooth functions of compact support). An heuristic (admittedly somewhat brutal) replacement of V(x) by $\lambda\delta(x)$ yields for the j-th term on the right of (2.1):

$$((G_zV)^jG_z\delta_y)(x) = \int G_z(x-x_1)\lambda\delta(x_1)G_z(x_1-x_2)\lambda\delta(x_2)\ldots$$
$$\ldots G_z(x_{j-1}-x_j)\lambda\delta(x_j)G_z(x_j-y) = \lambda^jG_z(x)G_z(y)G_z(0)^{j-1}.$$

Hence heuristically the right side of (2.1) becomes

$$G_z(x,y) - \lambda\sum_{j=1}^{\infty}\lambda^{j-1}G_z(0)^{j-1}G_z(x)G_z(y) =$$
$$= G_z(x,y) + G_z(x)\{\tfrac{1}{\lambda} + G_z(0)\}^{-1}G_z(y). \qquad (2.2)$$

However we have to remark that $G_z(x,y) \equiv G_z(x-y) = -\frac{1}{2\pi}\ln|x-y| + C_z(x-y)$

with $C_z(x-y)$ uniformly bounded. Hence in particular $G_z(0)$ is $+\infty$ for all z, but $G_z(0)-G_0(0) = C_z(0)-C_0(0)$ is finite. So obviously the above computation was really only formal, $G_z(0)$ being infinite. If we believe, nevertheless, that the final formula has heuristic meaning then we see, noting that $G_z(0)$ is infinite, that $(H-z)^{-1}(x,y) \neq G_z(x-y)$ (hence the potential V is effectively "felt") if we choose $\frac{1}{\lambda}$ so as to compensate the divergence of $G_z(0)$. Of course λ should be independent of z, hence we set

$$\frac{1}{\lambda} = -G_0(0) - (C_z(0)-C_0(0)) + \alpha, \quad \text{with } \alpha \text{ independent of } z.$$

This computation of the point interaction Hamiltonian H following a "don't worry" principle is heuristic but correct. How can we make it rigorous if not by nonstandard analysis?

Let us start with the hypercontinuous approach. The basic idea of this method is to replace the singular interaction, formally taken to be (infinitesimal) × (δ-function at 0), by a smooth nonstandard realization of the interaction, namely $V_\varepsilon(x) = \lambda_\varepsilon\delta_\varepsilon(x)$, with δ_ε a smooth realization of the δ-function at 0, for example

$$\delta_\varepsilon(x) = (\tfrac{4}{3}\pi\varepsilon^3)^{-1} \chi_\varepsilon(|x|), \tag{2.3}$$

with χ_ε the characteristic function of the unit ball with center at 0 and radius ε, $\varepsilon > 0$ infinitesimal.

Can we compute the resolvent $(-\Delta + V_\varepsilon - z)^{-1}$ and then take the standard part and get the resolvent of an operator realizing the formal operator H described above? The answer is yes. For this, one makes a computation like the one which we would do if ε were real positive and we wanted to compute the Green's function of the rotation invariant potential V (of course we exploit here the transfer principle). We give more details below; let us first state the theorem one arrives at:

2.4 Theorem

Let $\varepsilon > 0$ be infinitesimal and let χ_ε be the characteristic function of a ball of radius ε in ${}^*\mathbb{R}^3$, with center at the origin. For $\alpha,\beta \in \mathbb{R}$, and $\gamma \in \mathbb{N}_0 \equiv \mathbb{N} \cup \{0\}$, define

$$\lambda_\varepsilon(\alpha,\beta,\gamma) \equiv -(\gamma + \tfrac{1}{2})^2 \tfrac{4}{3}\pi^3\varepsilon + \frac{32}{3}\pi^2\alpha\varepsilon^2 + \tfrac{4}{3}\pi\beta\varepsilon^3.$$

Then $H_{\lambda_\varepsilon}(\alpha,\beta,\gamma) \equiv -\Delta + \lambda_\varepsilon\delta_\varepsilon(\cdot)$, with δ_ε given by (2.3), is a well defined *self-adjoint operator in* $^*L^2(\mathbb{R}^3)$, *on the domain of* $-\Delta$, *with lower bound* 0 *if* $\alpha \geq 0$ *(resp.* $-(4\pi a)^2$ *if* $\alpha < 0$*). The resolvent* $(H_{\lambda_\varepsilon} - z)^{-1}$ *(for* $z \in \mathbb{C}$*, with* $\text{Im } z \neq 0$*) has a kernel* $(H_{\lambda_\varepsilon} - z)^{-1}(x,y)$ *which is near standard and S-continuous, for finite* $x,y \in {}^*\mathbb{R}^3$, $x \neq y$. *In fact, for such* x,y, ${}^\circ(H_{\lambda_\varepsilon} - z)^{-1}(x,y)$ *is the kernel (evaluated at* ${}^\circ x$, ${}^\circ y$*) of the resolvent of a self-adjoint lower bounded operator* $H(\alpha)$ *in* $L^2(\mathbb{R}^3)$. $H(\alpha)$ *is independent of* β,γ. *One has, for* $x,y \in \mathbb{R}^3$:

$$(H(\alpha)-z)^{-1}(x,y) = G_z(x-y) - G_z(x)\left[\frac{ik}{4\pi} - \alpha\right]^{-1} G_z(y),$$

with

$$G_z(x) \equiv \frac{e^{iz^{1/2}x}}{4\pi|x|} = (-\Delta-z)^{-1}(x,x).$$

Remark. This in particular implies that the formal Hamiltonian given by $H = -\Delta + \lambda\delta$ is actually realized by a 1-parameter family of self-adjoint lower bounded operators $H(\alpha)$, the different $H(\alpha)$ being characterized by an "effective" coupling constant α (also called a renormalized coupling constant). In terms of the original coupling constant λ we see that we are taking λ as λ_ε, negative infinitesimal. No other choice of λ leads to a well defined self-adjoint operator that is different from the trivial ones 0 and $-\Delta$.

Sketch of the proof. We exploit the rotational symmetry of $V_\varepsilon(|x|)$ (in ${}^*\mathbb{R}^3$) by decomposing ${}^*L(\mathbb{R}^3)$ orthogonally in the direction of a rotation symmetric subspace ${}^*L^2_S(\mathbb{R}^3)$ and its orthogonal complement. $V_\varepsilon(|x|)$ only influences $-\Delta$ on the subspace ${}^*L_S(\mathbb{R}^3)$. Exploiting the unitary equivalence of $L^2(\mathbb{R}_+,r^2dr)$ with $L^2(\mathbb{R}_+)$ given by $f(r)/r \to f(r)$, we see that

$-\Delta + V_\varepsilon$, $V_\varepsilon \equiv \lambda_\varepsilon\delta_\varepsilon$, becomes $A \equiv -\dfrac{d^2}{dr^2} + V_\varepsilon(r)$ in ${}^*L^2(R_+)$ (defined by closure and transfer from twice differentiable functions of compact support, vanishing at the origin).

By Sturm Liouville theory and transfer again we get that $G_z^A \equiv (A-z)^{-1}$ has the kernel

$$G_z^A(x-y) = (2ia_+ z^{1/2})^{-1}\sin (z-\tilde{\lambda}_\varepsilon)^{1/2}\{c_+ e^{i(z-\tilde{\lambda}_\varepsilon)^{1/2}(x-y)} + c_- e^{-i(z-\tilde{\lambda}_\varepsilon)^{1/2}(x-y)}\}$$

for $x \le y \le \varepsilon$,

$$G_z^A(x-y) = (2ia_+ z^{1/2})^{-1}(a_+ e^{z^{1/2}x} + a_- e^{-z^{1/2}x})e^{-iz^{1/2}x} \quad \text{for } \varepsilon \le x \le y,$$

$$G_z^A(x-y) = (2ia_+ z^{1/2})^{-1}\sin(z-\tilde{\lambda}_\varepsilon)^{1/2} e^{-iky} \quad \text{for } x \le \varepsilon \le y,$$

where $a_\pm$, $c_\pm$ are the functions of z, ε, $\tilde{\lambda}_\varepsilon \equiv \frac{3}{4\pi\varepsilon^3}\lambda_\varepsilon$ given by

$$a_\pm \equiv \frac{1}{2}\, e^{\mp iz^{1/2}\varepsilon}[\sin(z-\tilde{\lambda}_\varepsilon)^{1/2}\varepsilon \pm \left[\frac{z-\tilde{\lambda}_\varepsilon}{z}\right]^{1/2} \cos(z-\tilde{\lambda}_\varepsilon)^{1/2}\varepsilon\ ],$$

$$c_\pm \equiv \frac{1}{2} \exp(-i(z^{1/2} \pm (z-\tilde{\lambda}_\varepsilon))^{1/2}\varepsilon)(1 \mp \left[\frac{z}{z-\tilde{\lambda}_\varepsilon}\right]^{1/2}).$$

The question is now: for which λ do we have $G_z^A(x,y)$ near standard for x,y finite and with non trivial (i.e. $\neq 0$, $\neq(-\frac{d^2}{dr^2} - z)^{-1}({}^\circ x, {}^\circ y)$) standard part? For the above formulae we see that we have to avoid $a_+/a_- \approx -1$ (where $\approx$ means equality modulo an infinitesimal). If $\tilde{\lambda}_\varepsilon$ is finite, recalling that $\varepsilon > 0$ is infinitesimal, we see that $a_+/a_- \approx -1$, hence we certainly need $\tilde{\lambda}_\varepsilon$ infinite. Also in this case, if the term $\pm\left[\frac{z-\tilde{\lambda}_\varepsilon}{z}\right]^{1/2}\cos((z-\lambda)^{1/2}\varepsilon)$ in $a_\pm$ turns out to be infinite, we again have $a_+/a_- \approx -1$. So we need λ infinite and $\left[\frac{z-\tilde{\lambda}_\varepsilon}{z}\right]^{1/2}\cos((z-\tilde{\lambda}_\varepsilon)^{1/2}\varepsilon)$ finite, hence $(-\tilde{\lambda}_\varepsilon)^{1/2}\cos((-\tilde{\lambda}_\varepsilon)^{1/2}\varepsilon)$ finite. This then means that $(-\tilde{\lambda}_\varepsilon)^{1/2}\varepsilon = (\gamma + \frac{1}{2})\pi + \rho\varepsilon$, for some $\gamma \in \mathbb{N}_0$, $\rho \in \mathbb{R}$. But this leads us then easily to the choice $\lambda_\varepsilon(\alpha,\beta,\gamma)$ of the theorem.◂

Remark. A form of this theorem was found originally by Nelson (1977). For closely related work see also Friedman (1972) (who uses standard methods) and Alonso y Coria (1978). The result was extended in Albeverio, Fenstad & Høegh-Krohn (1979). See also Albeverio, Fenstad, et al.(1986) and Albeverio, Gesztesy, et.al (1988) for details.

The idea of the *fine discrete* (*hyperfinite*) method as applied to the realization of the point interaction Hamiltonian H is very simple. Instead of enlarging $\mathbb{R}^3$ to $^*\mathbb{R}^3$ as in the hypercontinuous method and smoothing the singular V, we rather shrink $\mathbb{R}^3$ to $\varepsilon\mathbb{Z}^3$, $\varepsilon > 0$ infinitesimal, and replace the δ-function by a suitably scaled "Kronecker δ-function" and the differential operator by a hyperdiscrete difference operator, namely $H_\varepsilon = -\Delta_\varepsilon + \varepsilon^{-3}\lambda_\varepsilon\delta_{0,\varepsilon n}$ in $^*\ell^2(\varepsilon\mathbb{Z}^3, \mathbb{R}^3)$, with Δ_ε the discrete Laplacian on $\varepsilon\mathbb{Z}^3$; i.e.

$$(\Delta_\varepsilon f)(n\varepsilon) = \varepsilon^{-2} \sum_{|n-n'|=1} [f(n\varepsilon)-f(n'\varepsilon)].$$

A calculation of the resolvent along the lines of the above formal computation then yields the resolvent of H_ε and one realizes again that λ_ε has to be chosen in a one-parameter family of infinitesimal functions, as in Theorem 2.4, in order to get a non-trivial standard part of $(H_\varepsilon - z)^{-1}$ for Im z ≠ 0, defining the resolvent of a standard self-adjoint operator (coinciding with the operator H(α) of Theorem 2.4). We shall not do the computations in detail here since there is another version of the fine discrete method we want to discuss in greater generality here and in subsequent sections. By going from $L^2(\mathbb{R}^3)$ to $L^2(\mathbb{R}^3, d\mu(x))$, where $d\mu(x) \equiv (\exp(4\pi\alpha|x|)/|x|)dx$, and $\alpha\in\mathbb{R}$, H(α) can be seen to be unitary equivalent to the unique positive self-adjoint operator associated with the closed quadratic form

$$f \to \int_{\mathbb{R}^3} |\nabla f|^2(x)d\mu(x), \tag{2.5}$$

with

$$\nabla: L^2(\mathbb{R}^3) \to L^2(\mathbb{R}^3,\mathbb{R}^3), \text{ given by } (\nabla f)_i(x) \equiv \frac{\partial f}{\partial x_i}(x),\ i=1,2,3,$$

defined by closure from the same form defined on smooth functions with compact support.

Forms such as (2.5) are particular cases of Dirichlet forms. Dirichlet forms are basic objects of potential theory and are in one-to-one correspondence with symmetric (i.e. time reversible) Markov processes (cf. Fukushima (1980)). It is a major merit of nonstandard analysis that it provides a very convenient tool for describing and studying Dirichlet forms and symmetric Markov processes. This is the consequence of recent results by T. Lindstrøm, (see Albeverio, Fenstad, et al. (1986)), which we shall describe shortly, because of their relevance in many domains, including applications, especially in quantum physics. The basic idea is to "fine discretize" the Dirichlet forms involved by replacing the continuum $\mathbb{R}^d$ by its hyperfinite version $\varepsilon\mathbb{Z}^d$. Let us first do some simple functional analysis, using nonstandard tools (cf. also Henson's paper in this volume). Many applications of nonstandard analysis in physics are made precisely by applying nonstandard analytic tools to Hilbert space theory; see also, as a complement to this paper the references Albeverio, Fenstad, et al. (1986), and Todorov (1985) for example.

Other work concerning applications of nonstandard analysis to quantum mechanics includes the following:

(a) Callott (1983a,b). Here the concept of *visible* L^2-solutions of the stationary Schrödinger equation with infinitesimal Planck constant is analysed. This gives an original approach to the study of the classical limit for quantum mechanics.

(b) In quantum field theory multiplication of distribution valued fields is essential; hyperfinite methods give a tool to do this (c.f. Section 7). A discussion of multiplication of certain distributions using nonstandard (hypercontinuous) tools is in Li & Li (1985). Hyperfinite methods are developed in Kessler (1984).

(c) A problem in optics, somewhat mathematically related to problems in quantum mechanics, is the *moiré problem*. For beautiful discussions of this by nonstandard analytic tools see Harthong (1981) and Li (1986).

(d) For other topics in quantum field theory, in particular fermionic methods, discussed by nonstandard analytic tools see for example Nagamachi & Mugibayashi (1986), Nagamachi & Nishimura (1984), and Keleman & Robinson (1972); see also Section 7.

3. NONSTANDARD THEORY APPLIED TO CLOSED BILINEAR FORMS ON HILBERT SPACES

As we recalled in Section 2, a basic quantity of quantum mechanics is the Hamilton operator. In the case of a particle moving in $\mathbb{R}^d$ under the action of a (real-valued) potential $V(x)$, $x \in \mathbb{R}^d$, the Hamiltonian is a self-adjoint realization H of $-\Delta + V$ in $L^2(\mathbb{R}^d)$. In the case of a bounded V one can indeed define $-\Delta + V$ as the sum of $-\Delta$ (the

Laplacian, a pure differential operator) and the multiplication operator V, in the sense that the domain $D(-\Delta)$ of $-\Delta$ (the vectors to which $-\Delta$ can be applied) contains that of V; hence for $\psi \in D(-\Delta)$, $H\psi = -\Delta\psi + V\psi$ is well defined. In more general situations, such as the point interaction potential discussed in Section 2, such a simple definition of H is not possible, and H has to be obtained in a more indirect way, as we saw in Section 2. There are intermediate cases where, although $-\Delta$ and V are not defined simultaneously on any vector $\psi \not\equiv 0$, at least the associated quadratic forms $\psi \to \varepsilon_0(\psi,\psi) \equiv \int|\nabla\psi|^2 dx$ and $\psi \to \varepsilon_V(\psi,\psi) \equiv \int \psi V \psi dx$ are well defined, for a dense set of ψ. Note that for $\psi \in C_0^\infty(\mathbb{R}^d)$, $\int|\nabla\psi|^2 dx = \int\psi(-\Delta\psi)dx$, as seen by partial integration, and it is well known that the quadratic forms ε_0 and ε_V (for V locally integrable and bounded below) give a complete characterization of the operators $-\Delta$ and V respectively; see, for example Kato (1966), Reed & Simon (1972, Vol.II).

Moreover, associated with the sum of the forms $\varepsilon_0 + \varepsilon_V$, defined pointwise by $(\varepsilon_0 + \varepsilon_V)(\psi,\psi) = \varepsilon_0(\psi,\psi) + \varepsilon_V(\psi,\psi)$, there is a lower bounded self-ajoint operator H. If for simplicity we assume $V \geq 0$ (which we can always obtain by shifting V by a finite constant), then H is characterised by $(H^{1/2}\psi, H^{1/2}\psi) = \varepsilon_0(\psi,\psi) + \varepsilon_V(\psi,\psi)$ (with $H^{1/2}$ being well defined since $H \geq 0$). (Much less stringent conditions on V are possible; see for example Reed & Simon (1972, Vol.II)). One calls H the *form sum* of $-\Delta$ and V. The mapping $(\varphi,\psi) \to (H^{1/2}\varphi, H^{1/2}\psi)$ defines a positive, symmetric, bilinear form on the Hilbert space $L^2(\mathbb{R}^d)$. Every interesting quantum mechanical Hamiltonian can be associated with a positive, symmetric closed bilinear form on a Hilbert space, and for this reason we shall take a look at such forms. It turns out that they can always be obtained as standard parts of corresponding *hyperfinite* forms. Now hyperfinite forms can be thought of as forms associated with hyperfinite matrices. In this way nonstandard analysis gives an implementation of physicists thinking of operators as matrices. (Quantum mechanics used to be called matrix mechanics, because through work initiated by Heisenberg and pursued among others by Dirac and Jordan the first realization of the formalism of quantum mechanics was by operators represented by matrices - see, for example Jammer (1974)).

So let us now give some details of how one realizes bilinear forms by nonstandard analysis. Let $\mathcal{H}$ be an internal, hyperfinite dimensional linear space, with inner product $\langle , \rangle$ and norm $\|\cdot\|$ (with values in $*\mathbb{R}_+$) (in particular there exist internal e_i, $i \le \eta$, $\eta \in *\mathbb{N}$ such that $\mathcal{H} = \{ \sum_{i=1}^{\eta} \alpha_i e_i : \alpha_i \in *\mathbb{R}\}$. Let $\mathrm{Fin}(\mathcal{H}) = \{x \in \mathcal{H}: \|x\| < \infty\}$ be the set of vectors in $\mathcal{H}$ with finite norm. Let

$$\hat{\mathcal{H}} \equiv \mathrm{Fin}(\mathcal{H})/\approx \quad (\text{where } x \approx y \iff \|x-y\| \approx 0 \text{ in } *\mathbb{R} \text{ for } x, y \in \mathcal{H}),$$

with inner product $(\hat{x},\hat{y}) \equiv {}^\circ\langle x,y\rangle$, where $\hat{x}$ is the $\approx$ -equivalence class of x. $\hat{\mathcal{H}}$ is thus the *nonstandard hull* of $\mathcal{H}$ (see the paper by Henson, in this volume).

Let $\mathcal{E}: \mathcal{H} \times \mathcal{H} \to *\mathbb{R}$ be a positive, symmetric, bilinear form (defined by transfer from the corresponding standard concepts; see, for example, Kato (1966)). We call $\mathcal{E}$ a *hyperfinite form.* $\mathcal{E}$ is called *S-bounded* if there exists $K \in \mathbb{R}$ such that $|\mathcal{E}(u,v)|_{*\mathbb{R}} \le K \|u\|_{\mathcal{H}} \|v\|_{\mathcal{H}}$. In this case we can immediately associate with the hyperfinite form $\mathcal{E}$ a standard form E by

$$E(\hat{u},\hat{v}) \equiv {}^\circ\mathcal{E}(u,v), \qquad (u,v \in \mathcal{F}in(\mathcal{H})).$$

E is then a bounded positive, symmetric, bilinear form on a standard Hilbert space $\mathcal{X}$. But since in interesting applications $\mathcal{E}$ comes from a differential operator, the assumption that $\mathcal{E}$ is S-bounded is not satisfied. Yet it is still possible to associate with *any* hyperfinite form $\mathcal{E}$ a standard closed form, and vice-versa, but the association requires precision regarding domains; this we shall now discuss.

We want to define a suitable domain for $\mathcal{E}$. For this we use a connection between $\mathcal{E}$ and operators, obtained by transfer of the usual connection between positive closed bilinear symmetric forms and positive symmetric operators: to a general hyperfinite form $\mathcal{E}$ there exists a unique symmetric, positive operator A such that

$$\mathcal{E}(u,v) = \langle Au, v\rangle_{\mathcal{H}} \qquad \text{for all } u,v \in \mathcal{H}.$$

We shall now see that A generates a semigroup. Fix an infinitesimal Δt, $0 < \Delta t \le 1/\|A\|$, where $\|A\|$ is the norm of A (which is well defined in $*\mathbb{R}$). Define $Q^{\Delta t} \equiv 1 - A\Delta t$ (the *infinitesimal semigroup* for A). Then we have $Q^{\Delta t} \ge 0$ (as an operator), $\|Q^{\Delta t}\| \le 1$ and

$$\mathcal{E}(u,v) = \langle (1 - Q^{\Delta t})u, v\rangle/\Delta t.$$

Let $T \equiv \{k\Delta t: k \in {}^*\mathbb{N}\}$ be a hyperfinite time line and define $Q^t \equiv (Q^{\Delta t})^k$ for $t = k\Delta t \in T$. Then $(Q^t, t\in T)$ is a semigroup in $\mathcal{H}$ associated with $\mathcal{E}$ and temporal scale Δt. $A = (1-Q^{\Delta t})/\Delta t$ appears then as the generator of this semigroup.

We shall now make precise a domain $D(\mathcal{E})$ of $\mathcal{E}$:

$$D(\mathcal{E}) \equiv \{u\in\mathcal{H} : {}^\circ\mathcal{E}_1(u,u) < \infty;\ \mathcal{E}(Q^t u, Q^t u) \approx \mathcal{E}(u,u) \quad \text{whenever } t \approx 0\},$$

with $$\mathcal{E}_\alpha(u,v) \equiv \mathcal{E}(u,v) + \alpha\langle u,v\rangle,\ \alpha\geq 0.$$

It is not difficult to see that:

(a) $D(\mathcal{E})$ is linear (with respect to finite scalars in ${}^*\mathbb{R}$);

(b) $D(\mathcal{E})$ is $\mathcal{E}_1$-closed, in the sense that if

$${}^\circ\mathcal{E}_1(u_n - u_m, u_n - u_m) \to 0 \quad \text{as } n,m \to \infty,$$

then there is $u \in D(\mathcal{E})$ such that $\mathcal{E}_1(u_n-u, u_n-u) \to 0$ as $n \to \infty$;

(c) $u \to \mathcal{E}(u,u)$ is continuous in the sense that if $u,v \in D(\mathcal{E})$ and $u \approx v$, then $\mathcal{E}(u,u) \approx \mathcal{E}(v,v)$.

The definition of a standard, closed, positive, symmetric, bilinear form E on the standard Hilbert Space $\hat{\mathcal{H}}$, associated with the hyperfinite form $\mathcal{E}$, goes as follows:

i) $D(E) \equiv \{\hat{u} \in \hat{H} : \inf_{\substack{v\in\hat{u} \\ v\in\mathcal{H}}} \mathcal{E}_1(v,v) < \infty\}$

ii) for $\hat{x},\hat{y} \in D(E)$, $E(\hat{x},\hat{y}) \equiv {}^\circ\mathcal{E}(u,v)$, for any $u \in \hat{x}$, $v \in \hat{y}$, $u,v \in D(\mathcal{E})$.

It turns out that this association has also an inverse in the sense that given a standard Hilbert space $\mathcal{H}$ and a symmetric positive closed bilinear densely defined form F on it, there exists a hyperfinite dimensional subspace $\mathcal{H}$ of ${}^*\mathcal{X}$, S-dense in ${}^*\mathcal{X}$ (i.e. $\forall x \in \mathcal{X}\ \exists\ y \in \mathcal{H}$ such that $\|x-y\| \approx 0$) and a hyperfinite form $\mathcal{E}$ on $\mathcal{H}$ such that F is the restriction of the standard part of $\mathcal{E}$ to $\mathcal{X}$. Moreover the contraction semigroup P^t associated with F can be obtained as the standard part of the hyperfinite semigroup associated with $\mathcal{E}$, in the sense that ${}^\circ Q^s v = P^t u$, for all $t \in \mathbb{R}_+$ and $s \in T$ with ${}^\circ s = t$, and $u \in \mathcal{X}$, $v \in \mathcal{H}$ with $u \approx v$.

Remark. Proofs of the above are contained in Albeverio, Fenstad, et al. (1986). Of course these results are rather technical: the main point, however, is easy to grasp, namely that one can get any standard positive, symmetric, bilinear form as the standard part of a hyperfinite positive, symmetric bilinear form and vice-versa, and the same holds for the associated semigroups. The latter corresponds, in the standard world, to an approximation by discretization in time and space of the standard semigroup, a natural procedure used in numerical and some computational problems. As we shall see below, in the case of Markov semigroups associated with Markov processes this corresponds to approximating continuous time-continuous state space processes by discrete Markov chains.

Remark. The discussion at the beginning of this section concerned Hamilton operators for quantum mechanics, which are precisely given by closed bilinear forms, hence through a realization by hyperfinite bilinear forms. These are, of course, not the only objects of interest in quantum mechanics and this is not the only place where nonstandard (more precisely, hyperfinite) methods are useful in quantum mechanics. Another place is, for example, in the discussion of the spectral decompositions involved (which are necessary to associate numbers to operators and for comparison with experiments). Some references concerned with other relations between quantum mechanics and nonstandard analysis are Albeverio, Fenstad, et al. (1986), Callot (1983a,b), Todorov (1985, 1987), Harthong (1984), Sloan (1977, 1981), Keleman & Robinson (1972).

4. DIRICHLET FORMS AS STANDARD PARTS OF HYPERFINITE DIRICHLET FORMS, AND APPLICATIONS TO QUANTUM MECHANICS

As we saw at the beginning of Section 3 (and in Section 2) interesting quantum mechanical Hamiltonians can be represented as standard parts of closed bilinear hyperfinite forms. There is an important feature of such operators that relates them to objects studied in potential theory and in the theory of Markov processes, namely they generate symmetric Markov semigroups. By their association with hyperfinite forms it is natural to hope that those Markov processes and semigroups are associated with hyperfinite Markov processes and semigroups, i.e. hyperfinite Markov chains. This point of view, reducing the abstract and complex theory of continuous time, continuous state space Markov processes to the simpler theory of Markov chains, has had already many applications, also outside quantum mechanics; see Albeverio, Fenstad, et al. (1986) (and, for example, Lindstrøm's and Keisler's papers in this volume).

Let us now describe shortly how the relation Hamiltonian Dirichlet forms - Markov processes and its hyperfinite correlates comes about.

We recall that a standard bounded symmetric operator S on a $L^2(M,m)$-space is called *Markov* if $0 \le f \le 1$, $f \in L^2(M,m)$ implies that $0 \le Sf \le 1$. A semigroup $(S_t, t \ge 0)$ of such operators is called a *Markov semigroup*.

Markov semigroups are the transition semigroups of symmetric Markov processes; see for example Fukushima (1980), Silverstein (1974), Albeverio, Fenstad, et al. (1986). On the other hand there is a 1-1 correspondence of such semigroups with potential theoretical objects called Dirichlet forms. A positive symmetric bilinear closed and densely defined form $\mathcal{E}(f,f)$ (as we discussed in Section 3) is called a *Dirichlet form* if it has some suitable contraction property such as $E(f^{\#}, f^{\#}) \le E(f,f)$ for any f in its domain, where $f^{\#} = (f \vee 0) \wedge 1$.

The relation between Dirichlet forms and Markov semigroups follows from the fact that the infinitesimal generator of S_t is given by the unique positive self-adjoint operator $A_{\mathcal{E}}$ associated with $\mathcal{E}$ by $(A_{\mathcal{E}}^{1/2}f, A_{\mathcal{E}}^{1/2}f) = \mathcal{E}(f,f)$ for all f in the domain of $\mathcal{E}$.

This has been taken as the starting point of a theory of hyperfinite Dirichlet forms, yielding back, by standard parts, standard Dirichlet forms; see Albeverio, Fenstad, et al. (1986). So a hyperfinite Dirichlet form is defined as a positive bilinear symmetric form $\mathcal{E}$ on a hyperfinite dimensional Hilbert space $\mathcal{H}$ such that there exists $\Delta t > 0$, $\Delta t \approx 0$ such that $Q^{\Delta t} \equiv 1 - A_{\mathcal{E}}\Delta t$ is Markov (in the above sense), where $A_{\mathcal{E}}$ is the self-adjoint operator associated with $\mathcal{E}$.

Now the result of Section 3 giving a correspondence between standard and hyperfinite forms holds with the predicate Dirichlet added. In particular, to any standard Dirichlet form E on a L^2-space $L^2(M,m)$ there exists a hyperfinite Dirichlet form $\mathcal{E}$ on the L^2-space $L^2(Y,\mu)$ over a hyperfinite set Y, with μ an internal measure, such that E is obtained from $\mathcal{E}$ by taking standard parts (more precisely, we have $E(u,u) = \inf\{{}^{\circ}\mathcal{E}(v,v)\colon v \text{ a lifting of } u\}$. Let us now look at the relation between hyperfinite Dirichlet forms and hyperfinite Markov chains.

Let $X(t)$, $t \in T$ be a hyperfinite Markov chain, with some hyperfinite state space $S = \{s_0, \ldots, s_N\}$, $N \in {}^*\mathbb{N}\setminus\mathbb{N}$. Let q_{ij} be the transition probability from $X(t) = s_i$ to $X(t+\Delta t) = s_j$ and let s_0 be a trap so that $q_{oi} = 0$ if $i \neq 0$. Then $((q_{ij}))$, $0 \le i,j \le N$ is a stochastic matrix (i.e. $q_{ij} \ge 0$, $\sum_{j=0}^{N} q_{ij} = 1$). Let m be a hyperfinite measure on S, which we take as the starting measure for X. We assume the symmetry condition $q_{ij}m_i = m_j q_{ji}$ if $i,j \neq 0$ (such a condition, leading to symmetric processes, is related to time reversal invariance; it is often called the *detailed balance* condition, especially in the physics literature).

The Hilbert space for the Markov chain will be $\mathcal{H} = \{u:\ u: S\setminus\{s_0\} \to {}^*\mathbb{R}\}$, with scalar product $\langle u,u\rangle \equiv \sum_{i=1}^{N} u(s_i)u(s_i)m(s_i)\}$. Let $E^i(u(X(t)))$, $t \in T$, $u \in \mathcal{H}$, be expectation with respect to the Markov chain process started at s_i. Then $(Q^t u)(i) = E^i(u(X(t)))$, $t \in T$, is a symmetric hyperfinite Markov semigroup with generator

$$(Au)(i) = (u(i) - \sum_{i=1}^{N} q_{ij}u(j))/\Delta t.$$

The associated hyperfinite Dirichlet form is

$$\mathcal{E}(u,v) \equiv \langle Au,v\rangle_{\mathcal{H}}$$

$$= (\Delta t)^{-1} \sum_{i=1}^{N} (u(i)-u(j))(v(i)-v(j))q_{ij}m_i + \sum_{i=1}^{N} u(i)v(j)q_{i0}m_i.$$

Conversely, to any hyperfinite Dirichlet form on $\mathcal{H}$ there are Q^t, m as above. Such discrete forms (in the standard setting, with N finite) were originally studied by Beurling-Deny and are at the origin of the modern theory of Dirichlet forms. The passage from the discrete case to the continuous one (potential theory of elliptic operators over $\mathbb{R}^d$, for example) involves topological or measure theoretical tools of various kinds (for example Fukushima (1980), Silverstein (1974), Dynkin (1982), Bouleau & Hirsch (1986), Albeverio, Høegh-Krohn & Röckner (1988). The theory of hyperfinite Dirichlet forms (Albeverio, Fenstad, et al. (1986)) gives an alternative way of making this passage, by taking standard parts.

Remark. The potential theory (equilibrium potentials etc.) going with Dirichlet forms can also be deduced from the hyperfinite scheme; see Albeverio, Fenstad, et al. (1986). Also Fukushima's fundamental extension of the theory of stochastic differential equations can be obtained in this way; for such results see Albeverio, Fenstad, et al. (1986). One of the considerable advantages of the hyperfinite approach is that it unifies the theory of standard Dirichlet forms and associated processes in the cases where the state space is locally compact ($\mathbb{R}^d$, a finite dimensional manifold) as in non-relativistic quantum mechanics with scalar or spin particles; see Albeverio, Høegh-Krohn & Streit (1977); or in filter theory Mitter (1980), Arede (1986); or infinite dimensional (not locally compact) Banach, rigged Hilbert spaces, distribution spaces, etc. (with applications to quantum field theory Albeverio & Høegh-Krohn (1977a,b, 1985), Albeverio, Høegh-Krohn & Röckner (1988), Albeverio & Kusuoka (198?), Kusuoka (1982), Takeda (1985).

We close this section by mentioning a hyperfinite version of a fundamental formula, the Feynman-Kac formula, permitting us to construct by perturbations new Markov semigroups and symmetric processes starting from known ones. Such a formula is useful in quantum mechanics for constructing new Hamiltonians from known ones; we will have an application in Sections 6, 7, when discussing polymer measures and quantum fields. The formula is as follows: let X be a hyperfinite m-symmetric Markov chain, with time scale $\Delta t > 0$, $\Delta t \approx 0$, state space S and infinitesimal generator A as above.

Let $V: S\backslash\{s_0\} \to {}^*\mathbb{R}$ be an internal function such that $\|V\|_\infty/(\ln \Delta t) \approx 0$ and V is such that A + V is lower bounded; i.e. there is $\beta \in \mathbb{R}$ with $\langle(a + V)u, u\rangle \geq \beta\|u\|^2$ for all $u \in L^2(S,m)$. Then the semigroup $(P^t, t \in T)$ in $L^2(S,m)$ generated by A + V can be expressed by taking expectations with respect to the Markov chain as follows:

$$\|(P^t u)(\cdot)-E^{(\cdot)}(u(X(t))e^{-\int_0^t V(X(s)ds})\| \approx 0,$$

for all finite t and finite u. This is our hyperfinite Feynman-Kac formula. Perhaps it is instructive to give a sketch of the proof.

Lemma

Let $S^{\Delta t} \equiv (1-V\Delta t)(1-A\Delta t)$. *Then*

$$(S^t u)(s_i) = e^{-t\eta(t)}\, E_{s_i}(u(X(t))e^{-\int_0^t V(X(s))ds}).$$

Proof. Let Q^t be the semigroup generated by A; then

$$S^t = (1-V\Delta t)Q^{\Delta t}(1-V\Delta t)\dots(1-V\Delta t)Q^{\Delta t}.$$

By the probabilistic interpretation of the quantities involved we then get

$$(S^t u)(s_i) = E^{s_i}[u(X(t)\prod_{\substack{0\le s\le t\\ s\in T}}(1-V(X(s))\Delta t)].$$

But

$$\prod_{0\le s<t}(1-V(X(s))\Delta t) = \exp(\sum_{s=0}^{t}\ln[1-V(X(s))\Delta t]$$

$$= \exp(-\sum_0^t V(X(s))\Delta t - \tfrac{1}{2}\sum_0^t \theta(s)^2(\Delta t)^2),$$

with $0 \le \theta(s)^2 \le V(X(s))^2$. This then ends the proof of the lemma.◂

By the lemma it suffices then to show $P^t u \approx S^t u$, since by assumption $\|V\|_\infty^2 \Delta t \approx 0$. For $P^t u \approx S^t u$ we may observe that

$$S^t \approx S^{t-\Delta t}(1-(A+V)\Delta t + VA(\Delta t)^2)$$

$$= S^{t-\Delta t}P^{\Delta t} + S^{t-\Delta t}VA(\Delta t)^2$$

$$\approx S^{t-2\Delta t}P^{2\Delta t} + S^{t-2\Delta t}VAP^{\Delta t}(\Delta t)^2 + S^{t-\Delta t}VA(\Delta t)^2,$$

hence

$$S^t \approx P^t + \sum_{s=0}^{t} S^{t-s-\Delta t}VAP^s(\Delta t)^2.$$

But the second term on the right hand side is I - II,

with $II \equiv \sum_{s=0}^{t} S^{t-s-\Delta t}V^2P^s(\Delta t)^2 \approx 0$,

since $\|S^{t-s-\Delta t}\| \le e^{\|V\|_\infty(t-s)} \le e^{\|V\|t}$ (by the lemma) and $\|P^s\| \le e^{|\beta|t}$ (by the lower bound on A + V). Let us estimate

$$\|I\| \equiv \|\sum_{s=0}^{t} S^{t-s-\Delta t}V(A+V)P^s(\Delta t)^2\| \le e^{\|V\|_\infty t}\|V\|_\infty\sum_{s=0}^{t}\|(A+V)P^s\|(\Delta t)^2,$$

where we have again used the lemma.

An easy computation yields

$$\|(A+V)P^s\| \le M(s+\Delta t)^{-1},$$

where $M = \max\{|\beta|e^{|\beta|t},1\}$.

On the other hand $\sum_{s=0}^{t}\Delta t(s+\Delta t)^{-1} \le 2\ln(\Delta t)^{-1}$;

hence

$$\|I\| \leq \exp(\|V\|_\infty).\|V\|_\infty \, 2M \ln(\Delta t)^{-1}\Delta t \approx 0,$$

since $\exp(\|V\|_\infty)(\ln \Delta t) \not\approx 1$. This then shows $P^t u \approx S^t u$, which concludes the proof.◄

Remark. The Feynman-Kac formula gives a way to compute the heat semigroup e^{-tH}, $t \geq 0$, where $H = -\frac{1}{2}\Delta + V$; i.e. the solution of the heat equation $\frac{\partial}{\partial t} f = -Hf$ with given initial condition. It is well known that historically the Feynman-Kac formula was proved by Kac after stimulation by a lecture by Feynman where a corresponding formula for the unitary Schrödinger groups e^{-itH}, $t\in\mathbb{R}$, instead of the heat semigroup e^{-tH}, $t \geq 0$, was heuristically derived. The nonstandard, internal realization of Feynman's heuristic idea is as follows (see Albeverio, Fenstad, et al. (1986) for more details): the solution of the Schrödinger equation

$$ih \frac{\partial}{\partial t} \psi = H\psi, \quad \text{over } \mathbb{R}^d,$$

where $H = -\frac{h^2}{2}\Delta + V$, with initial condition $\psi(0) = \varphi$, is given by

$$^\circ\!\int_{*\mathbb{R}^{Nd}} \exp[\tfrac{i}{h} \; {}^*S_t(\gamma_N + x)]\,{}^*\varphi(\gamma_N(0) + x)\,{}^*d\gamma_N,$$

with

$$^*d\gamma_N = {}^*\prod_{j=0}^{N-1} \left(\frac{2\pi iht}{N}\right)^{-d/2} d\gamma_N \left(\frac{it}{N}\right),$$

for N a fixed infinite integer (i.e. $N \in {}^*\mathbb{N} \setminus \mathbb{N}$). $S_t(\gamma)$ is the (classical) action along the path γ; i.e.

$$S_t(\gamma) \equiv \tfrac{1}{2} \int_0^t \dot{\gamma}(\tau)^2 \, d\tau - \int_0^t \tilde{V}(\gamma(\tau))d\tau$$

(for $\gamma(\tau)$, $\tau \in [0,t)$ a piecewise linear path in $\mathbb{R}^d$); γ_N is a piecewise linear path through given points $\gamma_N(j\frac{t}{N})$ for $j=0,\dots,N$, and $\tilde{V}(\gamma_N(\tau)) = V(\gamma_N(j\frac{t}{N}))$, for $\tau \in [j\frac{t}{N}, (j+1)\frac{t}{N}]$; h is Planck's constant divided by 2π. From this picture it emerges clearly (as already remarked, in standard heuristic terms by Dirac) that $h \approx 0$ corresponds to passing from quantum mechanics to classical mechanics. For (partial) implementation of these ideas see Albeverio, Fenstad, et al. (1986), Harthong (1984); (for references to standard literature on the subject see also for example Albeverio (1986c), Albeverio & Høegh-Krohn (1976), Albeverio, Blanchard & Høegh-Krohn (1982a), Rezende (1985)).

5. HYPERFINITE ENERGY FORMS, DIFFUSIONS AND QUANTUM MECHANICS ON FRACTALS

Let us consider hyperfinite Dirichlet forms in $L^2(Y,m)$ in the case where the state space Y is the hyperfinite lattice $\{k\Delta x: k \in {}^*\mathbb{Z}^d, |k_i| \leq (\Delta x)^{-2}\}$, with $\Delta x > 0$, $\Delta x \approx 0$. Let $e = (0,\ldots, \pm 1,0,\ldots,0) \in \mathbb{R}^d$ with $\mathrm{sign}(e) = \pm$, according to the sign of the non-zero entry in e. For any internal map $f: Y \to {}^*\mathbb{R}$ we define $(D_e f)(y) \equiv (\mathrm{sign}(e)\Delta x)^{-1}[f(y + e\Delta x) - f(y)]$, for $y \in Y$. Let ν be an internal measure on Y. Then

$$(f,g) \to \mathcal{E}(f,g) \equiv \tfrac{1}{2} \sum_{x\in Y} \sum_e (D_e f)(y)(D_e g)(y)\nu(y)$$

defines a hyperfinite Dirichlet form which is called a *hyperfinite energy form*. The corresponding standard Dirichlet forms are called *energy forms* (Albeverio, Høegh-Krohn & Streit (1977), Fukushima (1985a)) and are of particular importance in potential theory and quantum physics; see for example Albeverio (1986a).

It is not difficult to show that the corresponding semigroup (Markov chain) is m-symmetric, with $m(y) \equiv \tfrac{1}{2}\nu(y) + \frac{1}{4d}\sum_e \nu(y + e\Delta x)$ and

$$q_{y,y+e\Delta x} = [4dm(y)]^{-1}[\nu(y) + \nu(y+e\Delta x)].$$

The Loeb measures $L(m)$ and $L(\nu)$ given by m and ν yield, by composition with st^{-1}, equal measures on $\mathbb{R}^d$. Under suitable assumptions on ν and domains, the generator $A_{\mathcal{E}}$ of $\mathcal{E}$ is a hyperfinite version of a diffusion operator of the form $-\Delta - 2\frac{\nabla\psi}{\psi}\nabla$ acting in $L^2(\mu)$, with $\mu = L(\nu)\circ\mathrm{st}^{-1}$ and $d\mu = \psi^2 dx$.

This yields in particular criteria for closability of standard densely defined energy forms; see Albeverio, Fenstad, et al. (1986) (which should be compared with those given in standard literature (Albeverio, Høegh-Krohn & Streit (1977), Röckner & Wielens (1985), Fukushima (1980), Silverstein (1974), Albeverio, Høegh-Krohn & Röckner (1988)). A particular result is the case where $\mu(dx) = \varphi_\alpha(x)^2dx$, and $\varphi_\alpha(x) = \frac{e^{4\pi\alpha|x|}}{|x|}$, in which case $\mathcal{E}$ is the hyperfinite version of the Dirichlet energy form associated with an operator unitary equivalent, by the map $L^2(\mu) \to L^2(\mathbb{R}^3)$ given by multiplication by $1/\varphi_\alpha$, to the operator $H(\alpha) + (4\pi\alpha)^2$ discussed in Section 2. Hyperfinite energy forms that yield diffusion processes by taking standard parts can also be used as tools to construct diffusion processes running on fractal sets. Fractals are certain geometrical structures of noninteger Hausdorff-dimension (known to mathematicians for a long time; a well known example is Peano's square filling curve); that such structures occur often in nature has been realized particularly through the stimulation of Mandelbrot's work, and so physical properties - heat diffusion, wave propagation, quantum mechanics - of fractals have been studied; well known examples of fractal media are porous media, galactic media, surfaces of turbulent clouds. See Mandelbrot (1977, 1982) and references therein. The use of nonstandard tools in this domain has been advocated in Notalle & Schneider (1984). Tom Lindstrøm has provided a detailed study of diffusions on certain fractals, using hyperfinite methods (Lindstrøm (1986, 1988)). Below we shall briefly give an idea of his recent work.

Almost everywhere in physics, differential operators play a role (wave equation, heat equation, Schrödinger equation, Navier-Stokes equations etc.). How can differential operators be defined on fractals, so that physics can be done on fractals? Since fractals do not have by their own nature a differentiable structure, the problem appears to be nontrivial and unnatural using classical analysis. Only probabilists (familiar anyway as they are with fractal structures like a Brownian path!) could, on the standard side, claim to have the means of providing a simple definition of (elliptic) differential operators on fractals, by defining diffusions on them (and this was effectively done by Kusuoka (1985), Goldstein (1986) and Barlow & Perkins (1987). But certainly nonstandard analysts should find themselves even more at home, living as they (possibly) do in a hyperfinite universe, which could also be a hyperfinite realization of a fractal. Indeed, this is the point of view taken by Lindstrøm, who combined elegantly and efficiently the nonstandard and probabilist's points of view. Lindstrøm considers a class of self-similar fractals that he calls nested, which includes such important cases as Koch curves and the Sierpiński gaskets, extensively studied by physicists. He constructs elliptic operators on fractals as self-adjoint operators on a L^2-space $L^2(S,m)$ over the fractal S with a probability measure m (the invariant measure for a diffusion on S). He proves among other things that Weyl's formula $n(E) \sim E^{d/2}$ as $E \to \infty$ for the number of eigenvalues of $-\Delta$ less than E, valid in d-dimensional manifolds, has to be modified on fractals to $n(E) \sim E^{d\log\nu/\log\lambda}$ as $E \to \infty$, where ν, λ are two parameters describing the fractal and d is the Hausdorff dimension (this should be compared with previous discussions Berry (1980), Brossard & Carmona (1986)). This kind of result depends on the following fact. For a Markov chain described by $Q = ((q_{ij}))$,

$$q_{ij} = \begin{cases} 0 & \text{if } i \text{ is not the nearest neighbour of } j \\ \tfrac{1}{2} & \text{if } i \neq 0,1, \text{ and } i \text{ is the nearest neighbour of } j \\ 1 & \text{if } i = 0,1, \text{ and } i \text{ is the nearest neighbour of } j \end{cases}$$

$$m(i) = \begin{cases} (\Delta x)^{-1}, & i \neq 0,1 \\ \tfrac{1}{2}(\Delta x)^{-1}, & i = 0,1, \end{cases}$$

one has to take $\Delta t = (\Delta x)^2$, $\Delta x > 0$, $\Delta x \approx 0$, for the hyperfinite version with state space $S \equiv \{k\Delta x: k \in {}^*\mathbb{N},\ k \leq 1/\Delta x\}$ to get a Brownian motion on [0,1] (reflected at the boundary). In order to get a diffusion process on a fractal one has to take $\Delta x = (\Delta t)^{\log\nu/\log\lambda}$. For the Sierpiński gasket (the limit of △ → ▲ → ▲ → ..., where black triangles are successively removed) we have $\nu = 2$, $\lambda = 5$ (we note that the Hausdorff dimension d is log 3/log 2). For reasons of time and space we cannot go into this brilliant work of Lindstrøm, but we urge the reader to go through his construction to get a better feeling for the way that nonstandard analysis, by allowing the limit object to be with us all the time (in its hyperfinite realization) encourages us to carry through constructions that otherwise would seem temerarious and bound not to lead anywhere.

Remark. Our last remark in this section concerns the fact that nonstandard analysis also provides a natural setting for infinite dimensional analysis, for example diffusion processes, Dirichlet forms, (and perhaps fractals) with infinite dimension. In fact in the above construction of Dirichlet forms and energy forms from hyperfinite ones nothing forbids us to take the dimension d of the state space to be a positive infinite (natural) number. For example, in the case of a standard state space that is a separable Hilbert space H with basis e_i, $i \in \mathbb{N}$, the hyperfinite realization of the state space is then some hyperfinite set Y such that $H \subset Y \subset {}^*H$. For this setting, not yet sufficiently exploited in applications, see Albeverio, Fenstad, et al. (1986); the standard theory of infinite dimensional Dirichlet forms has been constructed in Albeverio & Høegh-Krohn (1977a,b, 1985), Albeverio, Høegh-Krohn & Röckner (1988), Albeverio & Kusuoka (198?), Kusuoka (1982), Takeda (1985), Bouleau & Hirsch (1986).

6. POLYMER MEASURES

As a further application of the realization of diffusion processes by hyperfinite Markov chains, of energy forms by hyperfinite Dirichlet forms, and of the hyperfinite Feynman-Kac formula, let us look at the problem of giving a meaning to the heuristic operator

$$-\Delta + \lambda \int_0^t \delta(x - b(s))ds, \tag{6.1}$$

with $b(s)$, $s \in [0,t]$, a Brownian path in $\mathbb{R}^d$, describing the quantum mechanical Hamiltonian for a particle moving under the action of a potential created precisely on the path, with strength characterized by λ. Such a model has been considered for example by S.F. Edwards. A related problem is that of giving meaning to certain measures describing equilibrium statistical mechanical properties of polymers in solutions, described by heuristic measures of the forms

$$\exp\,[-\lambda \int_0^t \int_0^{t'} \delta(b(s)-b(s'))dsds']$$

times Wiener measure $dP(b)$ on $C_0([0,t];\ \mathbb{R}^d)$, and

$$\exp\,[-\lambda \int_0^t \int_0^{t'} \delta(b(s) - b'(s'))\, dsds']$$

times product Wiener measures $dP(b)dP(b')$, for two independent Brownian motions b,b'. Again a possible strategy (see Albeverio, Fenstad, et al. (1986)) is to put oneself in a hyperfinite setting. Hence $[0,t]$ is replaced by a hyperfinite time line $T \equiv \{k\Delta t,\ k \in {}^*\mathbb{N}\}$, $b(s)$ by Anderson's realization $B(s)$ of Brownian motion, $\mathbb{R}^d$ is replaced by a hyperfinite lattice Y, with spacing (or elementary lattice distance) $|\Delta x| = \Delta t^{1/2}$, $\Delta t > 0$, $\Delta t \approx 0$. Then $\int_0^t \delta(x-b(s))ds$ is replaced by the hyperfinite local time

$$L(t,x) \equiv \sum_{s<t} \delta_{x,B(s)}\, \Delta t^{1/2}$$

counting the number of visits to x up to time t.

Remark As shown by Perkins (1983) $l({}^\circ t, {}^\circ x) \equiv {}^\circ L(t,x)$ for $t \in T$, $x \in Y$, is the standard local time of Brownian motion in the case $d = 1$. In this case Perkins has also obtained, using the nonstandard representation, striking results such as the first proof of P. Lévy's formula

$$\lim_{\delta\downarrow 0} \sup_{\substack{{}^\circ t\le t_0\\ {}^\circ x\in\mathbb{R}}} |m({}^\circ t, {}^\circ x, \delta)\delta^{-\frac{1}{2}} - 2(2/\pi)^{\frac{1}{2}} l({}^\circ t, {}^\circ x)| = 0$$

a.s. for *each* $t_0 > 0$, where m is Lebesgue measure of the set of points whose distance to the level set $\{s \le t\colon b(s) = x\}$ is less than $\delta/2$. For $d > 1$ local time has been discussed by A Stoll (1985, 1986b). Of course $-\Delta + \lambda L(t,\cdot)$ in $L^2(Y,\mu)$, with μ counting measure on Y (corresponding to Lebesgue measure on $\mathbb{R}^d$) makes sense. Can we choose λ such that we can extract from it a nontrivial standard self-adjoint operator in $L^2(\mathbb{R}^d)$ (the Hamiltonian for the heuristic operator (6.1))?

It is useful to look at the problem in greater generality, replacing L by ρ/μ, with ρ some hyperfinite measure. The similarity with the simpler problem discussed in Section 2 is striking. Methods similar to those we used there, looking at the resolvent, lead us to analyse the nontrivial part of the resolvent, here given by

$$(\frac{1}{\lambda} + G'_z)^{-1},$$

with G'_z the operator $(-\Delta - z)^{-1}$ in $L^2(B,\rho)$, with (B,ρ) a hyperfinite realization of $[0,t]$ with Lebesgue measure. Using the precise infinite values of $G_z(x,y)$, $z \in \mathbb{C}$, $\operatorname{Im} z \neq 0$ on the diagonal, ($-\frac{1}{2\pi}\ln \|x-y\|$ for $d = 2$, and $\|x-y\|^{2-d}$ for $d \ge 3$, as $x \approx y$) we get that for $d \le 5$ the value of λ can be chosen independent of z, finite for $d \le 3$, infinitesimal negative for $d = 4,5$, in such a way that by standard parts one realizes (6.1) as a well defined and nontrivial self-adjoint operator (different from $-\Delta$ and zero!). These results are proven in Albeverio, Fenstad, et al. (1986) (see also Albeverio, Fenstad, et al. (1984)).

Remark. Many problems remain open; for example what is the spectrum and its nature, especially for $d = 4$? The case $d = 4$ has applications to the construction of quantum fields in space-time dimension 4, as we shall see in Section 7. This connection uses the construction of the above type of *polymer measures*. The nonstandard study of the *self-intersection* functional

$$\int_0^t \int_0^t \delta(b(s)-b(s'))dsds' = \int_0^t l(s,x)^2 ds \tag{6.2}$$

and the corresponding polymer measure has been done for d = 1,2, by Stoll (1985, 1986b), who extended results obtained by standard methods by Varadhan. Lawler (1980, 1986) has obtained partial related results by using nonstandard analytic tools for $d \geq 4$. It is a challenging open problem to complete these results and to provide an alternative nonstandard proof of the d = 3 result of Westwater (1980, 1982) (obtained by quite involved standard analytic tools). Edward's 2-polymers interaction (6.2) has been controlled by the above construction of (6.1) and a use of the hyperfinite Feynman-Kac formula, for $d \leq 5$, in Albeverio, Fenstad, et al. (1986). Heuristically, the one-polymer measure (6.2) is a limit of such 2-polymer measures, but no complete proof is as yet available, for $3 \leq d \leq 5$. Also much work remains to be done on studying the large time behaviour of the variance of the polymer process b(t); see references in Stoll (1985, 1986b); perhaps methods developing from Cutland (1987, 1988), could be useful.

7. QUANTUM FIELDS: HYPERFINITE MODELS AND CONNECTIONS WITH POLYMER MEASURES

It is well known that the problem of the mathematical construction of relativistic local quantum fields in 4 space-time dimensions (or relativistic strings) is perhaps the most challenging mathematical problem of modern theoretical physics. Important progress has been achieved in the last 20 years (see for example Albeverio, Fenstad, et al. (1986), Albeverio & Høegh-Krohn (1985), Glimm & Jaffe (1987). Such progress has been mainly obtained in the so called "Euclidean" approach, in which the solution is provided through the construction of suitable classical statistical mechanical systems. Nonstandard analytic tools have often been advocated to do this and solve the well known divergence problems. In our nonstandard language the problem can be formulated as follows. Take a $^*\mathbb{R}$-spin system on a hyperfinite d-dimensional lattice with infinitesimal spacing (d = 4 corresponds to the physical space-time dimension); of course we should specify interactions, but this is no problem, until one asks about standard parts. Can we take standard parts (to see that the model has some wanted properties)? The two main difficulties arising are the *infrared problem* (having to do with the infinite spatial extent of our hyperfinite lattice, in the large) and the *ultraviolet problem* (arising from removing the infinitesimal spacing; i.e. going to the continuum).

The infrared problem is similar to the problem of taking the thermodynamic limit in classical statistical mechanics (i.e. the construction of Gibbs states: for recent nonstandard methods, developed in connection with subtler questions like uniqueness and the global Markov property of these states, see Kessler (1984, 1985, 1986, 1987, 1988a,b)).

The real existence problem for quantum fields is, however, connected with the ultraviolet problem. Much heuristics and physics has been developed to discuss these problems, with two main strategies being the renormalization group method and perturbative renormalization. We would like to point out that nonstandard analysis gives other tools. The nonstandard solution approach is to set $\Lambda_{\delta,N} = \{\delta n\colon n\in{}^*\mathbb{Z}^d,\ |n|\le N\}$ and work with ${}^*\mathbb{R}^{\Lambda_{\delta,N}}$, the hyperfinite lattice with lattice spacing $\delta > 0$, $\delta \approx 0$ and length $N \in {}^*\mathbb{N} \setminus \mathbb{N}$. Take a suitable Loeb measure on ${}^*\mathbb{R}^{\Lambda_{\delta,N}}$ corresponding to an internal measure of the form $C\exp S_{N,\delta}(q)$ times Lebesgue measure $\prod_{n\delta\in\Lambda_{\delta,N}} dq_{n\delta}$ on ${}^*\mathbb{R}^{\Lambda_{\delta,N}}$, where $S_{N,\delta}(q)$ is the "Euclidean lattice action"

$$-\tfrac{1}{2}\sum_{n,n'} \delta^d(-\Delta_\delta + m^2)q_{n\delta}q_{n'\delta} - \lambda_\delta \sum_n \delta^d {}^*u_\delta q_{n\delta}$$

with m a positive constant ("the free mass"), $\lambda_\delta \in {}^*\mathbb{R}$, u_δ giving the *interaction term* and C being a normalisation constant. (The form of the choice of such measures is of course dictated by the physics involved; see for example, Glimm & Jaffe (1987)).

To understand this in a little more detail, let us first consider the case $u_\delta = 0$ (the *free* case).

In this case the above Loeb measure is the Loeb measure corresponding to the internal probability space $({}^*\mathbb{R}^{\Lambda_{\delta,N}},\ \mu_{0,\delta,N})$ with $\mu_{0,\delta,N}$ the normal distribution of mean 0 and covariance $\delta^{-d}(-\Delta_\delta + m^2)^{-1}|\Lambda_{\delta,N}$. The coordinate process $\Phi_\delta(\delta n)$ is then a realization of the free hyperfinite (Euclidean) field (of mass m). The standard part of $\sum_{n\delta\in\Lambda_{\delta,N}} \delta^d {}^*f(n\delta)\Phi_\delta(\delta n)$, for $f \in C_0^\infty(\mathbb{R}^d)$ is a.s. near standard with standard part the free Euclidean field $\Phi(f)$.

Remark. Whereas $\Phi(f)$ is not an ordinary random field, its hyperfinite counterpart $\Phi_\delta(\delta n)$ is pointwise defined, hence an ordinary random field. A more general discussion of hyperfinite random fields as realizations of generalized random fields is due to Kessler (1988a).

Interacting fields are defined as follows. Fix $g \in C_0^\infty(\mathbb{R}^d)$ and $u_\delta(\cdot) \in C(\mathbb{R})$ and consider

$$U_g^\delta(\Phi_\delta) \equiv \lambda_\delta \sum_{n\delta\in\Lambda_{\delta,N}} \delta^d * g(n\delta) * u_\delta(\Phi_\delta(n\delta)),$$

with $\lambda_\delta \in {}^*\mathbb{R}$ a *coupling constant*, and

$$d\mu_{g,\delta,N} \equiv \exp[-U_g^\delta]d\mu_{0,\delta,N} / \int \exp[-U_g^\delta]d\mu_{0,\delta,N}.$$

The Loeb measure given by $\mu_{g,\delta,N}$, and the corresponding coordinate process, are by definition the hyperfinite field with interaction given by λ_δ, u_δ (in a space-time region given by supp g). Crucial questions are now:

1. Does there exist a choice of λ_δ, u_δ such that one can extract from this hyperfinite field a *non-Gaussian* (generalized) random field?
2. Can one take g = 1 and thus obtain a Euclidean invariant (Euclidean stationary) (generalized) random field?

For d = 2 the answer is yes; for example $u_\delta(y) = e^{\alpha y}$, $y \in \mathbb{R}$, $\alpha \in \mathbb{R}$ gives Høegh-Krohn's or the exponential model (see Albeverio & Høegh-Krohn (1977b), Albeverio, Fenstad, et al. (1986)). In this case λ_δ is found by making sure that the random field obtained is non-trivial. A hyperfinite computation of $E_{\mu_{g,\delta,N}}((U_g^\delta)^2)$ shows that this is finite if $\lambda_\delta = \lambda \exp[-\frac{\alpha^2}{2}(-\Delta_\delta + m^2)^{-1}(0,0)]$, with $\lambda \in \mathbb{R}_+$. In this case $\exp[-U_g^\delta]$ is near standard, $\not\approx 1$. By monotonicity one can replace g by an internal g equal to 1 for all finite $n\delta$ and one obtains the so called $(\exp\varphi)_2$-model.

Remark. Such a model, for m = 0, has very interesting relations to quantum relativistic (bosonic) strings; see Albeverio, Høegh-Krohn, Paycha & Scarlatti (1988). By similar hyperfinite methods it is also possible to construct gauge fields, for d = 2; see Albeverio, Fenstad, et al. (1986).

Remark. Hypercontinuous methods have been applied to quantum electrodynamics in perturbation theory by Fittler (1984).

As a last topic in this section let us briefly show how by hyperfinite tools one can connect quantum fields with polymers. Following Symanzik (1968) and Dynkin (1982) we can realize the square $\Phi_\delta(\delta n)^2$ of the free hyperfinite field as a Poisson random field of local times of Brownian loops and if the interaction is given by an even function u_δ we obtain a gas of interacting local times of Brownian bridges; c.f. Albeverio, Fenstad, et al. (1986). In fact: let

$$\Lambda_\delta = \{n\delta \in \Lambda:\ n \in {}^*\mathbb{Z}^d,\ \delta > 0,\ \delta \approx 0\}$$

be a bounded subset of ${}^*\mathbb{R}^d$, and let $T \equiv \{k\Delta t:\ k \in {}^*\mathbb{N}\}$, $\Delta t > 0$, $\Delta t \approx 0$ be the hyperfinite time line. Assume $\delta \geq (2^{d-1}\Delta t/[1-m^2\Delta t])^{1/2}$. Let α be an internal measure on $\Lambda_\delta \times \Lambda_\delta \times T$ with $\alpha(\{i,j,t\}) \approx 0$. Now make the following definition; for $\sigma \in \{0,1\}^{\Lambda_\delta\times\Lambda_\delta\times T}$, let

$$q_{i,j,t}(\sigma) \equiv \begin{cases} \alpha(i,j,t) & \text{if } \sigma(i,j,t) = 1 \\ 1 - \alpha(i,j,t) & \text{if } \sigma(i,j,t) = 0 \end{cases}$$

Then (Q,σ), with $Q(\sigma) \equiv \prod_{(i,j,t)\in\Lambda_\delta\times\Lambda_\delta\times T} q_{i,j,t}(\sigma)$ is an internal Poisson random field with parameter α and its Loeb measure $L(Q)$ is such that

$$L(Q)\Big(\sum_{x\in C}\sigma(x) = m\Big) = (m!)^{-1}L(\alpha)(C)^m\exp[-L(\alpha)(C)], \quad \text{for all } m \in {}^*\mathbb{N}.$$

For internal $g\colon \Lambda_\delta \to {}^*\mathbb{R}_+$ let

$$T_\delta(g)(\delta) \equiv \sum_{(i,j,t)\in\Lambda_\delta\times\Lambda_\delta\times T} \sigma(i,j,t)\sum_{s=0}^{t} g(X^{i,j}(s))\Delta t,$$

where $X^{i,j}$ is the Markov process X generated by $\frac{1}{2}\Delta - m^2$, conditioned to start at i and be at j at time t.

Remark. Intuitively, $T_\delta(g)$ is a smoothed local time for Brownian bridges $X^{i,j}$, with (i,j,t) α-distributed and the distribution of the number of bridges being Poisson. If we assume $\sum_{i,j,t}(1 - e^{-t\|g\|_\infty}) < \alpha(i,j,t) \approx 0$, then an easy computation shows that

$$E(e^{-T_\delta(g)}) \approx \exp \int_{\Lambda_\delta\times\Lambda_\delta\times T} E_t^{i,j}(e^{-\int_0^t g(X_s)ds} - 1)d\alpha(i,j,t).$$

Theorem

For $\alpha(i,j,t) \equiv (2t)^{-1}\delta_{ij}(1 - m^2\Delta t)^{t/\Delta t}P_t^i(X(t) = j)\ \Delta t$, *with* P_t^i *the probability measure for the process* X *started at* i, *and if* $|\Lambda_\delta| \le (\Delta t)^{-1/2}$, *one has*

$$E(\exp(-\tfrac{1}{2}\,\Phi_\delta^2(g))) \approx E(e^{-T_\delta(g)}),$$

where $\Phi_\delta^2(g) = \sum_{i\in\Lambda_\delta} {}^*g(i)(\Phi_\delta(i))^2\delta^d$, *for all internal* $g \ge 0$ *with* $\|g\|_\infty \le (\ln 1/\Delta t)^{1/2}$. *Hence* $\Phi_\delta^2(g) \approx T_\delta(g)$ *in law.*

Proof. The proof is an easy computation, using among other things the hyperfinite Feynman-Kac formula. In fact the above left hand side, with $\Phi_\delta^2(g)$ being Gaussian, can be computed to be

$$(\mathrm{Det}\ C_\delta^{-1})^{1/2}\ [\mathrm{Det}\ (C_\delta^{-1} + \delta^{-d}g)]^{-1/2},$$

with

$$C_\delta \equiv \delta^{2d}\ (-\tfrac{1}{2}\Delta_\delta + m^2)^{-1}.$$

This is equal to

$$\exp\ [-\tfrac{1}{2}\mathrm{tr}[\ln(H_\delta + g) - \ln H_\delta]\},$$

with $H_\delta \equiv -\frac{1}{2}\Delta_\delta + m^2$ (where we used Det A = exp(tr(ln A)) for matrices, and transfer). Using $\ln x \approx -\sum_{k=1}^{\infty} \frac{1}{k}(1-x)^k$, for $x = (H_\delta + f)\Delta t$ and $x = H_\delta\Delta t$ (note that t drops out!) we get

$$\exp\ (\tfrac{1}{2}\mathrm{tr} \sum_{k=1}^{\infty} \frac{1}{k}\{(1 - (H_\delta + g)\Delta t)^k - (1 - H_\delta\Delta t)^k\}).$$

By the hyperfinite Feynman-Kac formula we have, with $e_i(j) \equiv \begin{cases}\delta^{-d/2} & i = j\\ 0 & i \ne j\end{cases}$,

$$(\sum_{k=1}^{\infty} \frac{1}{k}(1 - (H_\delta + g)\Delta t)^k e_i,\ e_i)$$

$$= \sum_{t=\Delta t}^{\infty} \delta^d\,\frac{\Delta t}{t}(1 - m^2\Delta t)^{t/\Delta t}\ P_t^i(X(t) = i)\ E_t^{i,i}(e^{-\sum_0^k f(X(s))\Delta t})$$

which is infinitely near to $E(e^{-T_\delta(g)})$. This concludes the proof.◂

We shall now use the above result to express the interacting fields, for u_δ even, by polymer measures. Let μ_δ be the interacting hyperfinite random field with interaction $U_\delta(\cdot)$ assumed to be an even function. We compute $E_{\mu_\delta}(e^{-\Phi_\delta(*g)})$ by realizing it has the form $kE_{\mu_{0,\delta}}(e^{-\Phi_\delta(*g)}F(½\Phi_\delta^2))$, ($k$ a constant) with $\mu_{0,\delta}$ the free hyperfinite field measure, F a function. Define $T_\delta(k)$ as $T_\delta(g)$ with g replaced by $\delta^{-d}\delta_k$. Define $T_{\delta,g}(k)$ similarly, with α replaced by $\alpha_g(i,j,t) \equiv ½g(i)g(j)(1-m^2\Delta t)^{t/\Delta t}\, P_t^i(X(t) = j)$. Then we have:

Proposition

There exists an $N \in {}^*\mathbb{N} \setminus \mathbb{N}$ *such that if* $|\Lambda_\delta| \le N$ *then for every* $F,g \in \ell^2(\Lambda_{\delta,N},\mathbb{R}^d)$, *with* $\|g\|_\infty < \infty$:

$$E_{\mu_{0,\delta}}(e^{-\Phi_\delta(*g)}F(½\,\Phi_\delta^2)) \approx E(F(T_\delta + T_{\delta,g})e^{½\langle g,H_\delta^{-1}g\rangle_{\ell^2}}).$$

Proof. One uses Dynkin's trick (Dynkin (1985)), to compute first for $F(\lambda) = e^{-\lambda}$ and use Stone-Weierstrass to extend to the general case. For $F(\lambda) = e^{-\lambda}$ we take $F(½\Phi_\delta^2) = F(½\Phi_\delta^2(f))$ for some $f \in C_0^\infty(\mathbb{R}^d)$ and the expectation on the left hand side in the proposition is equal to

$$(\mathrm{Det}\, C_\delta^{-1})^{½}\, [\mathrm{Det}\, (C_\delta^{-1} + \delta^{-2}f1)]^{-½}\, e^{½\langle g,(H_\delta+f)^{-1}g\rangle} \approx E(\exp[-T_\delta(f)]),$$

as computed before.

But $(H_\delta + f)^{-1} = \sum_{k=0}^{\infty} [1 - (H_\delta + f)\Delta t]^k\Delta t$; hence, by using the hyperfinite Feynman-Kac formula, we have

$$(g,(H_\delta + f)^{-1}g) \approx \sum_{t=0}^{\infty} E[g(X(t)g(X(0))e^{-\int_0^t f(X(s))ds}]\Delta t$$

$$= \sum_{(i,j,t)\in\Lambda_\delta\times\Lambda_\delta\times T} g(i)g(j)(1 - m^2\Delta t)^{t/\Delta t}\, P_t^i(X(t) = j)\, \delta^{2d}\Delta t e^{-\int_0^t f(X(s))ds},$$

where we introduced conditional measures. Using the definition of α this implies

$$e^{\langle g,(H_\delta + f)^{-1} g\rangle} \approx \exp\left(\int E_t^{i,j}\left(e^{-\int_0^t f(X(s))ds} - 1\right)d\alpha_g(i,j,t)\right)$$

where $E_t^{i,j}$ is expectation with respect to the Brownian bridge between $(t = 0,i)$ and (t,j). However, the right hand side is infinitely close to $E(e^{-T_g(f)})e^{-\frac{1}{2}\langle g,H_\delta^{-1}g\rangle}$, which is what we wanted to prove.◂

The so called $(\Phi^4)_d$-model of quantum fields corresponds to taking $u_\delta(\lambda)$ an even 4th order polynomial, with δ-depending coefficients. The above proposition gives the Laplace transform of the measure μ_δ in terms of expectations with respect to Brownian bridges and Poisson measures.

Recalling the expressions for $T_\delta(k)$ and $T_{\delta,g}(k)$ and inserting for $F(\Phi_\delta^2(f))$ we see easily that the right hand side in the above proposition contains terms involving densities of the type of those of 1 and 2-polymer measures.

These were given a non-trivial meaning by a suitable choice of strength, corresponding to a negative infinitesimal coupling constant in the leading term of the φ_d^4-interaction, for $d \le 5$. This makes it likely that $-|\lambda|(\varphi^4)_d$, with λ infinitesimal, exists and is non trivial (as opposed to the case $+|\lambda|(\varphi^4)_d$, which is likely to be trivial); this conjecture was made in Albeverio, Blanchard & Høegh-Krohn (1982b); see also Glimm & Jaffe (1987), Albeverio (1987a) for example.

Remark. Other types of interactions, reduced to the 2-polymer measures of Section 6 have been discussed in Albeverio, Fenstad, et al. (1986).

8. OTHER TOPICS AND CONCLUSIONS

We have tried to give an idea, not only of natural topics in physics which can and have been investigated using nonstandard tools, but also of some of the techniques involved. Of course this is far from being exhaustive. Fortunately, some complementary topics in the subject of interactions between physics and nonstandard analysis are discussed in other papers in this volume. Amongst these are the study of dynamical systems (ordinary differential equations) with nonstandard coefficients. This is a field where nonstandard analysis is most useful, helping to formulate problems conceptually and discover new phenomena. The Mulhouse-Strasbourg-Oran school has led in these investigations. For references see F. & M. Diener's paper in this volume and also, for example, Zvonkin & Shubin (1984), Albeverio, Fenstad, et al. (1986), Cartier (1982) and the references therein.

Another topic in physics where nonstandard methods have been applied successfully is non-equilibrium statistical mechanics. Here breakthrough results have been obtained by L. Arkeryd, and A.E. Hurd; cf. Arkeryd's paper (this volume), Albeverio, Fenstad, et al. (1986) and Hurd (1987a,b). More generally, nonstandard methods should prove useful in justifying finite elements, Galerkin and compactness methods, in mechanical and hydrodynamical problems for example; however, little has been done until now. See also for the study of partial differential equations Berger & Sloan (1983), and Kosciuk (1983) for example.

In equilibrium statistical mechanics the study of scaling phenomena, renormalization group methods, and phase transitions seems to be a promising domain for the application of nonstandard reasoning, modelling and tools; see for example Albeverio, Fenstad, et al. (1986) and references therein. Important results concerning Gibbs states have been obtained by Kessler (see Albeverio, Fenstad, et al. (1986) and Kessler (1984, 1985, 1986, 1987, 1988a,b)). Asymptotic phenomena (small time, large time, small diffusion, small parameters....), typically relevant in physical considerations seem also a potential field of fruitful applications of nonstandard analytic methods, and here tools of Cutland (1986a, 1987, 1988) should be useful. Let us also mention that the Efimov effect in nuclear physics seems to be a good candidate for such investigations; cf. Albeverio, Høegh-Krohn & Wu (1981). Let us mention finally a domain which seems ideal for nonstandard tools, but where almost nothing has been done up to now, namely the study of relativistic strings. Here heuristic computations involve measures on infinite dimensional non-flat manifolds and heuristic changes of variables, involving infinite determinants (for some partial justification see Albeverio, Høegh-Krohn, Paycha & Scarlatti (1988)). Could those computations of this type which are in a sense precise, although basically outside any rigorous mathematics, be justified by nonstandard analytic tools? Here as elsewhere problems arise. It is perfectly possible to do nonstandard analytic computations, but how do we interpret the result? How do we go back to the standard world, perhaps only for some suitable physical quantities? Or if one chooses to stay in the more comfortable nonstandard universe, how do we develop physics and interpretations within it? In our opinion the interaction of nonstandard analysis and physics can only get stronger by

1. a deeper use of nonstandard analysis as a way of thinking,
2. a development of better methods to recover standard results from nonstandard ones (a coming back to the standard world), in somewhat the same way as for the theory of generalized functions: generalized solutions of p.d.e.'s are particularly useful when they either can lead to classical solutions by some regularity theorem, or at least can be used to deduce some other results which have a convenient interpretation in classical terms.

Acknowledgements. It is a pleasure to thank N. Cutland and T. Lindstrøm for their very kind invitation to give the lectures which form the basis for this paper. I also take the opportunity to express my deep gratitude to my friends, J.-E. Fenstad, R. Høegh-Krohn, and T. Lindstrøm, with whom I wrote the book on nonstandard methods on which everything I have discussed in this paper is based. I also thank L. Arkeryd, N. Cutland, Ch. Kessler and A. Stoll for most useful discussions. I am also most grateful to Nigel Cutland for his patience and great help in the publication of this paper.

REFERENCES

Albeverio, S. (1984). Nonstandard analysis; polymer models, quantum fields, *Acta Phys. Austriaca, Suppl.* XXVI., 233-254.

Albeverio, S. (1986a) Some points of interaction between stochastic analysis and quantum theory; in *Stochastic Systems and Applications*, (eds. N. Christopeit, K. Helmes & M. Kohlmann), Lecture Notes on Control and Information Science **78**, Springer, Berlin, 1-26.

Albeverio, S. (1986b). Nonstandard analysis: applications to probability theory and mathematical physics; in *Mathematics and Physics, Vol. 2*, (ed. L. Streit), World Scientific Publications, Singapore, 1-49.

Albeverio, S. (1986c). Some recent developments and applications of path integrals; in *Path Integrals from mev to MeV*, (eds. M. C. Gutzwiller, A. Inomata, J. R. Klauder & L. Streit), World Scientific Publications, Singapore, 3-32.

Albeverio, S. (1987a). Some personal remarks on nonstandard analysis in probability theory and mathematical physics; in *Proc. VIIIth International Congress on Mathematical Physics* (eds. M.Mebkhout & R. Sénéor), World Scientific Publications, Singapore, 409-420.

Albeverio, S. (1987b). An introduction to nonstandard analysis and applications to quantum theory; in *Information Complexity and Control in Quantum Physics* (eds. A. Blaquière, S. Diner & G. Lochak), Springer, Vienna, 183-208.

Albeverio, S., Blanchard, Ph. & Høegh-Krohn, R. (1982a). Feynmann path integrals and the trace formula, *Comm. Math. Phys.* **83**, 49-76.

Albeverio, S., Blanchard, Ph. & Høegh-Krohn, R. (1982b). Some applications of functional integration; in *Mathematical Problems in Theoretical Physics* (eds. R. Schrader, R. Seiler & D. Uhlenbrock), Lecture Notes in Physics **153**, Springer, 265-275.

Albeverio, S., Fenstad, J.-E. & Høegh-Krohn, R. (1979). Singular perturbations and nonstandard analysis, *Trans. Amer. Math. Soc.* **252**, 275-295.

Albeverio, S., Fenstad, J.E., Høegh-Krohn, R., Karwowski, W. & Lindstrøm, T. (1984). Perturbations of the Laplacian supported by null sets, with applications to polymer measures and quantum fields, *Phys. Lett. A* **104**, 396-400.

Albeverio, S., Fenstad, J.E., Høegh-Krohn, R. & Lindstrøm, T. (1986). *Nonstandard Methods in Stochastic Analysis and Mathematical Physics*, Academic Press, New York.

Albeverio, S., Gesztesy, F., Høegh-Krohn, R. & Holden, H. (1988). *Solvable Models in Quantum Mechanics*, Springer, Berlin.

Albeverio, S. & Høegh-Krohn, R. (1974). The Wightman axioms and the mass gap for strong interactions of exponential type in two-dimensional space-time, *J. Funct. Anal.* **16**, 39-82.

Albeverio, S. & Høegh-Krohn, R. (1976). *Mathematical Theory of Feynman Path Integrals*, Lecture Notes in Mathematics **523**, Springer, Berlin.

Albeverio, S. & Høegh-Krohn, R. (1977a). Dirichlet forms and diffusion processes on rigged Hilbert spaces, *Z. Wahrsch. Th. verw. Geb* **40**, 1-57.

Albeverio, S. & Høegh-Krohn, R. (1977b). Hunt processes and analytic potential theory on rigged Hilbert spaces, *Ann. Inst. H. Poincaré B* **13**, 269-291.

Albeverio, S. & Høegh-Krohn, R. (1985). Diffusion fields,quantum fields and fields with values in Lie groups; in *Stochastic Analysis and Applications* (ed. M. Pinsky), M. Dekker, New York, 1-98.

Albeverio, S., Høegh-Krohn, R., Paycha, S. & Scarlatti, S. (1988). A probability measure for random surfaces of arbitrary genus and bosonic strings in 4 dimensions, Bochum Preprint; see also *Physics Letters B* **174**(1986), 81-86.

Albeverio, S., Høegh-Krohn, R. & Röckner, M. (1988). In preparation.

Albeverio, S., Høegh-Krohn, R. & Wu, T.T. (1981). A class of exactly solvable three-body quantum mechanical problems and the universal low energy behaviour, *Physics Letters A* **83**, 105-109.

Albeverio, S., Høegh-Krohn, R. & Streit, L. (1977). Energy forms, Hamiltonians and distorted Brownian paths, *J. Math. Physics* **18**, 907-917.

Albeverio, S. & Kusuoka, S. (1988). In preparation.

Alonso y Coria, A. (1978). Shrinking potentials in the Schrödinger equation, Ph.D. Dissertation, Princeton University.

Arede, T. (1986). A class of solvable non-linear filters, *Stochastics* **23**, 377-389.

Arkeryd, L. (1986). On the Boltzmann equation in unbounded space far from equilibrium and the limit of zero mean free path, *Commun. Math. Phys.* **105**, 205-219.

Barlow, M. T. & Perkins, E. (1987). Brownian motion on the Sierpiński gasket, Cambridge-Vancouver preprint.

van den Berg, I. (1987). *Nonstandard Asymptotic Analysis*, Lecture Notes in Mathematics **1249**, Springer, Berlin.

Berger, M. & Sloan, A. (1983). Explicit solutions of partial differential equations; in Hurd (1983), 1-14.

Berry, M. (1980). Some geometrical aspects of wave motion: wavefront dislocations, diffraction catastrophes, diffractals; in *Geometry of the Laplace Operator*, Proc. Symp. Pure Math **36**, Amer. Math. Soc, 13-38.

Birkeland, B. (1980). A singular Sturm-Liouville problem treated by nonstandard analysis, *Math. Scand.* **47**, 245-294.

Blanchard, Ph. & Tarski, J. (1978). Renormalisable interactions in two dimensions and sharp-time fields, *Acta Phys. Austr.* **49**, 129-152.

Bouleau, N. & Hirsch, F. (1986). Formes de Dirichlet générales et densité des variables aléatoires réelles sur l'espace de Wiener, *J. Funct. Anal.* **69**, 229-259.

Brasche, J. (1985). Perturbations of Schrödinger Hamiltonians by measures - self-adjointness and lower semiboundedness, *J. Math. Phys.* **26**, 621-626.

Brossard, J. & Carmona, R. (1986). Can one hear the dimension of a fractal?, *Comm. Math. Phys.* **104**, 103-122.

Callot, J. L. (1983a). Solutions visibles de l'equation de Schrödinger, Mulhouse preprint.

Callot, J. L. (1983b). Stroboscopie infinitésimale, in *Outils et modelès mathématiques pour l'automatique, l'analyse des systèmes et le traitement du signal*, CNRS **3**.

Cartier, P. (1982). Perturbations singulières des équations différentielles ordinaire et analyse nonstandard, *Sém. Bourbaki* **81-82**, *Astérisque* **92-93**, 21-44.

Cutland, N.J. (1983). Nonstandard measure theory and its applications, *Bull. London Math. Soc.* **15** (1983), 529-589.

Cutland, N.J. (1985). Simplified existence for solutions to stochastic differential equations, *Stochastics* **14**, 319-325.

Cutland, N.J. (1986a). Infinitesimal methods in control theory: deterministic and stochastic, *Acta Appl. Math.* **5**, 105-135.

Cutland, N. J. (1986b). Optimal controls for stochastic systems with singular noise, *Systems and Control letters* **7**, 55-59.

Cutland, N. J. (1987). Infinitesimals in action, *J. Lond. Math. Soc.* **35**, 202-216.

Cutland, N. J. (1988). An extension of the Ventcel-Freidlin large deviation principle, to appear in *Stochastics*.

Cutland, N. J. & Kendall, W. S. (1986). A nonstandard proof of one of David Williams' splitting-time theorems; in *Analytic and Geometric Stochastics*, supplement to *Adv. Appl. Prob.* 37-47.

Davis, M. (1977). *Applied Nonstandard Analysis*, Wiley, New York.

Dynkin, E. B. (1982). Green's and Dirichlet spaces associated with fine Markov processes, *J. Funct. Anal.* **47**, 381-418.

Dynkin, E. B. (1985). Random fields associated with multiple points of the Brownian motion, *J. Funct. Anal.* **62**, 397-434.

Edwards, S. F. (1965). The statistical mechanics of polymers with excluded volume, *Proc. Phys. Soc.* **85**, 613-624.

Farkas, E. J. & Szabo, M. E. (1984). On the plausibility of nonstandard proofs in analysis, *Dialectica* **38**, 297-310.

Farrukh, M. O. (1975). Applications of nonstandard analysis to quantum mechanics, *J. Math. Phys.* **16**, 177-200.

Fenstad, J.-E. (1987). The discrete and the continuous in mathematics and the natural sciences; in *L'Infinito nella scienza*, Istituto dell'Enciclopedia Italiana, 111-126.

Fittler, R. (1984) Some nonstandard quantum electrodynamics, *Helv. Phys. Acta* **57**, 579-609.

Fittler, R. (1987) More nonstandard quantum electrodynamics, *Helv. Phys. Acta*.

Francis, C. E. (1981). Applications of nonstandard analysis to relativistic quantum mechanics, *J. Phys. A* **14**, 2539-2551.

Friedman, C. N. (1972). Perturbations of the Schrödinger equation by potentials with small support, *J. Funct. Anal.* **10**, 346-360.

Fukushima, M. (1980). *Dirichlet Forms and Markov Processes*, North-Holland, Amsterdam

Fukushima, M. (1985). Energy forms and diffusion processes; in *Mathematics and Physics, Lectures on Recent Results, Vol I* (ed. L. Streit), World Scientific Publications, Singapore, 65-97.

Fukushima, M. (1986). On recurrence criteria in the Dirichlet space theory; in *From Local Times to Global Geometry, Control and Physics* (ed. K.D.Elworthy), Pitman Research Notes in Mathematics **150**, Longman, 100-110.

Giorello, G. (1973). Una rappresentazione nonstandard delle distribuzioni temperate e la transformazione di Fourier, *Boll. UMI* **IV**, Ser 7, 156-167.

Glimm, J. & Jaffe, A. (1987). *Quantum Physics* (2nd edn), Springer, Berlin.

Goldstein, S. (1986). Random walks and diffusions on fractals, Rutgers preprint.

Harthong, J. (1981). Le moiré, *Adv. Appl. Math.* **2**, 24-75.

Harthong, J. (1984). Études sur la méchanique quantique, *Astérisque* **111**.

Hejtmanek, J. (1986). Asymptotic behaviour of semigroups, Geerhart's theorem and nonstandard analysis, Wien preprint.

Helms, L.L. (1983). A nonstandard approach to the martingale problem for spin models; in Hurd (1983), 15-26.

Homer, B. J. & Thompson, C. L. (1986). Shadows and halos in nonstandard analysis with applications to topological dynamics, Southampton preprint.

Hurd, A.E. (ed.) (1983). *Nonstandard Analysis - Recent Developments*, Lecture Notes in Mathematics **983**, Springer-Verlag, Berlin and New York.

Hurd, A.E. (1987a). Global existence and validity for the BBGKY hierarchy, *Arch. Rational Mech. Anal.* **98**, 191-210.

Hurd, A.E. (1987b). Global existence and validity for the Boltzmann hierarchy, Victoria preprint.

Hurd, A.E. & Loeb, P.A. (1985). *An Introduction to Nonstandard Real Analysis*, Academic Press, New York.

Jammer, M. (1974). *The Philosophy of Quantum Mechanics*, Wiley, New York.

Kambe, R. (1974). On quantum field theory I, *Progr. Theor. Phys.* **52**, 688-706.

Kato, T. (1966). *Purturbation Theory for Linear Operators*, Springer, New York.

Keleman, P. J. & Robinson, A. (1972). The nonstandard $\varphi_2^4(x)$-model I .The technique of nonstandard analysis in mathematical physics , I & II, *J. Math. Phys.* **13**, 1870-1874, and 1875-1878.

Kessler, C. (1984). Nonstandard methods in the theory of random fields, Dissertation, Bochum.

Kessler, C. (1985). Examples of extremal lattice fields without the global Markov property, *Publ RIMS Kyoto Univ.* **21**, 877-888.

Kessler, C. (1986). Nonstandard conditions for the global Markov property for lattice fields, *Acta Applicandae Mathematicae* **7**, 225-256.

Kessler, C. (1987). Attractiveness of interactions for binary lattice systems and the global Markov property, *Stochastic Proc. App.*, **24**, 309-313.

Kessler, C. (1988a). Markov type properties of mixtures of probabilty measures, to appear in *Prob. Th. and Related Fields.*

Kessler, C. (1988b). On hyperfinite representations of distributions, to appear in *Bull. Lond. Math. Soc.*

Komkov, V. & McLaughlin, T. G. (1984). Local analysis of nonstandard C^∞ functions of predistributional type, *Ann. Pol Math.* **XLIV**, 15-38.

Kosciuk, S.A. (1983). Stochastic solutions to partial differential equations; in Hurd (1983), 113-119.

Krupa, A. & Zawisza, B. (1984). Applications of ultrapowers in analysis of unbounded selfadjoint operators, *Bull. Pol. Acad. Sci. Math.* **32**, 581-588.

Kusuoka, S. (1982). Dirichlet forms and diffusion processes on Banach spaces, *J. Fac. Sci. Univ. Tokyo* **29**, 79-95.

Kusuoka, S. (1985). A diffusion process on a fractal, *Taniguchi Symp. PMMP Karata*, 251-274.

Laugwitz, D. (1986). *Zahlen und Kontinuum*, Bibliographisches Inst., Mannheim.

Lawler, G.F. (1980). A self-avoiding random walk, *Duke Math. J.* **47**, 655-693.

Lawler, G.F. (1986). Gaussian behavior of loop-erased self-avoiding random walk in four dimensions, *Duke Math J.* **53**, 249-270.

Li Banghe (1986). On the moiré problem from the distributional point of view, *J. Sys. Sci. & Math. Sci.* **6**, 263-268.

Li Banghe & Li Yaqing (1985). Nonstandard analysis and multiplication of distributions in any dimension, *Scientia Sinica (Ser. A)* **27**, 716-726.

Li Yaqing (1985). The product of $x_+^\lambda \ln^p x_+$ and $x_-^N \ln^q x_-$, *J. Sys. Sci. & Math. Sci.* **5**, 241-250.

Lindstrøm, T. (1983). Stochastic integration in hyperfinite dimensional linear spaces; in Hurd (1983), 134-161.

Lindstrøm, T. (1986). Nonstandard energy forms and diffusions on manifolds and fractals; in *Stochastic Processes in Classical and Quantum Systems* (Albeverio, S. et al., eds.), Lecture Notes in Physics **262**, Springer-Verlag, 363-380.

Lindstrøm, T. (1988). Brownian motion on a class of self-similar fractals, Oslo preprint.

Loeb, P.A. (1982). A construction of representing measures for elliptic and parabolic differential equations, *Math. Ann.* **260**, 51-56.

Lutz, R. & Goze M. (1981). *Nonstandard Analysis*, Lecture Notes in Mathematics **881**, Springer-Verlag, Berlin and New York.

Mandelbrot, B. *B.* (1977). *Fractals*, Freeman, San Francisco.

Mandelbrot, B. *B.* (1982). *The Fractal Geometry of Nature*, Freeman, San Francisco.

Mitter, S. (1980). On the analogy between mathematical problems of nonlinear filtering and quantum physics, *Ric. di Automatica* **10**, 163-216.

Moore, S. (1980). Stochastic fields from stochastic mechanics, *J. Math. Phys.* **21**, 2102-2110.

Moore, S. (1982). Nonstandard analysis applied to path integral, *Nuovo Cim.* B, **70**, 227-290.

Murakami, H., Nakagiri, S. I. & Yeh, C. C. (1983). Asymptotic behaviour of solutions of nonlinear differential equations with deviating arguments via nonstandard analysis, *Ann. Pol. Math.* **41**, 203-208.

Nagamachi, S. & Mugibayashi, M. (1986). Nonstandard analysis of Euclidean Fermi fields, *BiBoS Preprint* **242**.

Nagamachi, S. & Nishimura, T. (1984). Linear canonical transformations on Fermion Fock space with indefinite metric.

Nelson, E. (1977). Internal set theory: a new approach to nonstandard analysis. *Bull. Amer. Math. Soc.* **83**,1165-1193.

Nelson, E. (1987a). *Predicative Arithmetic*, Princeton Univ. Press.

Nelson, E. (1987b). *Radically Elementary Probability Theory*, Princeton Univ. Press.

Nobis, K. (1984). On the application of nonstandard analysis in mechanics of porous media, *Bull. Pol. Acad. Sci. Tech. Sci.* **32**, 383-387.

Nobis, K., Wierzbicki, W. & Wozniak, C. (1984). On the physical integration of nonstandard methods in mechanics, *Bull. Pol. Acad. Sci. Tech. Sci.* **32**, 379-382.

Nottale, L. & Schneider, J. (1984). Fractals and nonstandard analysis, *J. Math. Phys.* **25**, 1296-1300.

Oikkonen, J. (1985a). Harmonic analysis and nonstandard Brownian motion in the plane, *Math. Scand.* **57**, 346-358.

Oikkonen, J. (1985b). The C^2 image of Brownian motion in the plane, Helsinki preprint.

Ostebee, A., Gambardella, P. & Dresden, M. (1976). A nonstandard approach to the thermodynamic limit II. Weakly tempered potentials and neutral Coulomb systems, *J. Math. Phys.* **17**, 1570.

Pecora, L. M. (1982). A nonstandard infinite dimensional vector space approach to Gaussian functional measures, *J. Math. Phys.* **23**, 969-982.

Perkins, E. (1983). Stochastic processes and nonstandard analysis; in Hurd (1983), 162-185.

Raskovic, M. (1985). An application of nonstandard analysis to functional equations, *Publ. Inst. Math. Nouv. Sci.* **37**, 23-24.

Reed, M. & Simon, B. (1972). *Methods of Modern Mathematical Physics I-IV*, Academic Press, New York.

Rezende, J. (1985). The method of stationary phase for oscillatory integrals on Hilbert spaces, *Comm. Math. Phys.* **101**, 187-206.

Richter, M.M. (1982). *Ideale Punkte, Monaden, und Nichtstandard-Methoden*, Vieweg, Wiesbaden.

Robert, A. (1985). *Analyse Non-Standard*, Presses Polytechnique Romandes, Lausanne.

Robert, A. (1988). *Nonstandard Analysis*, Wiley, New York & Chichester. (English translation of Robert (1985).)

Röckner, M. & Wielens, N. (1985). Dirichlet forms- closability and change of speed measure; in *Infinite Dimensional Analysis and Stochastic Processes* (ed. S.Albeverio), Pitman, London.

Silverstein, M. (1974). *Symmetric Markov Processes*, Lecture Notes in Mathematics **426**, Springer.

Sloan, A. D. (1977). An application of the nonstandard Trotter product formula, *J. Math. Phys.* **18**, 2495-2496.

Sloan, A. D. (1981). The strong convergence of Schrödinger propagators, *Trans. Amer. Math. Soc.* **264**, 557-570.

Stoll, A. (1985). Self-repellent random walks and polymer measures in two dimensions, Dissertation, Bochum.

Stoll, A. (1986a). A nonstandard construction of Lévy Brownian motion, *Prob. Th. and Related Fields* **71**, 321-334.

Stoll, A. (1986b). Invariance principles for Brownian intersection local time and polymer measures, to appear in *Math.Scand.*

Streit, L. (1981). Energy forms, Schrödinger theory, processes, *Phys. Reports* **77**, 363-375.

Streit, L. (1986). Quantum theory and stochastic processes - some contact points; in *Stochastic Processes and their Applications, Proc. Bernouilli Soc. Meeting, Nagoya* (eds. K.Ito & T.Hida), Lecture Notes in Mathematics **1203**, Springer, 197-213.

Stroyan, K.D. & W.A.J. Luxemburg. (1976). *Introduction to the Theory of Infinitesimals*, Academic Press, New York.

Takeda, M. (1985). On the uniqueness of Markovian self-adjoint extensions of diffusion operators on infinite dimensional spaces, *Osaka J. Math.* **22**, 233-242.

Tarski, J. (1976). Short introduction to nonstandard analysis and its physical applications; in *Quantum Dynamics* (ed. L.Streit), Springer, Vienna, 225-239.

Thomas, L. H. (1935). The interaction between a neutron and a proton and the structure of H^3, *Phys. Rev.* **47**, 903-909.

Todorov, T. D. (1985). Application of nonstandard Hilbert space to quantum mechanics, *Proc. IIIrd Int. Conf. Complex Anal. Varna*.

Todorov, T. D. (1987). Sequential approach to Colombeau's theory of generalised functions, Int. Centre Theor. Phys. Trieste preprint.

Tortorelli, V. M. (1987). *Γ-limits and infinitesimal analysis*, Scuola Norm. Sup. Pisa preprint.

Voros, A. (1973). Introduction to nonstandard analysis, *J. Math. Phys.* **14**, 292-296.

Wakita, H. (1962). On an extension of the mathematical framework of the quantum theory II, *Progr. Th. Phys.* **28**, 251-257.

Wakita, H. (1984). Mathematical framework of quantum electrodynamics, *Math. Jap.* **29**, 199-217.

Westwater, J. (1980, 1982). On Edwards' model for long polymer chains, *Comm. Math. Phys.* **72**, 131-174 & **84**, 459-470.

Woźniak, C. (1981). On the nonstandard analysis and the interrelation between mechanics of mass-point systems and continuum mechanics, *Mech. Teor. Stosow.* **19**, 511-525.

Zvonkin, A.K. & Shubin, M.A. (1984). Nonstandard analysis and singular perturbations of ordinary differential equations, *Russian Math. Surveys* **39**, 69-131

A LATTICE FORMULATION OF REAL AND VECTOR VALUED INTEGRALS

PETER A. LOEB

In classical analysis, the approach to general integration is either through measure theory or through functional analysis with a lattice formulation of the integral. Starting from the notion of "length", for example, one constructs Lebesgue measure and the class of Lebesgue measurable sets. Lebesgue measurable functions form an extension of the class of simple functions (linear combinations of characteristic functions of sets of finite measure), and the Lebesgue integral extends the obvious calculation for simple functions. Long before seeing this development, however, each student of mathematics has used the notion of length to obtain the Riemann integral. We wisely refrain from telling our calculus students that the Riemann integral is a positive linear functional on the space of continuous functions with compact support and is, therefore, represented by a measure. That measure is, of course, Lebesgue measure; it is obtained by extending the Riemann integral from the continuous functions with compact support to the class of measurable functions and then noting the action on the characteristic functions of measurable sets. Here, the Lebesgue integral is constructed before Lebesgue measure.

The major theorem generalizing the extension of "length" to Lebesgue measure is the Carathéodory Extension Theorem, and the major theorem associated with constructing a measure representing the Riemann integral (or any positive linear functional on continuous functions) is the Riesz Representation Theorem. These theorems have played an important role in the development of nonstandard measure theory. The Carathéodory theorem was used in a measure theoretic approach (Loeb (1975)) to extend internal measures to standard ones. The Riesz theorem was used to obtain standard measures on standard compact sets representing the action of internal measures (or functionals) on standard continuous functions. In the first case, the action of an internal measure on internal *simple functions* is extended; in the second case, the action of an internal measure on internal *continuous functions* is extended. These two processes will be combined and generalized here with the extension of a positive linear functional from an internal lattice of real valued functions to an external space of "measurable functions". Moreover, just as $\aleph_1$-saturation, can replace all uses of the Carathéodory theorem in the measure theoretic approach (see Stroyan & Bayod (1986)), $\aleph_1$-saturation will allow a development from first principles here with no use of the Riesz Representation Theorem. Indeed, the Riesz theorem will follow as a consequence of our construction.

We assume throughout this article that the constant function 1 is a member of the internal lattice and has a finite integral. This corresponds to starting with a finite measure space or a compact topological space. The original development of these results for arbitrary measure spaces and locally compact topological spaces can be found in Loeb (1984). We will conclude with some indications of recent work by Horst Osswald and the author extending the theory to the case of Banach-lattice valued functions and measures. For this extension, the external integrals are found in the nonstandard hull of the original Banach lattice. The interplay between nonstandard integration theory and the theory of nonstandard hulls is relatively unexplored and rich in possibilities for further research.

1. SCALAR FUNCTIONS AND MEASURES

Recall that a vector lattice of functions on a set X is a vector space with a pointwise ordering. That is, $(f+g)(x) = f(x) + g(x)$, $(\alpha f)(x) = \alpha(f(x))$ and $f \le g$ if $f(x) \le g(x)$ for all $x \in X$. A lattice is closed under the operations $\vee$ and $\wedge$; i.e., the function $f \vee g$ defined by setting $(f \vee g)(x) = f(x) \vee g(x)$ at each $x \in X$ is in the lattice; so is the function $f \wedge g = -(-f \vee -g)$. A linear mapping T from one lattice to another is called positive if $T(f) \ge 0$ whenever $f \ge 0$; T is called a functional if the range is in the scalar field.

In this section, X will denote an internal set in an $\aleph_1$-saturated enlargement of a structure containing the real numbers $\mathbb{R}$; L will denote an internal vector lattice of $^*\mathbb{R}$-valued functions on X, and

I will denote an internal positive linear functional on L. We will assume that $1 \in L$ and $I(1)$ is finite in $^*\mathbb{R}$. There are several examples to keep in mind here. First, X can be a hyperfinite set and L the space of all $^*\mathbb{R}$-valued internal functions on X. The functional I is determined by uniform counting measure; i.e., $I(f) = \frac{1}{|X|} \sum_{x \in X} f(x)$ where $|X|$ denotes the internal cardinality of X. A more general construction starts with an arbitrary internal probability space $(X, \mathcal{M}, P)$ (i.e. $\mathcal{M}$ is an internal σ-algebra and P an internal measure with $P(X) = 1$.) Here, L is the space of internal $\mathcal{M}$-measurable simple functions and $I(f) = \int f dP$ for all $f \in L$. A third example is constructed on the internal compact set $X = {}^*[0,1]$ with L the set of continuous functions on X. Here, I can be any positive linear functional on L; a prime example is the nonstandard extension of the Riemann integral: $I(f) = {}^*\int_0^1 f(x)dx$.

From the internal lattice L, we construct two external vector spaces L_0 and L_1 over the real numbers $\mathbb{R}$.

1.1 Definition

The class of *null functions* L_0 is the set of all internal and external $^*\mathbb{R}$-valued functions h on X such that for any $\varepsilon > 0$ in $\mathbb{R}$ there is a $\varphi \in L$ with $|h| \le \varphi$ and $I(\varphi) \le \varepsilon$. The class L_1 is the set of real valued functions f on X having representation $f = \varphi + h$ for some $\varphi \in L$ and $h \in L_0$. Given such a representation of an $f \in L_1$, we set $J(f) = {}^\circ I(\varphi)$.

First we need some preliminary results showing, among other things, that L_1 is a lattice and J is a well defined positive linear functional on L_1. Note that when $\varphi \in L \cap L_0$, $I(\varphi) \approx 0$. Also note that $1 \in L_1$.

1.2 Proposition

The sets L_0 *and* L_1 *are vector lattices over* $\mathbb{R}$. *If* f *is in* L_1 *with* $f = \varphi + h$ *for* $\varphi \in L$ *and* $h \in L_0$, *then* $I(|\varphi|)$ *is finite* *in* $^*\mathbb{R}$. *If* g *is also in* L_1 *with* $g = \psi + k$ *for* $\psi \in L$ *and* $k \in L_0$, *then* $(f \vee g) - (\varphi \vee \psi) \in L_0$ *and* $(f \wedge g) - (\varphi \wedge \psi) \in L_0$. *Moreover, if* $f = g$ *(that is* $f = \varphi + h = \psi + k$) *then* $\varphi - \psi \in L_0$, *whence* $J(f) = {}^\circ I(\varphi) = {}^\circ I(\psi)$. *It follows that* J *is a well-defined positive linear functional on* L_1.

Proof. It is easy to see that L_0 is a vector lattice and L_1 a vector space over $\mathbb{R}$. To show that $I(|\varphi|)$ is finite, fix γ in L with $|h| \le \gamma$ and $I(\gamma) \le 1$. Then $\varphi - \gamma \le f \le \varphi + \gamma$. Since f is real valued, the internal set $\{n \in N: \varphi - \gamma \le n\}$ contains every infinite element of $^*\mathbb{N}$ and thus some finite element. Similarly, $\varphi + \gamma \ge -n$ for some $n \in \mathbb{N}$. It follows that $I(|\varphi|) = I(\varphi \vee 0) + I(-\varphi \vee 0)$ is finite. Now fix $\varepsilon > 0$ in $\mathbb{R}$, and γ in L with $|h| + |k| \le \gamma$ and $I(\gamma) < \varepsilon$. From the arbitrary choice of ε and the inequality

$$(\varphi \vee \psi) - \gamma = (\varphi - \gamma) \vee (\psi - \gamma) \le (\varphi + h) \vee (\psi + k) = f \vee g$$
$$\le (\varphi \vee \psi) + \gamma$$

it follows that $(f \vee g) - (\varphi \vee \psi) \in L_0$. The rest is clear.◂

The reader should note that we have assumed no continuity properties for the internal functional I. Now, however, with $\aleph_1$-saturation we can establish continuity for the external functional J in the form of a monotone convergence property.

1.3 Theorem

If $\{f_n : n \in \mathbb{N}\}$ *is an increasing sequence in* L_1 *with real upper envelope* F *and* $\sup J(f_n) < +\infty$, *then* $F \in L$ *and* $J(F) = \lim J(f_n)$.

Proof. By replacing f_n with $f_n - f_1$, we may assume that each $f_n \ge 0$. By Proposition 1.2, we may fix $\varphi_n \in L$ and $h_n \in L_0$ for each $n \in \mathbb{N}$ so that $f_n = \varphi_n + h_n$ and $0 \le \varphi_n \le \varphi_{n+1}$. By the $\aleph_1$-saturation of our enlargement, there is a $\varphi_\omega \in L$ with $\varphi_\omega \ge \varphi_n$ for each $n \in \mathbb{N}$ and

${}^\circ I(\varphi_\omega) = \lim_{n\in\mathbb{N}} {}^\circ I(\varphi_n)$. We need only show that $F - \varphi_\omega \in L_0$. Fix $\varepsilon > 0$ in $\mathbb{R}$. Choose for each $n \in \mathbb{N}$ a $\psi_n \in L$ with $|h_n| \le \psi_n$ and $I(\psi_n) \le \varepsilon/2^n$. By $\aleph_1$-saturation, we may extend the sequence $\{\psi_n : n \in \mathbb{N}\}$ to an internal sequence $\{\psi_n : n \in {}^*\mathbb{N}\} \subset L$. We may also choose $\kappa \in {}^*\mathbb{N} - \mathbb{N}$ so that $0 \le \psi_n$ and $I(\psi_n) < \varepsilon/2^n$ when $1 \le n \le \kappa$ in ${}^*\mathbb{N}$. Setting $\psi = \sum_{n=1}^{\kappa} \psi_n$, we have $I(\psi) < \varepsilon$. Now for each $n \in \mathbb{N}$,

$$\varphi_n - \psi \le \varphi_n - \psi_n \le \varphi_n + h_n \le F \le (1 + \varepsilon)(\varphi_\omega + \psi),$$

so

$$(\varphi_n - \varphi_\omega) - \psi \le F - \varphi_\omega \le \varepsilon\,\varphi_\omega + (1 + \varepsilon)\psi.$$

The rest is clear.◂

The next two results exhibit the close relationship between the internal lattice L and the external lattice L_1.

1.4 Theorem

A real valued function f *on* X *is in* L_1 *if and only if for each* $\varepsilon > 0$ *in* $\mathbb{R}$ *there exist functions* ψ_1 *and* ψ_2 *in* L *with* $\psi_1 \le f \le \psi_2$ *and* $I(\psi_2 - \psi_1) < \varepsilon$, *in which case,* ${}^\circ I(\psi_1) \le J(f) \le {}^\circ I(\psi_1) + \varepsilon$.

Proof. First assume $f = \varphi + h \in L_1$ with $\varphi \in L$ and $h \in L_0$. Fix $\varepsilon > 0$ in $\mathbb{R}$. For each $n \in \mathbb{N}$, choose $\varphi_n \in L$ with $|h| \le \varphi_n$ and $I(\varphi_n) < \varepsilon/n$. Setting $\psi_1 = \varphi - \varphi_2$ and $\psi_2 = \varphi + \varphi_2$, we have $\psi_1 \le f \le \psi_2$ and $I(\psi_2 - \psi_1) < \varepsilon$. Moreover, given any ψ_1 and ψ_2 in L satisfying these conditions, we have for each $n \in \mathbb{N}$,

$$\psi_1 - \varphi_n \le \varphi \le \psi_2 + \varphi_n,$$

whence

$${}^\circ I(\psi_1) - \varepsilon/n \le {}^\circ I(\varphi) = J(f) \le {}^\circ I(\psi_2) + \varepsilon/n \le {}^\circ I(\psi_1) + \varepsilon + \varepsilon/n.$$

It follows that ${}^\circ I(\psi_1) \le J(f) \le {}^\circ I(\psi_1) + \varepsilon$.

Assume now that f is an arbitrary real valued function on X

for which there exists an increasing sequence $\{\varphi_n : n \in \mathbb{N}\} \subset L$ and a decreasing sequence $\{\psi_n : n \in \mathbb{N}\} \subset L$ with $\varphi_n \le f \le \psi_n$ and $I(\psi_n - \varphi_n) < 1/n$ for each $n \in \mathbb{N}$. By $\aleph_1$-saturation, we may extend both sequences to $^*\mathbb{N}$ and choose a $\psi_\omega \in L$ such that for each $n \in \mathbb{N}$, $\varphi_n \le \psi_\omega \le \psi_n$, whence $\varphi_n - \psi_n \le f - \psi_\omega \le \psi_n - \varphi_n$. It follows that $f - \psi_\omega \in L_0$ and thus $f \in L_1$.◂

Given an $^*\mathbb{R}$-valued function g on X, we will let $^\circ g$ denote the extended-real valued function on X defined by setting $^\circ g(x) = \mathrm{st}(g(x))$ for each x in X. Here, $\mathrm{st}(g(x))$ equals $+\infty$ or $-\infty$ if $g(x)$ is infinite in $^*\mathbb{R}$. For any function g taking values in $\mathbb{R}$ or $^*\mathbb{R}$, we set $g^+ = g \vee 0$ and $g^- = -g \vee 0$.

1.5 Proposition

If $\varphi \in L$ *and* $\varphi(x)$ *is finite in* $^*\mathbb{R}$ *for each* $x \in X$, *then* $^\circ\varphi \in L_1$ *and* $J(^\circ\varphi) = {^\circ I(\varphi)}$.

Proof. Recall that $^\circ I(1) < +\infty$. The proposition follows from the fact that for each $\varepsilon > 0$ in $\mathbb{R}$, $|^\circ\varphi - \varphi| < \varepsilon$, so $^\circ\varphi - \varphi \in L_0$.◂

1.6 Definition

Let M^+ denote the set of nonnegative, extended-real valued functions g on X such that for each $n \in \mathbb{N}$, $g \wedge n \in L_1$, and set $J(g) = \sup J(g \wedge n)$ for each g in M^+. Let $M = \{g : g^+ \in M^+$ and $g^- \in M^+\}$. For each g in M, set $J(g) = J(g^+) - J(g^-)$ if at least one of the right hand values is finite. Let $\mathcal{B} = \{A \subseteq X: \chi_A \in M^+\} = \{A \subseteq X: \chi_A \in L_1\}$, where χ_A denotes the characteristic function of A. For each A in $\mathcal{B}$ let $\mu(A) = J(\chi_A)$.

1.7 Theorem

The collection $\mathcal{B}$ *is a* σ-algebra *in* X *and* μ *is a complete, countably additive, finite measure on* $(X,\mathcal{B})$.

Proof. The completeness of μ follows from that fact that a real valued function $f = f + 0$ in L_0 is also in L_1. The rest is clear.◂

The next proposition contains preliminary results concerning the space M; from these results it will follow that $J(g) = \int_X g d\mu$ for each $g \in M^+$. Note that in Definition 1.6, we may replace the truncations $g \wedge n$ with truncations $g \wedge f$ for arbitrary elements $f \geq 0$ in L since $g \wedge f = \sup_n g \wedge f \wedge n$; therefore, $J(g) = \sup \{J(f): f \in L_1, f \leq g\}$.

1.8 **Proposition**

Fix $g \in M^+$, $p \in M^+$ *and* $\alpha \geq 0$ *in* $\mathbb{R}$. *Then* $g + p$, αg, $g \vee p$ *and* $g \wedge p$ *are in* M^+. *Moreover*, $J(g + p) = J(g) + J(p)$, $J(\alpha g) = \alpha J(g)$, *and if* $g \leq p$ *then* $J(g) \leq J(p)$. *If* $\{g_n : n \in \mathbb{N}\}$ *is an increasing sequence in* M^+ *with upper envelope* G, *then* $G \in M^+$ *and* $J(G) = \sup J(g_n)$.

Proof. For $n \in \mathbb{N}$, $(g + p) \wedge n = [(g \wedge n) + (p \wedge n)] \wedge n \in L_1$, and for $\alpha > 0$, $(\alpha g) \wedge n = \alpha(g \wedge n/\alpha) \in L_1$. It is easy to see that $J(g) + J(p) \leq J(g + p)$. The reverse inequality follows from the fact that if $f \in L_1$ and $f \leq g + p$ then $f \wedge g \leq g$ and $f - (f \wedge g) \leq p$. The rest is clear.◂

1.9 **Theorem**

A nonnegative extended-real valued function g *on* X *is* $\mathcal{B}$*-measurable if and only if* $g \in M^+$ *in which case*

$$J(g) = \int_X g \, d\mu.$$

Proof. Fix g in M^+ and let $A = \{g > 1\}$; we will show that $\chi_A \in L_1$. Let $f = (g \wedge 2) - (g \wedge 1)$. Then $f \in L_1$ and $\chi_A = \lim (1 \wedge nf) \in L_1$. Now for any positive α in $\mathbb{R}$, $\chi_A \in L_1$ when $A = \{g > \alpha\} = \{g/\alpha > 1\}$, and by Theorem 1.3, the same is true for $\alpha = 0$. Therefore, if $g \in M^+$ then g is $\mathcal{B}$-measurable. The converse and the equality $J(g) = \int g \, d\mu$ are obtained from the corresponding facts for $\mathcal{B}$-simple functions.◂

In light of Theorem 1.9, we may now call J an integral. If $\varphi \in L$ then by Proposition 1.5, ${}^\circ\varphi \in M$. It may not be the case, however, that $J({}^\circ|\varphi|) = {}^\circ I(|\varphi|)$; φ may take a large infinite value on an internal set A with $I(\chi_A) \approx 0$, i.e., $\mu(A) = 0$. Functions for which the standard part of the internal integral equals the integral of the standard part of the function form an important class in nonstandard integration theory. Following Anderson (1976), we say that $\varphi \in L$ is *S-integrable* if $I(|\varphi|)$ is finite and $J({}^\circ\varphi) = {}^\circ I(\varphi)$. Proposition 1.5 shows that a finite valued (and therefore bounded) function φ is S-integrable; the proof is much simpler than that for the measure-theoretic case (Loeb (1975)). The reader should view the criterion for the S-integrability of unbounded functions (Proposition 1.10) as an application of the usual procedure for extending integrals from bounded functions to unbounded ones, with the finiteness of the internal integral following as a bonus. A similar viewpoint (see the statement of Proposition 1.11) gives the existence of S-integrable "liftings" for those functions in M which have a finite integral when it is known that general liftings exist. Proposition 1.11 establishes the existence of a lifting for an arbitrary $g \in M$; i.e., there is a function $\varphi \in L$ with ${}^\circ\varphi = g$ μ-almost everywhere.

1.10 Proposition

A function φ *in* L *is S-integrable if and only if for each* $\omega \in {}^*\mathbb{N} - \mathbb{N}$, $I(|\varphi| - |\varphi| \wedge \omega) \approx 0$.

Proof. We may assume $\varphi \geq 0$. By the definition of the integral and Proposition 1.5,

$$J({}^\circ\varphi) = \sup_n J({}^\circ\varphi \wedge n) = \sup_n {}^\circ I(\varphi \wedge n).$$

Since ${}^\circ I(\varphi) = {}^\circ I(\varphi - \varphi \wedge n) + {}^\circ I(\varphi \wedge n)$ for each $n \in {}^*\mathbb{N}$, the proposition follows.◂

1.11 Proposition

Given $g \geq 0$ *in* M, *there is a* $\varphi \geq 0$ *in* L *such that for each* $n \in \mathbb{N}$, $(g \wedge n) - (\varphi \wedge n) \in L_0$, *whence*

$$J(g) = \sup_n J(g \wedge n) = \sup_n {}^\circ I(\varphi \wedge n) = {}^\circ I(\varphi \wedge \omega)$$

for some $\omega \in {}^*\mathbb{N} - \mathbb{N}$.

Proof. By Theorem 1.4, we may choose sequences $\{\varphi_n : n \in \mathbb{N}\}$ and $\{\psi_n : n \in \mathbb{N}\}$ in L so that $0 \leq \varphi_n \leq g \wedge n \leq \psi_n$, $\varphi_n \leq \varphi_{n+1}$ and

$I(\psi_n - \varphi_n) < 1/n$ for each $n \in \mathbb{N}$. Given $k \ge m \ge n$ in $\mathbb{N}$,

$$\psi_m \wedge n \ge g \wedge n \ge \varphi_k \wedge n \ge \varphi_m \wedge n.$$

By $\aleph_1$-saturation, we may choose a $\varphi \in L$ so that for every m and n with $m \ge n$ in $\mathbb{N}$, $\psi_m \wedge n \ge \varphi \wedge n \ge \varphi_m \wedge n$. Clearly, $(g \wedge n) - (\varphi \wedge n) \in L_0$ for each $n \in \mathbb{N}$.◂

We now return to the case of our first examples: an internal probability space $(X, \mathcal{M}, P)$, with L the space of internal $\mathcal{M}$-measurable simple functions and $I(f) = \int f dP$ for all $f \in L$.

1.12 **Example**

Let X be an internal set, $\mathcal{M}$ be an internal algebra on X, and ν an internal finitely additive measure on $(X,\mathcal{M})$ with $\nu(X)$ finite in ${}^*\mathbb{R}$; for example, $(X,\mathcal{M},\nu)$ may be an internal probability space. Let I be the ν-integral on the class L of internal $\mathcal{M}$-simple functions. For each A in $\mathcal{M}$, let $\nu_1(A) = \mathrm{st}(\nu(A))$. Then $\mathcal{M}$ is an algebra in the ordinary sense, and ν_1 is a finitely additive, real valued measure on $\mathcal{M}$. The construction of this section produces a standard measure space $(X,\mathcal{B},\mu)$ that extends the finitely additive space $(X,\mathcal{M},\nu_1)$; i.e. $\mathcal{M} \subset \mathcal{B}$ and $\mu(A) = \nu_1(A)$ for each $A \in \mathcal{M}$. If $B \in \mathcal{B}$ and $\varepsilon > 0$ in $\mathbb{R}$, then from the existence of functions φ and ψ in L with $\varphi \le \chi_B \le \psi$ and $I(\psi - \varphi) < \varepsilon$, we obtain sets $A_1 = \{\varphi > 0\}$ and $A_2 = \{\psi \ge 1\}$ in $\mathcal{M}$. Clearly $A_1 \subseteq B \subseteq A_2$ and $\nu(A_2 - A_1) \le I(\psi - \varphi) < \varepsilon$. This is the internal approximation result that characterizes externally measurable sets in the measure-theoretic approach to nonstandard integration theory.

2. INTERNAL FUNCTIONALS ON CONTINUOUS FUNCTIONS

The third example of Section 1 sets $X = {}^*[0,1]$; L is the set of internal, continuous functions on X, and I is a positive linear functional on L. A prime example of I is the nonstandard extension of the Riemann integral. Instead of just the interval $[0,1]$, this section deals with an arbitrary standard set Y supplied with a compact Hausdorff topology $\mathcal{T}$ in an enlargement of a structure containing Y and $\mathbb{R}$. We assume that the enlargement is κ-saturated with $\kappa \ge \aleph_1$ and $\kappa \ge \mathrm{Card}(\mathcal{T})$, so that in particular, the enlargement is $\aleph_1$-saturated.

Since the topology $\mathcal{T}$ is Hausdorff, each point x in $^*Y$ is in the monad $m(y)$ of a unique standard point y in Y; i.e. $y = st(x)$. With each extended-real valued function g on Y, we associate the extended-real valued function $\tilde{g}$ on $^*Y$ where $\tilde{g}(x) = g(st(x))$. With each subset A of Y we associate the subset $\tilde{A} = \cup\{m(y) : y \in A\}$ of $^*Y$, so that $\chi_{\tilde{A}} = \widetilde{(\chi_A)}$ on $^*Y$. The standard part map has played an important role in nonstandard measure theory; its inverse appears here in the form of the mapping $g \to \tilde{g}$. We now fix an internal positive linear functional I on $^*C(Y)$, with $I(1)$ finite in $^*\mathbb{R}$ (recall that $C(Y)$ denotes the space of continuous, real valued functions on Y.) We shall apply the results of Section 1 with $X = {}^*Y$ and $L = {}^*C(Y)$.

2.1 **Proposition**

For each compact $K \subseteq Y$, $\tilde{K} \in \mathcal{B}$, *and* $\mu(\tilde{K}) = \alpha_K$ *where* $\alpha_K = \inf\{{}^\circ I({}^*f) : f \in C(Y),\ \chi_K \le f \le 1\}$.

Proof. It follows from κ-saturation that there is a $\varphi \in L$ with $\chi_{*K} \le \varphi \le \chi_{\tilde{K}}$ and ${}^\circ I(\varphi) = \alpha_K$. Given any $f \in C(Y)$ with $\chi_K \le f \le 1$ and given $\varepsilon > 0$ in $\mathbb{R}$, we have $\varphi \le \chi_{\tilde{K}} \le (1+\varepsilon){}^*f$. It follows that $\chi_{\tilde{K}} - \varphi \in L_0$, $\chi_{\tilde{K}} \in L_1$, and $\mu(\tilde{K}) = J(\chi_{\tilde{K}}) = {}^\circ I(\varphi) = \alpha_K$.◂

2.2 **Theorem**

Let $\mathcal{B}_Y = \{B \subseteq Y : \tilde{B} \in \mathcal{B}\}$, *and let* $\mu_Y(B) = \mu(\tilde{B})$ *for each* $B \in \mathcal{B}_Y$. *Then* $\mathcal{B}_Y$ *is a* σ-algebra *in* Y *containing the Borel* σ-*algebra, and* μ_Y *is a complete, regular measure on* $(Y, \mathcal{B}_Y)$. *A function* g *on* Y *is* $\mathcal{B}_Y$-*measurable if and only if* $\tilde{g}$ *is* $\mathcal{B}$-*measurable on* $^*Y$, *in which case, if* $g \ge 0$ *then* $\int_Y g\,d\mu_Y = \int_{*Y} \tilde{g}\,d\mu$. *For each* $f \in C(Y)$, $\int_Y f\,d\mu_Y = {}^\circ I({}^*f)$.

Proof. To show that μ_Y is regular, choose $B \in \mathcal{B}_Y$ and $\psi \in L$ with $0 \le \psi \le \chi_{\tilde{B}}$. Let $K = \{st(x) : \psi(x) > 0\}$. By Theorem 3.4.2 of Luxemburg (1969), K is compact. Since $K \subseteq B$ and $\mu_Y(K) = \mu(\tilde{K}) \ge {}^\circ I(\psi)$, it follows from Theorem 1.4 that μ_Y is regular. If $f \in C(Y)$, then $*f \in L$, $\tilde{f} = {}^\circ(*f) \in L_1$, and $\tilde{f} - *f \in L_0$, whence $\int_Y f\,d\mu_Y = \int_{*Y} \tilde{f}\,d\mu = {}^\circ I(*f)$. The rest is clear.◂

2.3 Example

If $Y = [0,1]$ and I is the extension of the standard Riemann integral on Y, then a real valued g on Y is Lebesgue integrable if and only if $\tilde{g} = \varphi + h$ where $\varphi \in *C(Y)$ and $h \in L_0$. In this case, the Lebesgue integral of g equals the standard part of the internal Riemann integral of φ. A bounded g is Riemann integrable if and only if for any $\varepsilon > 0$, there are continuous functions p and q on $[0,1]$ with $p \le g \le q$ and $\int_0^1 (q - p)dx < \varepsilon$. It follows that g is Riemann integrable if and only if ${}^\circ(*g) \in L_1$ (result with A. Cornea.)

Numerous applications of nonstandard measure theory to problems in standard analysis have used the standard part map to convert an internal measure to a standard one. (See, for example, Anderson & Rashid (1978) and Loeb (1979), (1985).) Indeed, except for coin tossing, the first application of nonstandard measure theory used a construction of this type to obtain representing measures for harmonic functions in potential theory. The results stemming from this construction (Loeb (1976)) were later recast without nonstandard analysis and are now used in the work of a number of researchers investigating ideal boundaries and in particular the Martin boundary.

2.4 Example

If (Y,T) is a completely regular Hausdorff space then the results of Anderson & Rashid (1978) on tightness and weak convergence can be formulated in terms of the results of this section by extending internal functionals to $*C(Z)$, where Z is the Stone-Čech compactification of Y. Tightness corresponds to having $\chi_{\tilde{Z}-\tilde{Y}} \in L_0$ in which case, $\tilde{Y} \in L_1$ so $Y \in \mathcal{B}_Z$ and $\mu_Z(Y) = \sup_{\substack{K \text{ compact} \\ K \subset Y}} \mu_Z(K) = \mu_Z(Z)$.

3. VECTOR FUNCTIONS AND MEASURES

In this section, we outline some recent work of Horst Osswald and the author extending the results of Section 1 to vector lattices. We indicate the considerations that are involved in this research by describing, without proof, the results leading up to the generalizations of Theorems 1.3 and 1.4. As noted in Loeb (1984), the appropriate space for the standard integral is the nonstandard hull of the internal vector lattice. Here, however, we also want the functions of L_1 to take their values in a nonstandard hull. This means that we can not just modify an internal function φ with a null function h to obtain a function in L_1; we must take the projection of $\varphi + h$ in a nonstandard hull. Consequently, as we now show, the theory is a little more difficult than that of Section 1.

A vector space $\mathbf{B}$ over $\mathbb{R}$ equipped with a neighborhood base $\mathcal{U}$ of 0 and a partial ordering $\leq$ is called a *locally convex lattice* if the following conditions (A), (B), (C), (D) hold:

(A) The topology on $\mathbf{B}$ given by $\mathcal{U}$ makes $\mathbf{B}$ into a locally convex topological vector space.

(B) For all $x, y, z \in \mathbf{B}$ and all $\lambda \in \mathbb{R}$,

$$x \leq y \text{ implies } x + z \leq y + z \text{ and}$$
$$x \leq y \text{ implies } \lambda\cdot x \leq \lambda\cdot y \text{ if } 0 \leq \lambda.$$

(C) For all $x, y \in \mathbf{B}$, $\sup\{x, y\}$ (denoted by $x \vee y$) and $\inf\{x, y\}$ (denoted by $x \wedge y$) exist.

(D) The origin 0 has a neighborhood base consisting of *solid* sets.

Recall that a subset A of $\mathbf{B}$ is called *solid*, if for all $a, b \in \mathbf{B}$, $a \in A$ and $|b| \leq |a|$ implies $b \in A$. The binary operations sup and inf along with the corresponding unary operations $+$, $-$, and $|\cdot|$ defined on $\mathbf{B}$ by setting $x^+ = x \vee 0$, $x^- = -x \vee 0$, and $|x| = x \vee -x = x^+ + x^-$ are called the *lattice operations* in $\mathbf{B}$. It is well known that the lattice operations of a locally convex lattice are uniformly continuous.

In the following, let $(\mathbf{B},+,\cdot,\mathcal{U},\leq)$ be a locally convex lattice. We may assume that every set in the neighborhood base $\mathcal{U}$ of 0 is convex, circled and solid. (See Peressini (1967), p.104.) Following Henson & Moore (1972), we say an element $a \in {}^*\mathbf{B}$ is *S-finite* if for all $U \in \mathcal{U}$ there exists a standard $n \in \mathbb{N}$ with $a \in n{}^*U$. We let *fin* ${}^*\mathbf{B}$ be the set of all S-finite elements in $^*\mathbf{B}$. For $\alpha, \beta \in {}^*\mathbf{B}$, we write $\alpha \approx \beta$ when α is *infinitely close* to β in the topology on $\mathbf{B}$ given by $\mathcal{U}$ (not the order topology.) Given a, b S-finite elements of $^*\mathbf{B}$, $\lambda \in \mathbb{R}$, and $V \in \mathcal{U}$. We will use the following notation: $\overline{a} = \{ \beta \in {}^*\mathbf{B} : \beta \approx a\}$, $\hat{\mathbf{B}} = \{ \overline{\alpha} : \alpha \text{ is S-finite}\}$, $\hat{V} = \{ \overline{\alpha} \in \hat{\mathbf{B}} : \overline{\alpha} \cap {}^*V \neq \emptyset \}$, $\hat{\mathcal{U}} = \{ \hat{U} : U \in \mathcal{U} \}$, $\overline{a} \mathbin{\hat{+}} \overline{b} = \overline{a + b}$, $(\lambda \cdot \overline{b})^{\wedge} = \overline{\lambda \cdot b}$.

Fix a, b $\in {}^*\mathbf{B}$ and $U \in \mathcal{U}$. When a and b are finite, we will write $\overline{a} \mathrel{\hat{\leq}} \overline{b}$ if there are elements $\alpha \in \overline{a}$ and $\beta \in \overline{b}$ with $\alpha \leq \beta$. In general, we will write $a \mathrel{\tilde{\leq}} b$ if $a \vee b - b \in {}^*U$ for all $U \in \mathcal{U}$; i.e. $a \vee b \approx b$, which is equivalent to $a \wedge b \approx a$. It is easy to show that $a \mathrel{\tilde{\leq}} b$ if and only if for some $\alpha \approx a$ and $\beta \approx b$, $\alpha \leq \beta$.

3.1 Proposition

The relation $\hat{\leq}$ *is a partial ordering on* $\hat{\mathbf{B}}$ *and Properties* (A) *through* (D) *hold; i.e.* $(\hat{\mathbf{B}},\hat{+},\hat{\cdot},\hat{\mathcal{U}},\hat{\leq})$ *is a locally convex lattice.*

The lattice of Proposition 3.1 is called the *nonstandard hull* of $\mathbf{B}$. Let $(\mathbf{D},+,\cdot,\mathcal{D},\leq)$ be a second locally convex lattice with nonstandard hull $(\hat{\mathbf{D}},\hat{+},\hat{\cdot},\hat{\mathcal{D}},\hat{\leq})$. We will write $+$, $\cdot$ and $\leq$ for the operations and order relation in $\hat{\mathbf{D}}$ when there is no risk of confusion. Fix a nonempty internal set X. The ordering $\leq$ and the lattice operations can be applied to both internal and external functions mapping X into $^*\mathbf{B}$. For example, $f \vee g$ is defined pointwise; i.e. by setting $(f \vee g)(x) = f(x) \vee g(x)$. Fix an *internal vector lattice* L of functions mapping X into $^*\mathbf{B}$. That is, L is an internal vector space over $^*\mathbb{R}$ consisting of $^*\mathbf{B}$-valued functions such that for each $\varphi, \psi \in L$, the functions $\varphi \vee \psi$ and $\varphi \wedge \psi$ are again in L. We also fix an internal positive linear operator I mapping L into $^*\mathbf{D}$; we will call $I(\varphi)$ the *internal integral* of φ.

3.2 Definition

A internal or external function $h: X \to {}^*\mathbf{B}$ is called a *null function* if for any $V \in \mathcal{D}$ there is $\varphi \in L$ with $|h| \leq \varphi$ and $I(\varphi) \in {}^*V$. The set of null functions on X is denoted by L_0. We let L_1 denote the set of all $\hat{\mathbf{B}}$-valued functions f on X for which there exist $\varphi \in L$ and $h \in L_0$ such that $\varphi + h$ is (pointwise) S-finite and $f = \overline{\varphi + h}$.

3.3 Proposition

The sets L_0 *and* L_1 *are vector lattices over* $\mathbb{R}$.

At times, one needs to work with a weak notion of null function. An internal or external function $h: X \to {}^*B$ is called a *weak null function* if for any $V \in \mathcal{D}$ there is a $\varphi \in L$ with $|h| \lesssim \varphi$ and $I(\varphi) \in {}^*V$. We will denote the class of weak null functions by $\tilde{L}_0$. The next result shows that we cannot augment the class L_1 by replacing L_0 with $\tilde{L}_0$.

3.4 Proposition

For each $g \in \tilde{L}_0$, *there is an* $h \in L_0$ *with* $h \approx g$ *pointwise. Thus if* f *is a* $\hat{B}$*-valued function on* X *for which there exist* $\varphi \in L$ *and* $h \in \tilde{L}_0$ *such that* $\varphi + h$ *is (pointwise) S-finite and* $f = \overline{\varphi + h}$, *then* $f \in L_1$.

Given $f = \overline{\varphi + h}$ in L_1, we use $I(\varphi)$ to define an integral of f. To do so, we need to know that $I(\varphi)$ is S-finite and that $I(\varphi) \approx I(\psi)$ when f also equals $\overline{\psi + j}$, that is, when $\varphi - \psi = h - j + \varepsilon$ where $\varepsilon \approx 0$ pointwise on X. For the latter property, we need to know that $I(\lambda) \approx 0$ when $\lambda \in L \cap \tilde{L}_0$. We start with the following condition:

3.5 Definition

The operator I is called *S-continuous* if for each $\varphi \in L$ with $\varphi \approx 0$ (pointwise) we have $I(\varphi) \approx 0$.

3.6 Proposition

The following are equivalent:

(a) I *is S-continuous.*

(b) $I(\varphi)$ *is S-finite for every pointwise S-finite* $\varphi \in L$.

(c) *If* $\varphi \lesssim \psi$ *in* L, *then* $I(\varphi) \lesssim I(\psi)$ *in* ${}^*D$.

(d) $I(\varphi) \approx 0$ *for every* $\varphi \in L \cap \tilde{L}_0$.

We will assume hereafter that I is S-continuous. As we now note, this assumption implies the finiteness of the internal integrals of representatives of L_1.

3.7 Proposition

Given $\varphi \in L$, $h \in L_0$, *with* $\varphi + h$ *(pointwise)* S-*finite, it follows that* $I(\varphi)$ *is* S-*finite.*

3.8 Definition

Let J be the mapping from L_1 into $\hat{D}$ defined by setting $J(f) = \overline{I(\varphi)}$ for $f = \overline{\varphi + h}$ in L_1 (where $\varphi \in L$, $h \in L_0$).

Given the above results, it follows that the mapping J is well-defined, positive and linear. From these results, Horst Osswald and the author obtain the following monotone convergence theorem and generalization of Theorem 1.4.

3.9 Theorem

Let $\{f_n : n \in \mathbb{N}\}$ *be an increasing sequence in* L_1 *such that for each* $x \in X$, $\sup f_n(x) = f(x)$ *exists in* $\hat{B}$. *Then* $f \in L_1$ *if and only if* $\sup J(f_n)$ *exists in* $\hat{D}$, *in which case* $J(f) = \sup J(f_n)$.

3.10 Theorem

Fix an S-*finite* $g: X \to {}^*B$. *Then* $f = \overline{g}$ *is in* L_1 *if and only if for each* $V \in \mathcal{D}$ *there are functions* φ *and* ψ *in* L *with* $\varphi \tilde{\leq} g \tilde{\leq} \psi$ *and* $I(\psi - \varphi) \in {}^*V$. *In particular, if* $g \in L$, *then* $\overline{g} \in L_1$.

REFERENCES

Anderson, R.M. (1976). A nonstandard representation for Brownian motion and Itô integration. *Israel J. Math.*, **25**, 15-46.

Anderson, R.M. & Rashid, S. (1978). A nonstandard characterization of weak convergence, *Proc. Amer. Math. Soc.*, **69**, 327-32.

Henson, C.W. & Moore, L.C. Jr. (1972). The nonstandard theory of topological vector spaces, *Trans. Amer. Math. Soc.*, **172**, 405-35.

Henson, C. W. & Moore, L.C. Jr. (1983). The theory of Banach spaces; in *Nonstandard Analysis Recent Developments*, (ed. A. E. Hurd), 27-113. *Lecture Notes in Mathematics* **983**, Springer, Berlin.

Hurd, A.E. & Loeb, P.A. (1985). *An Introduction to Nonstandard Real Analysis*. Academic Press Series on Pure and Applied Mathematics.

Loeb, P.A. (1975). Conversion from nonstandard to standard measure spaces and applications in probability theory. *Trans. Amer. Math. Soc.*, **211**, 113-22.

Loeb, P.A. (1976). Applications of nonstandard analysis to ideal boundaries in potential theory. *Israel J. Math.*, **25**, 154-87.

Loeb, P.A. (1979). Weak limits of measures and the standard part map. *Proc. Amer. Math. Soc.*, **77**, 128-35.

Loeb, P.A. (1984). A functional approach to nonstandard measure theory. *Contemporary Mathematics*, **26**, 251-61.

Loeb, P.A. (1985). A nonstandard functional approach to Fubini's Theorem. *Proc. Amer. Math. Soc.*, **93**, 2, 343-46.

Luxemburg, W.A.J. (1969). A general theory of monads; in *Applications of Model Theory to Algebra, Analysis, and Probability* (ed. W.A.J. Luxemburg), 18-86. Holt, Rinehart, and Winston, New York.

Peressini, A. (1967). *Ordered Topological Vector Spaces*. New York: Harper and Row, New York.

Robinson, A. (1966). *Non-standard Analysis*. North-Holland, Amsterdam.

Stroyan, K.D. & Bayod, J.M. (1986). *Foundations of Infinitesimal Stochastic Analysis*. North-Holland Studies in Logic 119, Amsterdam.

AN APPLICATION OF NONSTANDARD METHODS TO COMPUTATIONAL GROUP THEORY

B. Benninghofen and M.M. Richter

INTRODUCTION

The methodological background of this paper is the theory of rewrite rules, although it is never mentioned explicitly. Finite complete sets of rewrite rules for a finitely presented algebra solve a number of decision problems for this algebra such as the word problem, the finiteness problem or the existence of a nilpotent subgroup of finite index. These problems remain solvable also for many infinite complete systems; what really matters is that the irreducible words form a regular set in the sense of automata theory. The regularity of the set of irreducible words can be regarded as an asymptotic aspect of the so-called Knuth-Bendix completion procedure. This is connected with an ordering on the words and some of these orderings are the central subject of our investigation. For more information on these topics we refer, for example to Benninghofen, Kemmmerich & Richter (1987) or to Richter (1987). In Richter (1987) one finds also a proof of the main result of this paper in a much more restricted case.

The main result can be found in section 4 and shows the non-regularity of the minimal words of the free nilpotent group of class 2 for a certain set of widely used orderings. We study a nonstandard model of this group (which was introduced in van den Dries & Wilkie (1984) in the context of Gromov's theorem) and its nonstandard hull and use a connection with the growth function of the group and with (nonstandard) automata theory.

1. GROUP THEORETIC PRELIMINARIES

The free nilpotent group of class two is defined by

$$FN = FN(a,b) = \langle a,b:\ [a,[a,b]] = [b,[a,b]] = 1\rangle$$

where

$[a,b] = aba^{-1}b^{-1}$ is the commutator of a and b.

We first summarize a number of results for this group needed later on. FN(a,b) has an isomorphic representation as

$$G = (\mathbb{Z}^3, *, -, E)$$

where the multiplication "*" is defined as the matrix multiplication of matrices of the form

$$\begin{bmatrix} 1 & x_1 & x_3 \\ 0 & 1 & x_2 \\ 0 & 0 & 1 \end{bmatrix}.$$

An isomorphism φ between G and FN is given by

$$\varphi(1,0,0) = a, \quad \varphi(0,1,0) = b, \quad \varphi(0,0,1) = c := [b,a].$$

Another way to express this is that FN is represented exactly by the words

$$W = \{a^n b^m c^k : n,m,k \in \mathbb{Z}\}.$$

These are not the shortest words which can denote the elements of FN. There is, however, another (partial) ordering $\sqsubseteq$ on the words such that W contains exactly the $\sqsubseteq$-minimal words denoting elements of FN.

We put

$$\Sigma = \{a,a^{-1},b,b^{-1},c,c^{-1}\}$$

as our alphabet and consider the set Σ^* of words over Σ.

1.1 **Definition**

Let $u = x_1 \ldots x_n$, $x_i \in \Sigma$. We put

(i) $K_x(u) = \mathrm{card}(\{i: x_i \in \{x,x^{-1}\}\})$, for $x \in \{a,b,c\}$;

(ii) $V_1(u) = \mathrm{card}(\{(i,j): i<j,\ x_i \in \{b,b^{-1}\},\ x_j \in \{a,a^{-1}\}\})$;

(iii) $V_2(u) = \mathrm{card}(\{(i,j): i<j,\ x_i \in \{c,c^{-1}\},\ x_j \in \{a,b,a^{-1},b^{-1}\}\})$

(iv) $u \sqsubseteq v$ iff

$$(K_a(u), K_b(u), V_1(u), K_c(u), V_2(u)) < (K_a(v), K_b(v), V_1(v), K_c(v), V_2(v)),$$

where "<" denotes the lexicographic ordering on 5-tuples of natural numbers.

The partial ordering is well-founded and satisfies

$$u \sqsubseteq v \Rightarrow sut \sqsubseteq svt.$$

We need the following further notation:

1.2 Definition

(i) $K_x^+(u) = \mathrm{card}(\{i: x_i = x\})$,

(ii) $K_x^-(u) = \mathrm{card}(\{i: x_i = x^{-1}\}) = K_x(u) - K_x^+(u)$,

(iii) $K_x^0(u) = K_x^+(u) - K_x^-(u)$,

(iv) $V^+(u) = \mathrm{card}(\{(i,j): i<j,\ [(x_i = b \ \&\ x_j = a)$
$\text{or } (x_i = b^{-1} \ \&\ x_j = a^{-1})]\})$,

(v) $V^-(u) = \mathrm{card}(\{(i,j): i<j,\ [(x = b \ \&\ x_j = a^{-1})$
$\text{or } (x_i = b^{-1} \ \&\ x_j = a)]\})$,

(vi) $V(u) = V^+(u) - V^-(u)$.

With this notation a computation verifies the following relation: for each word $u \in \Sigma^*$ the $\sqsubseteq$-minimal word $v = a^n b^m c^k$ for which $u \equiv v \pmod{FN}$ holds is given by $n = K_a^0(u)$, $m = K_b^0(u)$, $k = -V(u)$. Another way to express this is to state that the isomorphism φ: $G \to FN$ can be described by

$$\varphi^{-1}(K_a^0(u),\ K_b^0(u),\ -V(u)) = g$$

where $g \in FN$ is the element denoted by $u \in \Sigma^*$.

As in any group the powers g^n, $g \in G$, $n \in \mathbb{Z}$ are defined in $G \cong FN$. They are computed as

$$(x_1,x_2,x_3)^n = (nx_1,nx_2,nx_3 + \tfrac{1}{2}n(n-1)x_1x_2).$$

Later on we need:

1.3 Definition

$$G_\infty = \{(x_1,x_2,\ \tfrac{1}{2}\, x_1\cdot x_2): x_i \in \mathbb{Z},\ 1 \le i \le 2\}.$$

FN is also a torsion free polycyclic group as the central series $G_1 = GN$, $G_2 = \langle b,c\rangle$, $G_3 = \langle c\rangle$, $G_4 = \{1\}$ shows. The lower central series of FN is given by $G_1 = GN$, $G_2 = \langle c\rangle$, $G_3 = \{1\}$. This allows us to compute the growth function γ: $\mathbb{N} \to \mathbb{N}$ of FN. It is defined as

$$\gamma(n) = \mathrm{card}(\{g \in FN: g \text{ is denoted by some } u \in \Sigma^*,\ |u| \le n\})$$

where $|u|$ is the length of u. The growth function of FN is polynomial (see section 2), its degree d is computed by the Bass formula (cf. Bass (1975)):

$$d = \sum_{k\geq 1} k\cdot r_k = 1\cdot 2 + 2\cdot 1 = 4$$

where r_k is the rank of the free abelian group in G_k/G_{k+1} of the lower central series. The polycyclic series also gives rise to the Malcev-coordinates. For FN the Malcev-basis is (a,b,c) and the coordinates of $a^n b^m c^k$ are (n,m,k). In general (in a torsion free polycyclic group) the coordinates of products and inverses can be computed in terms of polynomials:

if $g = a_1^{n_1} \ldots a_m^{n_m}$, $h = a_1^{k_1} \ldots a_m^{k_m}$, then $gh = a_1^{l_1} \ldots a_m^{l_m}$ where

$$(l_1, \ldots ,l_m) = (f_1(n_1, \ldots k_m), \ldots ,f_m(n_1, \ldots ,k_m))$$

and the f_i are polynomials in 2m variables. The operations in G are a special case of this. The polynomials f_i allow us to embed the original group as a discrete subgroup in a Lie group: one defines on $\mathbb{R}^m$ a multiplication (and inversion) using the same f_i. In our case we get $(\mathbb{R}^3,*)$ as this extension. The exponentiation g^n $(n \in \mathbb{Z})$ of $\mathbb{Z}^3$ can be extended to an operation g^x, $g \in \mathbb{R}^3$, $x \in \mathbb{R}$, using the same formula. For our purposes, however, we need more insight into the relationship between $(\mathbb{Z}^3,*)$ and $(\mathbb{R}^3,*)$ than can be obtained from the way the extension was defined. Therefore we will construct the extension a second time by using nonstandard hulls.

2. THE GROWTH FUNCTION AND AUTOMATA

In section 1 we introduced the partial ordering $\sqsubseteq$ on Σ^*. The general definition of a word ordering is:

2.1 Definition

A partial ordering $<$ on Σ^* is a *word ordering* iff

(i) $u < v$ implies $w_1uw_2 < w_1vw_2$,

(ii) $<$ is well-founded,

(iii) $e < u$ for $u \neq e$.

2.2 Definition

For a word ordering $<$ the set *Min*($<$) is the set of $<$-minimal words.

Let $F(\Sigma)$ be the free group generated by Σ and let $N \subseteq F(\Sigma)$ be a normal subgroup and $E \subseteq F(\Sigma) \times F(\Sigma)$ a set of equations.

2.3 Definition

(i) A word ordering $<$ is *compatible with* N iff each coset of N contains exactly one $<$-minimal word,

(ii) $<$ is *compatible with* E iff $<$ is compatible with the normal subgroup N induced by E.

If $<$ is compatible with N then there is a one-one correspondence between Min($<$) and $F(\Sigma)/N$. In the above case $\sqsubseteq$ is compatible with the equations defining FN(a,b). A more popular set of orderings was introduced in Knuth & Bendix (1970).

2.4 Definition

(i) A *weight function* is a mapping $g: \Sigma \to \mathbb{N}$, such that $g(a) \geq 1$ for $a \in \Sigma$; the weight function is extended to Σ^* by

$$g(\varepsilon) = 0 \quad (\text{where } \varepsilon = \text{empty word})$$

$$g(a_1 \ldots a_n) = \sum_{i=1}^{n} g(a_i), \quad a_i \in \Sigma.$$

Here we assume that $a \in \Sigma$ implies that $a^{-1} \in \Sigma$ and $g(a) = g(a^{-1})$.

(ii) The *KB-ordering* $<_g$ relative to a weight function g is defined by

$$u <_g v \text{ iff } g(u) < g(v), \text{ or } g(u) = g(v)$$
and u is less than v in the lexicographic ordering with respect to $\Sigma = \{a , \ldots ,a \}$.

The KB-orderings $<_g$ are always total. For a word ordering $<$ the set Min($<$) is always closed under the formation of subwords. Min ($<$)

is recursively enumerable for KB-orderings; we are interested in studying additional properties.

2.5 Definition

< is *regular* iff Min(<) is regular (i.e. accepted by a finite automaton).

2.6 Definition

(i) For regular < the transition graph of the minimal automaton accepting Min(<) is called the *word graph* Gr(<) of <.

(ii) A *simple path* in Gr(<) starts at the initial node and visits each node at most once.

The elements of Min(<) correspond to the (labelled) paths in Gr(<) beginning at the starting node. A word ordering < also gives rise to a growth function.

2.7 Definition

The <-growth function $\gamma(<)$ is defined by

$$\gamma(<)(n) = \text{card } (\{u \in \text{Min}(<) : |u| \leq n\}).$$

If $H = F(\Sigma)/N$ and < is compatible with N, then $\gamma(n) \leq \gamma(<)(n)$ for all $n \in \mathbb{N}$. For regular < we not only have a finiteness test for Min(<); the word graph also provides us with information about $\gamma(<)$.

2.8 Definition

A function $\gamma\colon \mathbb{N} \to \mathbb{N}$ is called

(i) *polynomial* iff $\gamma(n) \leq p(n)$ for some polynomial p; the minimal possible degree for p is called $\deg(\gamma)$;

(ii) *exponential* iff $\gamma(n) \geq \exp(n)$ where exp is some exponential function.

The relation between $\gamma(<)$ and Gr(<) is given by:

(1) $\gamma(<)$ is exponential iff there are at least two cycles in Gr(<) which have a node in common (in this case the graph Gr(<) is also called exponential);

(2) $\gamma(<)$ is polynomial iff it is not exponential.

In this latter case Gr(<) is called polynomial. The degree deg(γ(<)) is given by the maximum number of cycles intersecting a branch of Gr(<). For the proof we refer to Gilman (1979). In the polynomial case the degree deg(γ($<_g$)) is independent of g; if $<_g$ is compatible with N then it also equals deg(γ) where γ is the growth function of $F(\Sigma)/N$. For general < we may have deg (γ) $\neq$ deg(γ(<)), however. An example is the group FN(a,b) where deg(γ) = 4 $\neq$ 3 = deg(γ($\sqsubseteq$)). The finitely presented polynomial groups have been characterized by M. Gromov as those which have a nilpotent subgroup of finite index (cf. for example van den Dries & Wilkie (1984)).

3. THE NONSTANDARD HULL OF FN(a,b)

We consider a fixed nonstandard model and assume for simplicity that it is a model of ZFC that satisfies Nelson's axioms of internal set theory (see the article by Diener & Stroyan (this volune), or Nelson (1977) or Richter (1982)). In the sequel $\omega \in {}^*\mathbb{N}$ will be some nonstandard natural number.

In this model FN(a,b) is defined as above. Our alphabet is

$$\Sigma = \{a, b, c_1, \dots, c_n, a^{-1}, b^{-1}, \dots, c_n^{-1}\}$$

and the c_i are governed by equations $c_i = u_i$ where u_i is a word in a, b, a^{-1}, b^{-1}. Let < be a fixed KB-ordering defined by a weight function g. For the rest of the paper $\omega \in {}^*\mathbb{N}$ is a nonstandard integer.

For x, y $\in$ G $\cong$ FN we put:

3.1 Definition

(i) $\|x\|_\omega = \|x\|_{g,\omega} = \frac{1}{\omega}{}^*g(u)$, where u is the <-minimal word denoting x;

(ii) $d_\omega(x,y) = \|x \cdot y^{-1}\|_\omega$;

(iii) $x \approx_\omega y$ iff $d_\omega(x,y) \approx 0$;

(iv) ${}^\omega G = \{x \in {}^*G: d_\omega(e,x) = \|x\|_W \lessapprox \omega\}$ where $s \lessapprox t$ iff s/t is a finite real number.

(v) $\mu_\omega = \{x \in {}^*G: x \approx_\omega 0\}$

It is easy to see that $\approx_\omega$, ${}^\omega G$, u_ω do not depend on the particular weight function g. For $s, t \in {}^*\mathbb{R}$, $t \neq 0$, we put $s \ll t$ iff $s/t \approx 0$.

3.2 Proposition

$$(n, m, l) \approx_\omega (n', m', l') \quad \textit{iff} \quad \frac{1}{\omega}|n - n'| \approx \frac{1}{\omega}|m - m'| \approx \frac{1}{\omega^2}|l-l'| \approx 0$$

Proof. Without loss of generality we put $n' = m' = l' = 0$ and restrict ourselves to the alphabet $\Sigma = \{a,b,a^{-1},b^{-1}\}$ and the weight function $g(a) = g(b) = 1$. Let the minimal word denoting (n,m,l) be the word $w = a^{s_0}b^{t_0}\ldots a^{s_k}b^{t_k}$; from $(n,m,l) \approx_\omega (0,0,0)$ we obtain

$$\frac{1}{\omega}\Big(\sum_{i=0}^{k}(|s_i| + |t_i|)\Big) \approx 0,$$

which implies

$$|n| = |{}^*K_a^0(\omega)| \leq |\sum_{i=0}^{k} s_i| \ll \omega,$$

$$|m| = |{}^*K_b^0(\omega)| \leq |\sum_{i=0}^{k} t_i| \ll \omega$$

and

$$|l| = |{}^*V(\omega)| \leq |\sum_{i=0}^{k} t_i| \ll \omega^2.$$

If on the other hand

$$\frac{1}{\omega}\, n \approx \frac{1}{\omega}\, m \approx \frac{1}{\omega^2}\, l \approx 0$$

the word

$$u = a^n b^m [a^k, b^k]\cdot[a,b]^{1-k^2}, \quad k = [\surd l],$$

then denotes (n,m,l) and has weight ${}^*g(u) \leq n + m + 6k \ll \omega$ which implies $(n,m,l) \approx_\omega 0$.◂

Similarly we see that ${}^\omega G = \{(n,m,l) \in {}^*G\colon n,m \lesssim \omega,\ l \lesssim \omega^2\}$. This in turn implies that ${}^\omega G \subseteq {}^*G$ is a subgroup. A straightforward computation shows that the group operations in ${}^\omega G$ are S-continuous with respect to $\approx_\omega$.

An alternative way to prove the last proposition is to derive the inequality

$$\omega.\|(n,m,l)\|_\omega \le |n| + |m| + 12. [\sqrt{l}]$$

This is done for $l \ge 0$ (in an anlogous way for $l < 0$) by considering the equation

$$a^n b^m [a,b]^l = a^n b^m [a^k, b^k][a,b]^{l-k^2}$$

for $k = \sqrt{l}$.

3.3 Definition

The standard part mapping θ_ω: ${}^\omega G \to \overline{G} = (\mathbb{R}^3, *)$ is obtained by

$$\theta_\omega(n,m,l) = (st(\tfrac{n}{\omega}), st(\tfrac{m}{\omega}), st(\tfrac{l}{\omega^2}))$$

where st is the usual standard part mapping.

From the definition and proposition it follows immediately that θ_ω is an epimorphism with μ_ω as its kernel. The metric d_ω induces a metric $\overline{d}_\omega$ on $\overline{G}$ in the usual way:

$$\overline{d}_\omega(\theta_\omega(x), \theta_\omega(y)) = st(d_\omega(x,y)).$$

Hence we obtain

$$(\overline{G}, \overline{d}_\omega) \qquad (\cong {}^\omega G/\mu_\omega)$$

as the nonstandard hull of $({}^*G, d_\omega)$. As d_ω is left invariant, $\overline{d}_\omega$ is left invariant, too. The corresponding norm on $\mathbb{R}^3$ is denoted by $\|x\|_0$. Finally we put:

3.4 Definition

$$\overline{G}_\infty = \{(x_1, x_2, \tfrac{1}{2} x_1 x_2): x_i \in \mathbb{R}, 1 \le i \le 2)\}$$

We note that G is embedded in $\mathbb{R}^3$ via ${}^\omega G$, but that this embedding is very different from the one mentioned in section 1. The metric $\overline{d}_\omega$ induces a uniformity and a topology on $\mathbb{R}^3$. We will see that the topology is the usual one. By the remarks above we first obtain for $(x,y,z) \in \overline{G}$ and $r = \|(x,y,z)\|_0$ the inequalities

$$|x| \le r, \ |y| \le r, \ |z| \le r^2.$$

Now we consider $(\overline{G}, \overline{d}_\omega)$ as an ordinary (standard) metrizable group in its own right. In the next proposition we use nonstandard analysis in order to argue about $(\overline{G},\overline{d}_\omega)$. For this purpose we use an ad hoc notion and put $\simeq$ for the infinitesimal relation induced by the metric $\overline{d}_\omega$.

3.5 Proposition

If $u_1 = (x_1,y_1,z_1)$, $u_2 = (x_2,y_2,z_2)$ *and* $u_1, u_2 \in \text{fin}(^*\mathbb{R}^3)$, *then* $u_1 \simeq u_2$ iff $u_1 \approx u_2$.

Proof. First we get

$$\begin{aligned} u_1 \simeq u_2 \quad &\text{iff} \quad u_1^{-1}u_2 \simeq (0,0,0) \\ &\text{iff} \quad \|(x_2 - x_1,\ y_2 - y_1,\ z_2 - z_1 + y_1\ (x_2 - x_1))\|_0 = 0 \\ &\text{iff} \quad x_1 \approx x_2,\ y_1 \approx y_2 \text{ and } \surd|z_2 - z_1 + y_1(x_2 - x_1)| \approx 0 \\ &\text{iff} \quad x_1 \approx x_2,\ y_1 \approx y_2 \text{ and } z_2 - z_1 + y_1(x_2 - x_1) \approx 0. \qquad (1) \end{aligned}$$

If now $u_1 \approx u_2$ and $y_1 \in \text{fin}\ (^*\mathbb{R})$, then (1) is immediate; i.e. $u_1 \simeq u_2$ holds. If on the other hand (1) holds and $y_1 \in \mathbb{R}$, then $y_1(x_2 - x_1) \approx 0$, therefore $z_2 \approx z_1$ and $u_1 \approx u_2$ is true.◂

A consequence is:

3.6 Corollary

$\overline{d}_\omega$ *induces the usual topology on* $\mathbb{R}^3$.

We remark that $\overline{d}_\omega$ does not induce usual the uniformity on $\mathbb{R}^3$. This can be shown by considering the points

$$(0,n,0) \text{ and } (\tfrac{1}{n},n,-1) \text{ with } n \in {}^*\mathbb{N} \text{ nonstandard.}$$

Next we turn our attention to integration theory. We define an internal function Λ on the (internal) subsets of $^*G$ by putting

$$\Lambda(X) = \frac{1}{\omega^4}\ {}^*\text{card}(X), \qquad X \subseteq {}^*G \text{ internal.}$$

It is immediate that Λ is a left and right invariant *-finitely additive

*-measure on *G. For $I = \{(n,m,l) \in {}^*G: 0 \le n,\ m \le \omega,\ 0 \le l \le \omega^2\}$ we get $\Lambda(I) \approx 1$.

3.7 Definition

$\bar{\Lambda}$ is the Loeb measure corresponding to Λ restricted ${}^{\omega}G$ and $\Lambda_0 = \theta_\omega(\bar{\Lambda})$ is its image on $\mathbb{R}^3$.

It follows that $\bar{\Lambda}$ is left and right invariant too, and $\bar{\Lambda}(I) = 1$ holds.

3.8 Proposition

Λ_0 is the Lebesgue measure on $\mathbb{R}^3$.

Proof. For $\varepsilon \in \mathbb{R}$, $\varepsilon > 0$ we put

$$I(\pm\varepsilon) := \{(n,m,l) \in {}^*G: \pm\,\varepsilon \le \frac{1}{\omega}\,n,\ \frac{1}{\omega}\,m \le 1 \mp \varepsilon,\ \pm\,\varepsilon \le \frac{1}{\omega^2}\,l \le \mp\,\varepsilon\}$$

and get

$$I(+\varepsilon) \subseteq \theta_\omega^{-1}([0,1]^3) \subseteq I\,(-\varepsilon).$$

This implies

$$\bar{\Lambda}(\theta_\omega^{-1}([0,1]^3)) = 1 \text{ and } \Lambda_0\ ([0,1]^3) = 1.$$

Next we will show that Λ_0 is left invariant. For this purpose we take some Λ_0-measurable set $X \subseteq \mathbb{R}^3$ and $(x_0,y_0,z_0) \in \mathbb{R}^3$ and put $X' = (x_0,y_0,z_0)*X$. Then

$$M = \theta_\omega^{-1}(X) \text{ and } M' = (n_0,m_0,l_0)*M \text{ where}$$

$$\theta_\omega(n_0,m_0,l_0) = (x_0,y_0,z_0) \text{ are } \bar{\Lambda}\text{-measurable and } \bar{\Lambda}(M) = \bar{\Lambda}(M') \text{ holds.}$$

It remains to show $M' = \theta_\omega^{-1}(X')$. We consider some $(x,y,z) = \theta_\omega(n,m,l)$ and obtain the result from

$$(n,m,l) \in \theta_\omega^{-1}(X') \text{ iff } (x,y,z) \in X' \text{ iff } (\overline{x_0,y_0,z_0})*(x,y,z) \in X$$
$$\text{iff } (n_0,m_0,l_0)*(n,m,l) \in M \text{ iff } (n,m,l) \in M'.$$

So far we have seen that Λ_0 is the normalized Haar measure on G; the proposition will be proved if we can show that the Lebesgue measure λ on $\mathbb{R}^3$ is also left invariant for $G = (\mathbb{R}^3,*)$. For fixed (u,v,w) we consider the transformation

$$F(x,y,z) = (u,v,w)*(x,y,z).$$

The result now follows from the fact that the value of the Jacobian determinant is

$$\det(J(F(x,y,z))) = 1.◂$$

3.9 Corollary

If $A \subseteq \Sigma^*$ *is a set of words the growth function of which has degree* ≤ 3 *and* $U \subseteq {}^*G$ *is the set of group elements denoted by* A, *then* $\theta_\omega(U \cap {}^\omega G)$ *has Lebesgue measure* 0.

Proof. $\frac{1}{\omega^4}\cdot{}^*\mathrm{card}(U \cap {}^\omega G)$ is infinitesimal; hence the proposition gives the result.◂

4. NON-REGULARITY RESULTS

We consider FN(a,b) with finitely many new constants denoting words in a, b, a^{-1}, b^{-1} as in section 3. Let < be a KB-ordering with Min(<) as the set of minimal words denoting elements of FN. Our main result is

4.1 Theorem

Min (<) *is not regular for any KB-ordering.*

The usual way to show that some set of words is not regular employs the pumping lemma of automata theory. In the case that no additional constants are allowed such a proof is given in Richter (1987). In our more general situation this method of proof breaks down, however. To establish the proof we need to study the situation in the nonstandard model.

We will proceed indirectly and assume that Min(<) is regular with Gr(<) as its word graph. As the degree deg(γ) of the growth function γ(<) is 4 we can select a simple path of Gr(<) which intersects (the maximal number of) 4 cycles. Together with these cycles and its labels it has the form shown in the following diagram

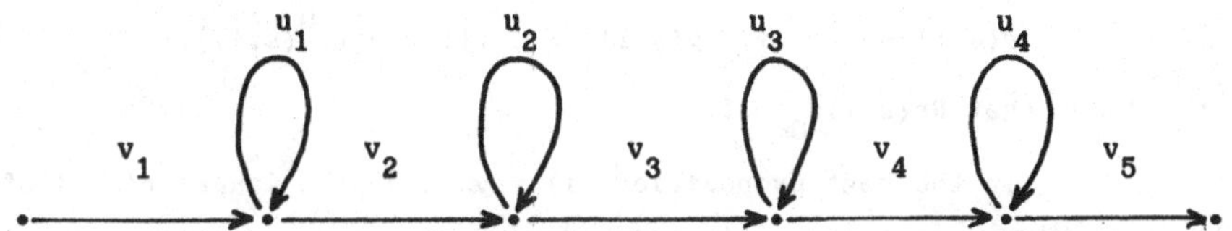

where v_2, v_3 and v_4 are not the empty word.

Each path s in Gr(<) determines in this way a word $v_1(s)u_1^{k_1}(s)v_2(s)u_2^{k_2}(s) \ldots v_l(s)$ where $l \le 5$ and the $u_i(s)$ label the cycles of s. For a simple path we introduce the following notation:

$f(s,i) := (n(s,i), m(s,i), l(s,i)) \in G$ is the group element denoted by $u_i(s)$;

$h(s,i) := (n'(s,i), m'(s,i), l'(s,i) \in G$ is the group element denoted by $v_i(s)$.

By definition of Gr(<) each element of G has a unique representation of the form

$$(n,m,l) = h(s,1)f^{k_1}(s,1) \ldots f^{k_{l-1}}(s,l-1)h(s,l)$$

where $k_i \ge 0$. Furthermore we have

$$\|(n,m,l)\|_\omega = \sum_{i=1}^{l} \|h(s,i)\|_\omega + \sum_{i=1}^{l-1} k_i \|f(s,i)\|_\omega$$

$$\approx_\omega \sum_{i=1}^{l-1} k_i \|f(s,i)\|_\omega .$$

Now we put

$$k(s,i) := \left[\frac{1}{\|f(s,i)\|_\omega}\right] \in {}^*\mathbb{N}$$

and normalize the f(s,i) as

$$f^0(s,i) := (f(s,i))^{k(s,i)}$$

which gives

$$\|f^0(s,i)\|_\omega \approx 1.$$

Finally we put

$$r(s,i) := (x(s,i),\ y(s,i),\ z(s,i)) = \theta_\omega(f^0(s,i)).$$

It follows that $\|r(s,i)\|_0 = 1$.

In the next proposition (i) says that the images r(s,i) of the infinite powers of the cycle's words depend on two coordinates only; (iii) expresses a fact difficult to formulate in a standard approach: small words (between the cycles) can asymptotically be neglected.

4.2 Proposition

(i) $r(s,i) \in \overline{G}_\infty$,

(ii) $x(s,i) \neq 0$ *or* $y(s,i) \neq 0$,

(iii) $\theta_\omega(h(s,1)f(s,1)^{k_1} \ldots h(s,l)) = r^{p_1}(s,1)*\ldots*r^{p_l}(s,l)$,

where $p_i = \text{st}\,(\frac{k_i}{k(s,i)})$;

(iv) $\|r^{p_1}(s,1)*\ldots*r^{p_{l-1}}(s,\ l-1)\|_0 = \sum_{i=1}^{l-1} p_i$

Proof. (i) For $(n,m,l) \in {}^\omega G$ in the case that $\frac{p}{\omega^2} \approx_\omega 0$ we have $(n,m,l)^p \approx_\omega (np, mp, \frac{1}{2}npmp) \in {}^\omega G_\infty$.

(ii) Suppose $x(s,i) = y(s,i) = 0$. From $\frac{k(s,i)}{\omega} \approx 1$ we obtain $n(s,i) \approx m(s,i) \approx 0$ and therefore $n(s,i) = m(s,i) = 0$. We get a contradition from

$$1 = \|r(s,i)\|_0 = \|\theta_\omega(0,0,l(s,i))^{k(s,i)}\|_0$$

$$= \|\theta_\omega(0,0,l(s,i)\cdot k(s,i))\|_0 = 0.$$

The last equation holds because $(0,0,l(s,i)k(s,i))$ can be denoted by a word w with the length |w| in the magnitude of

$$\sqrt{(l(s,i)k(s,i))};\quad \text{i.e. } |w| \ll \omega.$$

(iii) We have

$$\theta_\omega(f(s,i)^{k_i}) = (\text{st}\left[\frac{k_i}{\omega}n(s,i)\right],\ \text{st}\left[\frac{k_i}{\omega}m(s,i)\right],\ \tfrac{1}{2}\text{st}\left[\frac{k_i}{\omega}n(s,i)\right]\text{st}\left[\frac{k_i}{\omega}m(s,i)\right]).$$

From

$$\text{st}\left[\frac{k_i}{\omega}n(s,i)\right] = \text{st}\left[\frac{k_i}{k(s,i)}\cdot\frac{k(s,i)}{\omega}\cdot n(s,i)\right] = p_i x(s,i)$$

and the analogous equation for m(s,i) we infer that

$$\theta_\omega(f(s,i)^{k_i}) = r(s,i)^{p_i}.$$

The assertion then follows from $\theta_\omega(h(s,i)) = 0$.

(iv) This is now immediate from the definition of $\|\cdot\|_0$ and

$$\|h(s,i)\|_\omega \approx 0 \text{ and } \|f(s,i)^{k_i}\|_\omega \approx p_i. ◂$$

So far we have considered an arbitrary simple path s. We will now see that it is sufficient to restrict ourselves to those s which intersect exactly 4 cycles.

4.3 Proposition

For each $(x,y,z) \in \mathbb{R}^3$ *there is a path* s *that intersects 4-cycles such that* $(x,y,z) = r(s,1)^{p_1} * \dots * r(s,4)^{p_4}$.

Proof. Let $W_4 \subseteq \mathrm{Min}(<)$ be the set of words labelling a path which intersects 4-cycles and let G_4 be the set of group elements donoted by W_4. Then $(*G_4, d_\omega)$ is a *-metric space with $(\theta_\omega(*G_4 \cap {}^\omega G, \bar{d}_\omega)$ as its nonstandard hull. Therefore $\theta_\omega(*G_4 \cap {}^\omega G)$ is complete with respect to $\bar{d}_\omega$ and hence a closed subspace of $\mathbb{R}^3$. Consequently $T := \mathbb{R}^3 \backslash \theta_\omega(*G_4 \cap {}^\omega G)$ is open. It suffices to show that the Lebesgue measure $\Lambda_0(T)$ is 0. As the degree of the growth function of $\mathrm{Min}(<)\backslash W_4$ is less than 4, corollary 3.9 gives the desired result.◂

4.4. Definition

For each path s intersecting 4-cycles $F_s: (\mathbb{R}^+)^4 \to \mathbb{R}^3$ is defined by

$$F_s(p_1,\dots,p_4) = r(s,1)^{p_1} * r(s,2)^{p_2} * r(s,3)^{p_3} * r(s,4)^{p_4}$$

The range of F_s is a closed subspace of $\mathbb{R}^3$ because it is again the nonstandard hull of a suitable subset of *G.

4.5 Proposition

If s *intersects 4-cycles then for* i < j

$$\begin{vmatrix} x(s,i) & x(s,j) \\ y(s,i) & y(s,j) \end{vmatrix} = 0$$

implies (x(s,j),y(s,j)) = (x(s,i),y(s,i)).

***Proof*.** As (x(s,j),y(s,j)) ≠ (0,0) there is some $\lambda \in \mathbb{R}$ such that (x(s,j),y(s,j)) = $\lambda\cdot$(x(s,i),y(s,i)). From the definition of the exponentation and the fact that r(s,i), r(s,j) $\in \overline{G}_\infty$ we conclude $r(s,j) = r(s,i)^\lambda$. The fact that $\|r(s,i)^\lambda\|_0 = |\lambda|$ then gives $|\lambda| = 1$. From $\lambda = -1$ we get the following contradiction: put $k_v = 1$ for $v \in \{i,j\}$ and $k_v = 0$ otherwise and consider

$$2 = \|r(s,1)^{k_0}*\ldots*r(s,4)^{k_4}\|_0 = \|r(s,i)*r(s,j)\|_0 = \|r(s,i)*\overline{r(s,i)}\|_0$$
$$= \|(0,0,0)\|_0 = 0. ◀$$

The next lemma will immediately lead to the proof of the theorem. It uses heavily that the range of the F_s is essentially two-dimensional.

4.6 Lemma

Suppose s *intersects 4-cycles. Then there are* α, $\beta \in \mathbb{R}$ *such that for all* $(u,v,w) \in \mathrm{range}(F_s)$, $\|(u,v,w)\|_0 \le \alpha\cdot u + \beta\cdot v$.

***Proof*.** We will distinguish several cases. In each case we will define, using the r(s,i), some continuous automorphism φ: $\overline{G} \to \overline{G}$, $\varphi(\overline{G}_\infty) \subseteq \overline{G}_\infty$. Each such φ defines a linear transformation $\overline{\varphi}$: $\mathbb{R}^2 \to \mathbb{R}^2$ given by the matrix

$$M = \begin{bmatrix} x_1 & x_2 \\ y_1 & y_2 \end{bmatrix}$$

where $\varphi(1,0,0) = (x_1,y_1,z_1)$, $\varphi\,(0,1,0) = (x_2,y_2,z_2)$. Finally α and β are obtained as $(\alpha,\beta) = (1,1)\cdot M$. The cases are divided into two groups. We put

$$A_{i,j} = \begin{vmatrix} x(s,i) & x(s,j) \\ y(s,i) & y(s,j) \end{vmatrix}$$

(I) $A_{1,4} = 0$.

(a) $A_{1,i} = 0$ for $1 \le i \le 4$. This gives $(x(s,i),y(s,i)) = (x(s,j),y(s,j))$ for $1 \le i,j \le 4$ by the previous proposition and therefore $r(s,1) = r(s,2) = r(s,3) = r(s,4)$. Some computation shows the existence of φ such that $\varphi(1,0,0) = r(s,i)$.

Now take some $(u,v,w) = F_s(k_1,\dots,k_4) \in \text{range}(F_s)$. We get $t := \sum_{i=1}^{4} k_i = \|(u,v,w)\|_0$. For $(x,y,z) := \varphi^{-1}(u,v,w)$ we obtain

$$(x,y,z) = (\varphi^{-1}(r(s,1))^t = (1,0,0)^t = (t,0,0)$$

and $$t = \|\varphi(x,y,z)\|_0 = \|\varphi(t,0,0)\|_0.$$

Therefore we have $(t,0) = (x,y) = \bar{\varphi}^{-1}(u,v)$ and finally get $\|(u,v)\|_0 = \alpha u + \beta v$ from $t = (1,1)\cdot\begin{bmatrix} t \\ 0 \end{bmatrix} = (1,1)\cdot M\cdot\begin{bmatrix} u \\ v \end{bmatrix}$.

(b) $A_{1,3} = A_{1,4} = 0$, $A_{1,2} \ne 0$. Similarly to (a) we take φ such that $\varphi(1,0,0) = r(s,1)$ and $\varphi(0,1,0\} \ne r(s,2)$ and proceed analogously to (a).

(c) $A_{1,4} = 0$, $A_{1,3} \ne 0$. We choose φ such that

$$\varphi(1,0,0) = r(s,1) = r(s,4) \text{ and } \varphi(0,1,0) = r(s,3).$$

Because $\varphi(\overline{G_\infty}) \subseteq G_\infty$ there are $d,e \in \mathbb{R}$ such that $\varphi^{-1}(r(s,2)) = (d,e,\tfrac{1}{2}de)$. With the same notation as in (a) a computation gives

$$(x,y,z) = (k_1 + dk_2 + k_4,\ ek_2 + k_3,\ \tfrac{1}{2}dek_2^2 + k_4(ek_2 + k_3)).$$

If $d + e = 1$ we would be done because $(1,1)\begin{bmatrix} x \\ y \end{bmatrix} = (x + y) = t$.

It remains to show that $d + e = 1$; we consider again different cases. One easily verifies that $\bar{\varphi} \in GL(2,\mathbb{R})$. Then $d = 0$ gives $e \ne 0$ and $A_{1,2} = 0$ which implies

$$(x(s,2),\ y(s,2)) = (x(s,3),\ y(s,3))$$

and hence $(0,e) = (0,1)$; i.e. $e = 1$. The case $e = 0$ is treated in the same way.

Now assume that $d \neq 0$, $e \neq 0$. For $y = 1$, $z = -1$ the conditions on the k_i are rewritten as

$$k_1 = x -(dx_2 + k_4), \quad k_3 = 1 - ek_2, \quad k_4 = 1 - \tfrac{1}{2}dek_2^2.$$

There is some $\delta > 0$ such that for all $k_1 \in [0,\delta]$ we have

$$1 -ek_2 \geq 0 \text{ and } 1 - \tfrac{1}{2}dek_2^2 \geq 0.$$

If we choose x sufficiently large there are for all $k_2 \in [0,\delta]$ real numbers k_1, k_3, $k_4 \geq 0$ with

$$\varphi(x,1,-1) = r(s,1)^{k_1} * \ldots * r(s,4)^{k_4}.$$

We obtain $\|\varphi(x,1,-1)\|_0 = (1+x) + (1 - (d+e))k_2$; because the last summand does not depend on x we get $1 - (d+e) = 0$, i.e. $d + e = 1$.

(II) $A_{1,4} \neq 0$.

We choose φ such that $\varphi(1,0,0) = r(s,1)$, $\varphi(0,1,0) = r(s,4)$.

(a) $A_{i,i+1} \neq 0$ for $1 \leq i < 4$. We put

$$(a_1, b_1, \tfrac{1}{2}a_1b_1) := \varphi^{-1}(r(s,2))$$

$$(a_2, b_2, \tfrac{1}{2}a_2b_2) := \varphi^{-1}(r(s,3))$$

and get from the assumptions on the $A_{i,i+1}$

$$\begin{vmatrix} 1 & a_1 \\ 0 & b_1 \end{vmatrix} \neq 0, \quad \begin{vmatrix} a_1 & a_2 \\ b_1 & b_2 \end{vmatrix} \neq 0, \quad \begin{vmatrix} a_2 & 0 \\ b_2 & 1 \end{vmatrix} \neq 0.$$

Therefore the quadratic form Φ given by the matrix

$$\begin{bmatrix} a_1b_1 & a_2b_1 \\ a_2b_1 & a_2b_2 \end{bmatrix}$$

has rank 2. From $(x,y,z) = \varphi^{-1}(F_s(k_1,\ldots,k_4))$ we get

$$\tfrac{1}{2}\, a_1b_1k_2^2 + \tfrac{1}{2}\, a_2b_2k_3^2 + a_2b_1k_2k_3 = z,$$

or equivalently $\Phi(k_2,k_3) = 2z$. We define

$$K_z := \{(k_2,k_3) \in \mathbb{R}^2 : \Phi(k_2,k_3) = 2z, \quad k_2,k_3 \geq 0\}.$$

As our quadratic form is not degenerate we can always find $z \neq 0$ (depending on the different possibilities for Φ) such that K_z is not contained in any straight line. The remaining arguments are now similar to those in (Ic). We take a compact subset $K \subseteq K_z$ and sufficiently large $x, y, \in \mathbb{R}$. Next we chose $k_2, k_3 \in K$ and put

$$k_1 := x - (a_1k_2 + a_2k_3) \geq 0$$

$$k_4 := y - (b_1k_2 + b_2k_3) \geq 0$$

which implies

$$\|\varphi(x,y,z)\|_0 = (x + y) + (1 -(a_1 + b_1))\, k_2 + (1 - (a_2 + b_2))k_3.$$

Now $\|\varphi(x,y,z)\|_0$ should be independent of the choice of k_2 and k_3; therefore the linear function

$$f(k_2,k_3) = (1 - (a_1+b_1))k_2 + (1- (a_2+b_2))k_3 = \|\varphi(x,y,z)\|_0 - (k_2 + k_3)$$

has to be constant on K which implies $f \equiv 0$. This gives $a_1 + b_1 = 1$ and $a_2 + b_2 = 1$; the rest is as in (Ic).

(b) $A_{i,i+1} = 0$ for $1 \leq i < 4$. This is impossible because it implies $r(s,i) = 0$ for $1 \leq i \leq 4$ and therefore $A_{1,4} = 0$.

(c) For exactly one i, $1 \leq i < 4$, $A_{i,i+1} \neq 0$. We consider only $A_{1,2} \neq 0$; the other cases are similar. Now $A_{2,3} = A_{3,4} = 0$ implies $r(s,2) = r(s,3) = r(s,4)$ and one can proceed as in (Ia).

(d) For exactly one i, $1 \leq i < 4$, $A_{i,i+1} = 0$. Then exactly two consecutive $r(s,i)$ are equal; hence we have some function $t(s,i)$ such that $\{t(s,i): 1 \leq i \leq 3\} = \{r(s,i): 1 \leq i \leq 4\}$ and

$$r(s,1)^{k_1} * \ldots * r(s,4)^{k_4} = t(s,1)^{p_1} * t(s,2)^{p_2} * t(s,3)^{p_3}$$

where

$$p_i = k_i + k_{i+1} \quad \text{iff} \quad A_{i,i,+1} = 0 \text{ and } p_i = k_i$$

otherwise. If $\pi: \mathbb{R}^3 \to \mathbb{R}^2$ denotes the projection onto the first two coordinates then $\pi(t(s,1))$ and $\pi(t(s,2))$ are linearly independent, from which we conclude $d \neq 0 \neq e$ for $(d, e, \frac{1}{2}de) = \varphi^{-1}(t(s,2))$. Analogously to

(Ic) one gets for $(u,v,w) = \varphi(x,y,z) \in$ range (F_s):

$$\|(u,v,w)\|_0 \le p_1 + (d+e)p_2 + p_3 = x + y$$

and one can proceed as in (Ic), provided that $d + e \ge 1$. It remains to confirm this last inequality. There are four cases according to whether d and e are positive or negative.

(1) $d > 0$ and $e < 0$. We consider

$$(4d,\ e,\ ½de) = (d,\ e,\ ½de)*(1,0,0)^{3d}$$

and get

$$\|\varphi(4d,\ e,\ ½de)\|_0 = 3d + 1$$

which implies for all $n \in \mathbb{N}$

$$\|\varphi(4d,\ e,\ ½\ de)^n\| \le n\ (3d + 1).$$

Considering $(4d,\ e,\ ½\ de)^n = (4nd,\ ne,\ ½\ de(4n^2 - 3n))$ and observing that the $t(s,i)$ are from the same set as the $r(s,i)$ one verifies that $\varphi(4d,\ e,\ ½de)^n \in$ range (F_s). This gives

$$\|\varphi(4d,\ e,\ ½de)\|_0 = p_1 + p_2 + p_3 = (4d + e)n + (1 - (d+e))p_2$$
$$\le n\ (3d + 1).$$

The last inequality holds for all $n \in \mathbb{N}$; therefore we have $d + 3 \ge 1$.

(2) $d < 0$ and $e > 0$. This case is similar to (1) and will be omitted.

(3) $d < 0$ and $e < 0$. The choice of $p_1 = -2d > 0$, $p_2 = 1$, $p_3 = -2e > 0$ shows that $\varphi(-d,\ -e,\ ½de) \in$ range (F_s). We get a contradiction from $1 = \|\varphi(-d,-e,\ ½de)\|_0 = p_1 + p_2 + p_3 > 1$; hence this case is impossible.

(4) $d > 0$ and $e > 0$.

By a computation in $\overline{G}$ one verifies directly

$$1 = \|\varphi(d,\ e,\ ½de)\|_0 \le d(1-\tfrac{1}{\sqrt{2}}) + \tfrac{e}{\sqrt{2}} + \tfrac{d}{\sqrt{2}} + e(1 - \tfrac{1}{\sqrt{2}}) = d + e$$

Therefore all cases are considered and this finishes the proof of the lemma.◄

Now we are ready for the

Proof of Theorem 4.1. We proceed indirectly and assume that Min(<) is regular. Then there are some 4-cycles intersecting s such that $(0,1,1) \in \text{range}(F_s)$. The lemma gives us α and $\beta \in \mathbb{R}$ such that

$\|(u,v,w)\|_0 \leq \alpha \cdot u + \beta \cdot v$ for all $(u,v,w) \in$ range (F_s). We obtain a contradiction from

$$0 < \|(0,0,1)\|_0 \leq \alpha \cdot 0 + \beta \cdot 0 = 0. \blacktriangleleft$$

REFERENCES

Bass, H. (1975). The degree of polynomial growth of finitely generated nilpotent groups, *Proc. London Math. Soc.* **25**, 603-614.

Benninghofen, B. Kemmmerich S. & Richter, M.M. (1987). *Systems of Reductions*, to appear in Springer Lecture Notes in Computer Science.

van den Dries, L. & Wilkie, A.J. (1984). On Gromov's theorem on groups of polynomial growth and elementary logic, *Journal of Algebra* **89**, 349-374.

Gilman, R.H. (1979). Presentations of groups and monoids, *Journal of Algebra* **57**, 544-554.

Knuth, D.E. & Bendix, P.B. (1970). Simple word problems in universal algebra; in *Computational Problems in Abstract Algebra*, (ed. J. Leech), Pergamon Press, 263-297.

Nelson, E. (1977). Internal set theory, *Bull Amer. Math. Soc.*, 1165-1198.

Richter, M.M. (1982). *Ideale Punkte, Monaden und Nichtstandard Methoden*, Vieweg Verlag.

Richter, M.M. (1987). The Knuth-Bendix completion procedure, the growth function and polycyclic groups; to appear in *Proc. Logic Colloquium* '86 (ed. F. Drake & J. Truss), North-Holland, Amsterdam.

SYNTACTICAL METHODS IN INFINITESIMAL ANALYSIS

FRANCINE DIENER and KEITH D. STROYAN

1. INTRODUCTION

In (1977) Edward Nelson gave a new formulation of Abraham Robinson's Theory of Infinitesimals known as Internal Set Theory. In (1980) Nelson refined this to give a solution to Robinson's (1973) Metamathematical Problem 11.

Nelson's approach to infinitesimal analysis has been taken up by a large number of workers in various fields. Lawler (1980) has obtained interesting results on a kind of self-avoiding random walk. The article by the Dieners below describes some of the many applications of Internal Set Theory to the study of differential equations. The article by Stroyan describes some extensions of Nelson's methods which are useful in topology and functional analysis. This article is an introduction to the ones by the Dieners and Stroyan.

This article is also a description of the common ground shared by the two approaches to Robinson's theory. We hope that our short presentation of IST and its interpretation in a superstructure will help those familiar with superstructures and those familiar with IST to understand each other.

In section 2 we give a brief introduction to IST (see references at the end for more complete introductions). In section 3 we restrict Nelson's methods to a superstructure as described above by Lindstrøm. This means that we only consider predicates from Nelson's formal language whose quantifiers are bounded by a standard entity. With this restriction, we shall prove in section 4 that Nelson's axiom schemes (I), (S) and (T) hold in a superstructure. Sections 5, 6 and 7 point out that (I), (S) and (T) can also be considered as quantifier manipulation rules. These rules allow us to study formal properties of some external sets.

2. IST: AN ALTERNATE AXIOMATIZATION OF SET THEORY

Zermelo-Fraenkel set theory with the axiom of choice (ZFC) is cast in a formal language that uses only one non-logical binary predicate $\in$. (The language also contains connectives, $\neg$ (not), $\wedge$ (and), $\vee$ (or), quantifiers, $\forall$, $\exists$, equality, $=$, and countably many variables, x,y,....) The properties of $\in$ are given by seven axioms. For example, the first two are:

1. $(\exists x)(\forall y)[\neg\, y \in x]$ (existence of the empty set)
2. $(\forall x)(\forall y)(\forall z)[[z \in x \Leftrightarrow z \in y] \Rightarrow x = y]$

Most mathematicians accept ZFC as a good formalization of the set theoretical principles they use.

Theorems – of set theory or mathematics – are just those statements of the language that can be proved from the axioms using well defined rules of proof. In theory at any rate, all theorems are established purely syntactically without reference to any real world of sets. In practice, however, most mathematicians have in mind a world of sets which is described, in part at least, by ZFC. Another aspect of mathematical practice should be mentioned: Extensive use is made of abbreviations for things that could be expressed in the formal language. In particular, constants such as $\emptyset$, $\cup$, $\cap$, 0, 1, π, e, $\mathbb{R}$, $\sin(\cdot)$, etc. can be regarded as abbreviations for the formal statements asserting their unique existence in the theory.

2.1 Description of IST

Nelson's formalization of (part of) infinitesimal analysis simply extends ZFC by adding a new (undefined, non-logical) unary predicate $st(\cdot)$. Formulas of the language that contain $st(\cdot)$ are called *external formulas*, others are called *internal*. The most important way that $st(\cdot)$ occurs is as bounds on quantifiers. (We abbreviate $(\forall x)[st(x) \Rightarrow F]$ by $(\forall^{st} x)F$ and $(\exists x)[st(x) \wedge F]$ by $(\exists^{st} x)F$. Also, $(\forall_{fin} x)$ and $(\exists_{fin} x)$ are abbreviations for 'for all finite x' or 'there exists finite x', where 'finite' is a formula stating that every injection is a surjection.) We think of $st(x)$ as meaning "x is standard", but three new axioms governing its use are added to those of ZFC in the official syntactical theory. These axioms are as follows:

Transfer. For any internal formula F containing at most the free variables $x, t_1, \cdots, t_m$

(T′) $(\forall^{st} t_1)\cdots(\forall^{st} t_m)[(\forall^{st} x)F(x,t_1,\cdots,t_m) \Rightarrow (\forall x)F(x,t_1,\cdots,t_m)]$.

Idealization. For any internal formula F with free variables x, y and possibly other free variables

(I′) $(\forall^{st}_{fin} z)(\exists x)(\forall y \in z)[F(x,y)] \Leftrightarrow (\exists x)(\forall^{st} y)[F(x,y)]$

Standardization. For any (internal or external) formula F with free variable z (and possibly others)

(S′) $(\forall^{st} x)(\exists^{st} y)(\forall^{st} z)[z \in y \Leftrightarrow z \in x \wedge F(z)]$.

The resulting formal set theory (IST) was shown to be a conservative extension of ZFC in Nelson (1977); i.e. IST is consistent if ZFC is and moreover any formula which can be written in the language of ZFC (i.e. without st) is a theorem in ZFC if and only if it is a theorem in IST.

2.2 Infinite Sets Contain Nonstandard Elements

In IST any infinite standard set (such as $\mathbb{N}$, $\mathbb{R}$, $\mathscr{C}[0,1]$) contains nonstandard members. This follows immediately from the idealization axiom (I′). In contrast to Robinson's original approach, it is not necessary to construct a new set $^*\mathbb{R}$ in order to obtain infinitesimals: all nonstandard objects are already in the universe of IST. The standard predicate allows us to distinguish standard elements of a set from the rest. The set $\mathbb{R}$ in IST (which is an abbreviation for statement in the formal language asserting the unique existence of a complete ordered field) corresponds to the set $^*\mathbb{R}$ in a superstructure. Similarly, other objects of IST such as $\mathbb{N}$, $\sin(\cdot)$, etc., correspond to $^*\mathbb{N}$, $^*\sin(\cdot)$, etc. in a superstructure. We denote the collection of all standard elements of $\mathbb{R}$ by $^\sigma\mathbb{R}$ (also sometimes denoted $\underline{\mathbb{R}}$). In the pure syntactical theory, $^\sigma\mathbb{R}$ is not a set since it is illegally formed, but rather is an abbreviation for the formula $x \in \mathbb{R} \wedge \mathrm{st}(x)$. This corresponds to the external set $^\sigma\mathbb{R} = \{^*r : r \in \mathbb{R}\}$ in a superstructure.

The distinction in notation, $\mathbb{R}$ vs. $^*\mathbb{R}$, reflects a philosophical difference. Mathematics is intended to model reality as well as possible. The nonstandard universe gives a richer model of the reals than the standard universe, so it is natural to identify the real line with $\mathbb{R}$ from IST. From this point of view the standard objects are merely "shadows" of the real objects: taking standard parts loses information.

2.3 Internal and External Sets in IST

In ZFC, there is a set that corresponds to each formula: the set of all elements satisfying the formula. In IST, this still remains true for internal formulas but becomes false for external ones, because the rules of set formation are contained in the first 7 axioms and, thus, are the same as in ZFC. This means that any variable in a formula must be interpreted as an internal set. For example, if A is a set, the formula $[\forall B, B \subset A, \cdots]$ means in fact: for all *internal* subsets B of $A, \cdots$, because we introduced $\subset$ above as a syntactical abbreviation in IST.

Nevertheless, many external sets can be easily treated in IST as abbreviations. Suppose one has to consider the halo of a real number x (see section 7), $hal(x) = \{y \in \mathbb{R},\ y \approx x\}$, in a formula such as:

$$\forall x \in \mathbb{R},\quad hal(x) \cap \mathbb{Q} \neq \emptyset.$$

This is not, properly speaking, a formula of IST, but one can use it as an abbreviation of the following external formula of IST:

$$(\forall x \in \mathbb{R})(\exists q \in \mathbb{Q})(\forall^{st}\varepsilon > 0)[\ x-\varepsilon \leq q \leq x+\varepsilon\ ]$$

In general, one can consider any external set which can be defined by a formula just as an abbreviation. For special external sets, this turns out to be convenient. Finite unions and intersections, inclusions, etc. all work naturally. Difficulties do occur when the reasoning involves quantification on external sets.

We conclude this section with some useful definitions and theorems of IST.

For nonstandard mathematicians who do not know IST, the most unusual axiom is the standardization axiom (see section 4 for a proof of its validity in the superstructure). This axiom formalizes the following idea: any collection C of standard objects, for example the standard objects satisfying an external formula Φ of IST, defines a unique standard set denoted by ${}^{s}C$. The standard elements of ${}^{s}\{x : \Phi(x)\}$ are precisely those which satisfy Φ, but its nonstandard elements may or may not satisfy Φ. (In a superstructure ${}^{s}C$ is the star of the discrete standard part of the external set C, ${}^{s}C = {}^{*}D$.)

The standardization axiom is useful in making definitions. For example, one may define the set $\mathcal{C}(\mathbb{R},\mathbb{R})$ of all continuous functions from $\mathbb{R}$ to $\mathbb{R}$, by

$$\mathcal{C}(\mathbb{R},\mathbb{R}) = {}^{s}\{f : \mathbb{R} \longrightarrow \mathbb{R}: \ \forall^{st}x \ \forall y \ (x \approx y \Rightarrow f(x) \approx f(y))\}.$$

Other examples are given after Proposition 4.4 below.

Applying the normal induction principle to the standard set ${}^{s}\{x : \Phi(x)\}$ one obtains the useful principle:

2.4 External Induction Principle

For any external or internal formula Φ, *one has*

$$(\ \Phi(0) \ \wedge \ [\forall^{st}n(\Phi(n) \Rightarrow \Phi(n+1))]\ ) \ \Rightarrow \ (\forall^{st}n \ \Phi(n)).$$

One of the most important notions for the users of IST is the shadow of a set or a function. It is often denoted by $°A$ or $°f$, but it is not obtained by taking the standard part of the elements or values (a notation sometimes used in superstructures).

Let X be a standard topological space, $A \subset X$ and $x \in X$ and let $hal(x)$ be the intersection of all standard neighborhoods of x. The *shadow* of A is defined by $°A = {}^{s}\{x \in X : hal(x) \cap A \neq \emptyset\}$. For standard A, this is just the closure. In a superstructure, $°A = {}^*[st(A)]$. According to C. L. Thompson & B. J. Homer, the interior shadow or *soul* of A is defined by ${}^{s}\{x \in X: \ hal(x) \subset A\}$. For standard A, this is just the interior. In Robinson's terminology, the soul of A is the nonstandard extension of the set of standard points in its S-interior.

Now let X and Y be standard metric spaces and A be an internal subset of X. A function f from X to Y is called *of class* S^0 *on* A (or just S^0 on A) if for all $x \in A$, near standard in A, $f(x)$ is near standard and

$$y \approx x \Rightarrow f(y) \approx f(x).$$

A standard function $°f$ is the *shadow of* f *on* A if it is infinitely close to f at all near the standard points of A. In a superstructure, $°f = {}^*g$, where $g(x) = st(f(x))$, for standard x. Any function has at most one shadow. One can show (Continuous Shadow Theorem) that f actually has a shadow if and only if it is of class S^0. This is a nonstandard version of the Ascoli theorem (see Lindstrøm's lecture III, 2.7).

Let X and Y be standard normed spaces and A an internal subset of X. A function f from X to Y is called *of class* S^1 *on* A if for all $x \in A$, near standard in A, f(x) is near standard and there exists a standard linear mapping $L : X \longrightarrow Y$ such that: $(y \approx x, y' \approx x, y \neq y') \Rightarrow ([f(y)-f(y')]/\|y-y'\| \approx L[y-y']/\|y-y'\|)$.

For an internal differentiable function, being of class S^1 just means that f and f′ are both of class S^0. And in that case, one has $(°f)' = °(f')$. (Also see the description of $\mathscr{C}^1$ for standard f in the examples following 4.4.)

One can prove the following result along the lines of Behrens (1974) or Stroyan & Luxemburg (1976, 5.7.11).

2.5 Near Standard Local Inverse Function Theorem

Let X *be a standard Banach space,* x *a near standard point of* X *and let* $f : X \longrightarrow X$ *be an internal function of class* S^1 *at* x. *If the derivative of the shadow of* f *at* x *is invertible then* f *is invertible on a standard open neighborhood of* x *and* f^{-1} *is of class* S^1 *at* f(x).

3. BOUNDED INTERNAL SET THEORY

Let V(S) denote the superstructure over a set of atoms $S \subseteq \mathbb{R}$ as defined in the article by Lindstrøm. Let V(*S) be the full superstructure over the atoms *S of a polysaturated nonstandard extension. Recall that an entity $b \in V(*S)$ is called *standard* if there is an $a \in V(S)$ such that b = *a. Everything in V(S) is "standard", it is only important in V(*S) to know which entities come from V(S). Also recall that an entity c is called *internal* if $c \in b$ for some standard b.

3.1 Definition

The *language of Bounded Internal Set Theory* L(*V(S),st(·)) contains variables, the predicates $\in$, =, st(·), logical connectives, parentheses, the quantifiers $\forall, \exists$ and

(a) constants for every internal (and standard) entity of V(*S)

(b) function symbols for every tame standard function, together with Lindstrøm's above rules for forming terms, his rules for forming formulas plus the additional rule

(iii′) if t is a term $st(t)$ is a formula.

We wish to stress that we require Lindstrøm's bounded quantifier rule. Unbounded quantifiers are NOT allowed. Bounded quantifiers may be thought of as abbreivations for formulas $(\exists v)[v \in t \wedge \varphi]$ and $(\forall v)[v \in t \Rightarrow \varphi]$. Later we will want to incorporate the standard predicate in another quantifier abbreviation.

3.2 Definition

A formula φ of $L({}^*V(S),st(\cdot))$ is called *internal* if it does not contain the standard predicate $st(\cdot)$. An internal formula is called *standard* if it at most contains standard constants. A formula which contains $st(\cdot)$ is called an *external* formula.

A standard formula of $L({}^*V(S),st(\cdot))$ is also a formula of Lindstrøm's $L^*(V(S))$ and an internal formula φ equals $\psi(\bar{v},\bar{c})$ where $\psi(\bar{v},\bar{w})$ is a formula of $L^*(V(S))$ with some free variables $\bar{w}$ replaced by internal constants $\bar{c}$.

The importance of the standard predicate $st(\cdot)$ is that it permits us to make formal statements about certain *external* sets. For example, the set of *inifinitesimal* *reals is described by

$$a \approx 0 \iff (\forall v \in {}^*\mathbb{R})[(st(v) \wedge v > 0) \Rightarrow |a| < v].$$

The set of *limited* *reals $\mathbb{G}$ is defined by

$$b \in \mathbb{G} \iff (\exists v \in {}^*\mathbb{R})[st(v) \wedge |b| < v].$$

The standard part function is given by

$$w = st(v) \iff [v \in \mathbb{G} \wedge w \in {}^*\mathbb{R} \wedge st(w) \wedge w \approx v]$$

where "$v \in \mathbb{G}$" and "$w \approx v$" are replaced by their formal equivalents from above.

More generally, if $B = {}^*A$ is an infinite standard set, the external set consisting of only its standard elements, ${}^{\sigma}B = \{b \in B : (\exists a \in A)[b = {}^*a]\}$ is described in $L({}^*V(S),st(\cdot))$ by

$$b \in {}^{\sigma}B \iff b \in B \wedge st(b).$$

For example, the set of standard subsets of ${}^*\mathbb{R}$ is ${}^{\sigma}\mathcal{P}(\mathbb{R})$, where $\mathcal{P}(\mathbb{R})$ denotes the power set.

Recall that a sentence is a formula with no free variables (see Lindstrøm's article). Only *sentences* have a truth value in $V(*S)$. We may assume that the only way external sets enter the *sentences* of $L(*V(S),st(\cdot))$ is as bounds of the form $^{\sigma}B$ in quantifiers. First, in any sentence we may replace the expressions $st(t)$ by $(\exists v \in B)[st(v) \wedge v = t]$ where v does not occur in t and B is a bound high enough in $*V(S)$ so that $t \in B$. Next, the quantifiers $(\forall v \in B)[st(v) \Rightarrow \cdots]$ and $(\exists v \in B)[st(v) \wedge \cdots]$ may be abbreviated $(\forall v \in {^{\sigma}B})[\cdots]$ and $(\exists v \in {^{\sigma}B})[\cdots]$. Notice that if a term t contains free variables, replacement of $st(t)$ by $(\exists v)[st(v) \wedge v = t]$ is illegal unbounded quantification and not a formula of $L(*V(S),st(\cdot))$. However, if $b_1, \cdots, b_n$ are internal constants, replacement of the free variables by these constants makes $\varphi(b_1, \cdots, b_n)$ a *sentence* of $L(*V(S),st(\cdot))$. The sentence $\varphi(\overline{b})$ is equivalent to one in which $st(\cdot)$ only occurs in bounds on quantifiers.

Also notice that term bounds $(\forall v \in t)\cdots$ and $(\exists v \in t)\cdots$ may be replaced by standard constant bounds $(\forall v \in b)[v \in t \Rightarrow \cdots]$ and $(\exists v \in b)[v \in t \wedge \cdots]$ in every *sentence* of $L(*V(S),st(\cdot))$. Again, each term has some maximum height, $t \in *V_m$, because sentences have no free variables.

3.3 Definition

Quantifiers of the forms $(\forall v \in b)[\cdots]$ and $(\exists v \in b)[\cdots]$ where b is a standard constant $(b = *a,$ for some $a)$ are called *bounded internal quantifiers*. Quantifiers of the forms $(\forall v \in {^{\sigma}b})[\cdots]$ and $(\exists v \in {^{\sigma}b})[\cdots]$ which are abbreviations for $(\forall v \in b)[st(v) \Rightarrow \cdots]$ and $(\exists v \in b)[st(v) \wedge \cdots]$, are called *bounded external quantifiers.*

The point of the preceding discussion is:

3.4 Proposition

Every sentence of $L(*V(S),st(\cdot))$ *is equivalent to one which only has bounded internal and external quantifiers and at most has* $st(\cdot)$ *appearing in external quantifier bounds.*

Notice that the variables in quantifiers only take internal values because $v \in b$ or $v \in {^{\sigma}b}$ means v belongs to a standard entity.

3.5 Definition

A set $X \in V(*S)$ is called a *definable external set* (or function) if it is not internal, but there is a formula $\chi(v)$ of $L(*V(S),st(\cdot))$ with only v free such that for some $*V_n$, $X = \{v \in *V_n : \chi(v)\}$. A set $X \in V(*S)$ is called an *undefinable external set* if there is no formula $\chi(v)$ in $L(*V(S),st(\cdot))$ such that $X = \{v \in *V_n : \chi(v)\}$.

In many syntactical calculations it is convenient not to explicitly write bounds on quantifiers. In these cases the internal quantifers $(\forall v \in b)$ and $(\exists v \in b)$ will be written $(\forall v)$ and $(\exists v)$ where the bounds are understood to exist, while the external quantifiers $(\forall v \in {}^{\sigma}b)$ and $(\exists v \in {}^{\sigma}b)$ will be written $(\forall^{st} v)$ and $(\exists^{st} v)$. We stress that unbounded quantifiers are not permitted in Bounded Internal Set Theory because they would take us outside $V(*S)$.

4. THE BOUNDED FORMULAS (T), (I), (S)

Here is a way to formulate transfer in $L(*V(S),st(\cdot))$. (We have the constants b, so we do not need the t_i quantifiers of (T′) in section 2.)

4.1 Proposition (Transfer Axiom (T))

If $\varphi(v_1,\cdots,v_m;b_1,\cdots,b_n)$ *is a standard formula of* $L(*V(S),st(\cdot))$ *with only the free variables* $v = (v_1,\cdots,v_m)$ *and only the standard constants* $b = (b_1,\cdots,b_n)$ *and if* C *is a standard constant, then the following holds in* $V(*S)$:

$$\text{(T)} \qquad (\forall v \in {}^{\sigma}C)[\varphi(v;b)] \Leftrightarrow (\forall v \in C)[\varphi(v;b)]$$

or abbreviated, $(\forall^{st} v)[\varphi] \Leftrightarrow (\forall v)[\varphi]$.

Proof. Only the $\Rightarrow$ implication is nontrivial, since more is better. Since $\varphi(v,b)$ only contains standard constants, it is the $*$-transfer of a formula in $L(V(S))$, $\varphi(v,b) = \varphi(v,*a)$. Also $C = *D$. Given any $d \in D$, $c = *d$ is standard, $c \in {}^{\sigma}C$. Since we know $\varphi(c,b)$ from the formula (T), we know $\varphi(*d,*a)$. Hence, by Lindstrøm's Transfer Principle IV.2.4, we know that for each $d \in D$, $\varphi(d,a)$, that is, $V(S) \models (\forall d \in D)[\varphi(d,a)]$.

This is equivalent to $*V(S) \models (\forall d \in *D)[\varphi(d,*a)]$ by Lindstrøm's Transfer Principle IV.2.4. This is the same formula as $(\forall c \in C)[\varphi(c,b)]$.

All we have done is to re-phrase Lindstrøm's Transfer Principle in $L(*V(S),st(\cdot))$. Notice that we cannot allow $\varphi(v,b)$ to contain external predicates, free variables other than v, or internal nonstandard constants in our proof.◂

The finite power set operator is a tame function in V(S):

$$FP(A) = \{B \subseteq A : B \text{ is finite}\}.$$

"B is finite" means every injection of B into itself is onto; equivalently, that there is a bijection from an initial segment of $\mathbb{N}$ onto B, or any of the other formal equivalents. If B is an internal set, *FP(B) consists of all the "hyperfinite" subsets of B as described by Lindstrøm. This *finiteness property may be expressed by transfer of any of the formal equivalents of finiteness, for example, $C \in *FP(B)$ if and only if there is an $n \in *\mathbb{N}$ and an internal bijection from $\{k \in *\mathbb{N} : 1 \le k \le n\}$ onto C.

Nelson's axiom of the ideal point is as follows for V(*S).

4.2 Proposition (Saturation Axiom (I))

Let $\varphi(v;b)$ *be an internal formula of* $L(*V(S),st(\cdot))$ *with only* $v = (v_1,\cdots,v_m)$ *free and only internal constants* $b = (b_1,\cdots,b_n)$. *Let* B *and* C *be standard constants. Then*

(I) $$(\forall Y\in {}^{\sigma}FP(B))(\exists x\in C)(\forall y\in Y)[\varphi(x,y;b)] \Leftrightarrow (\exists x\in C)(\forall y\in {}^{\sigma}B)[\varphi(x,y;b)]$$

or abbreviated, $(\forall^{st}_{fin} Y)(\exists x)(\forall y \in Y)[\varphi] \Leftrightarrow (\exists x)(\forall^{st} y)[\varphi]$.

Proof. If Y is *finite and standard, there is a standard $m \in \mathbb{N}$ and a standard function $f : \{k \in *\mathbb{N} : 1 \le k \le m\} \longrightarrow Y$ which is 1-1 and onto, so Y is really a finite set. The family of sets of the form $\Phi(Y) = \{x \in C: (\forall y \in Y)[\varphi(x,y;b]\}$ has the finite intersection property by the left side of (I). These are internal sets by Lindstrøm's Internal Definition Principle IV.2.5., so

$$\bigcap_{Y\in {}^{\sigma}FP(B)} \Phi(Y) \neq \emptyset$$

by Lindstrøm's saturation principle III.1.1, since $B \in *V_m$, for some m, and $\operatorname{card}({}^{\sigma}FP(B)) < \operatorname{card}(V(S))$. Any $x \in \cap \Phi(Y)$ satisfies the right hand side of (I).

The opposite implication is trivial since every element of a standard finite set is standard, so we have proved (I) in $V(*S)$.◀

Axiom (I) is not equivalent to Lindstrøm's saturation principle; the family of internal sets $\Phi(Y)$ must be indexed by standard finite subsets of some C and described by an internal property only depending on finitely many other internal constants, b. In some sense it is exactly the amount of saturation you can detect from within $L(*V(S), st(\cdot))$.

A positive infinitesimal number is an ideal point. For any standard finite set of positive reals, $Y = \{\varepsilon_1, \varepsilon_2, \cdots, \varepsilon_m\} \in {}^{\sigma}FP((0,\infty))$, there is a $\delta > 0$ such that $(\forall i)[\delta < \varepsilon_i]$. Thus, by the right side of (I) there is a $\delta > 0$ such that for all $\varepsilon \in {}^{\sigma}\mathbb{R}$ with $\varepsilon > 0$, $0 < \delta < \varepsilon$, i.e., $0 < \delta \approx 0$. This property has no internal constants. Internal constants let us make even "more ideal" points.

Let $C_0 = *D_0$ be the standard set of all positive *real-valued internal functions defined on a neighborhood of zero in $*\mathbb{R}$. Let δ be a positive infinitesimal number. If $\varepsilon \in C_0$ is standard, $\varepsilon \in {}^{\sigma}C_0$, then ε is defined on a standard neighborhood of zero, so $\varepsilon(\delta)$ is defined and $\varepsilon(\delta) > 0$. For any standard finite set $\{\varepsilon_1, \cdots, \varepsilon_m\} \in {}^{\sigma}FP(C_0)$, $\bigcap_{j=1}^{m} (0, \varepsilon_j(\delta)) \neq \emptyset$, hence by axiom (I)

$$(\exists \iota > 0)(\forall \varepsilon \in {}^{\sigma}C_0)[0 < \iota < \varepsilon(\delta)]$$

that is, ι is smaller than any standard function at δ, for example, $0 \approx \delta > \delta^2 > \delta^3 > \cdots > e^{-1/\delta} > \iota > 0$ and each ratio in the inequalities above is unlimited δ/δ^2, $\delta^3/e^{-1/\delta}$, $e^{-1/\delta}/\iota$. In this example, the intervals $(0, \varepsilon(\delta))$ depend on the internal nonstandard constant $\delta \approx 0$.

Consider the property $\varphi(F,x) = [x \in F \wedge$ "F is *finite"] where "F is *finite" is an abbreviation for one of the formal equivalents of finiteness such as "there exists an internal bijection from an initial segment of $*\mathbb{N}$ onto F". Whenever B is an internal set we know that

$$(\forall X \in {}^{\sigma}FP(B))(\exists F \in *FP(B))(\forall x \in X)[\varphi(F,x)]$$

therefore by axiom (I), there is a *finite set F containing all the standard points of B, ${}^{\sigma}B \subseteq F \subseteq B$.

Nelson applies (I) to the φ formula above without quantifier bounds to obtain one *finite set containing all standard sets. This result is false in $^*V(S)$.

The most important form of "standardization" is:

4.3 Proposition (Standardization Axiom (S))

Let $\Phi(x,y)$ *be any formula of* $L({}^*V(S),st(\cdot))$ *with only* x *and* y *free. Let* A *and* B *be standard sets. Then*

$$\text{(S)}\qquad (\forall x \in {}^{\sigma}A)(\exists y \in {}^{\sigma}B)[\Phi(x,y)] \Rightarrow (\exists y \in {}^{\sigma}[B^A])(\forall x \in {}^{\sigma}A)[\Phi(x,y(x))]$$

abbreviated, $(\forall^{st}x)(\exists^{st}y)[\Phi(x,y)] \Rightarrow (\exists^{st}y(\cdot))(\forall^{st}x)[\Phi(x,y(x))]$.

Proof. Let $A = {}^*C$ and $B = {}^*D$ and assume that the left side of (S) holds. Then the sets $D_c = \{d \in D : \Phi({}^*c,{}^*d)\}$ are nonempty for each $c \in C$. The axiom choice in $V(S)$ says there is a choice function $z \in D^C$. Let $y = {}^*z$.◂

Notice that $\Phi(x,y)$ in (S) may be external and contain both internal and standard constants.

Another form of "standardization" is:

4.4 Proposition

Let $\Phi(x)$ *be any formula from* $L({}^*V(S),st(\cdot))$ *with only the free variable* x. *Let* $B \in {}^*V_{n+1}$ *be a standard set, then* $(\exists y \in {}^{\sigma}V_{n+1})(\forall x \in {}^{\sigma}V_n)[x \in y \Leftrightarrow (x \in B \wedge \Phi(x))]$ *in other words, there is a standard subset of* B *whose standard elements are determined by* $\Phi(x)$. *This set is denoted*

$${}^{s}\{x \in B : \Phi(x)\}.$$

Proof. Define a set in $V(S)$ by $D = \{d \in V_n : {}^*d \in B \wedge \Phi({}^*d)\}$ then ${}^*D = {}^{s}\{x \in B : \Phi(x)\}$. This ends the proof.◂

This may seem a little awkward when there is a standard model $V(S)$ around, but this is the place where we can discard the standard model altogether. For example, we may define the weak derivative of a standard function $f : {}^*\mathbb{R} \longrightarrow {}^*\mathbb{R}$ by

$$Df = {}^{s}\{(x,y) : x \in {}^{\sigma}\mathbb{R} \wedge y \in {}^{\sigma}\mathbb{R} \wedge (\forall\delta\neq 0)[\delta\approx 0 \Rightarrow y \approx \frac{f(x+\delta)-f(x)}{\delta}]\}.$$

The set of functions which have a weak derivative is

$$D^1 = {}^{s}\{f \in {}^*(\mathbb{R}^{\mathbb{R}}) : (\forall x\in{}^{\sigma}\mathbb{R})(\exists y\in{}^{\sigma}\mathbb{R})(\forall\delta\neq 0)[\delta \approx 0 \Rightarrow y \approx \frac{f(x+\delta)-f(x)}{\delta}]\}$$

In other words, $f \in D^1$ if for each standard x, $st(\frac{f(x+\delta)-f(x)}{\delta})$ is the same for all nonzero $\delta \approx 0$. The external definition of Df need not apply at nonstandard points. If f is always differentiable, but Df is discontinuous at a standard x, there is a $\xi \approx x$ and a $\delta \approx 0$ such that $Df(\xi) \not\approx \frac{f(\xi+\delta)-f(\xi)}{\delta}$. This is because even if x is standard and $\xi = x - \delta$, $Df(x) \approx (f(\xi+\delta) - f(\xi))/\delta = -(f(x-\delta) - f(x))/\delta \not\approx Df(\xi)$.

The set of continuously differentiable functions is given by

$$\mathscr{C}^1 = {}^{s}\{f \in {}^*(\mathbb{R}^{\mathbb{R}}) : (\exists f'\in{}^{\sigma}\mathbb{R}^{\mathbb{R}})(\forall x\in\mathbb{G})(\forall\delta\neq 0)[\delta \approx 0 \Rightarrow f'(x) \approx \frac{f(x+\delta)-f(x)}{\delta}]\}$$

The formulas (I), (S), (T) may be used to re-write these standard definitions in their familiar pointwise and uniform $\varepsilon - \theta$ definitions. The next section gives rules for systematic application of (I), (S) and (T) that always result in a canonical reduced form. For now we just apply rules unsystematically. Ordinary quantifier rules apply to both internal and external quantifiers because these rules apply to bounded quantifiers and the $st(\cdot)$ predicate is part of the bound. Recall how to pull quantifiers out of an implication: if x does not occur in ψ and y does not occur in φ,

$$\left.\begin{aligned} &[(\forall x)[\varphi(x)] \Rightarrow (\forall y)[\psi(y)]] \\ \Longleftrightarrow\ &(\forall y)(\exists x)[\varphi(x) \Rightarrow \psi(y)] \\ \Longleftrightarrow\ &(\exists x)(\forall y)[\varphi(x) \Rightarrow \psi(y)] \end{aligned}\right\} \quad (\forall\exists)$$

Also note that the negations of both sides of each of (I), (S), (T) yields a quantifier rule, for example,

$$(\neg I) \qquad (\exists^{st}_{fin}E)(\forall\delta)(\exists\varepsilon \in E)[\varphi(\varepsilon,\delta)] \Longleftrightarrow (\forall\delta)(\exists^{st}\varepsilon)[\varphi(\varepsilon,\delta)].$$

The definition of f' for $f \in \mathcal{C}^1$ replacing the abbreviations for $\mathbb{G}$ and $\approx$ is:

$$(\forall x)[((\exists^{st} b)[|x|<b] \wedge (\forall^{st}\varepsilon)[|\delta|<\varepsilon]) \Rightarrow (\forall^{st}\theta)[|f'(x) - \frac{\delta f}{\delta x}| < \theta]]$$

$$\Leftrightarrow \qquad (\forall\exists)$$

$$(\forall x)(\forall^{st} b)(\forall^{st}\theta)(\exists^{st}\varepsilon)[(|x| < b \wedge |\delta| < \varepsilon) \Rightarrow |f'(x) - \frac{\delta f}{\delta x}| < \theta]$$

$$\Leftrightarrow \qquad (\forall\forall \Leftrightarrow \forall\forall)$$

$$(\forall^{st} b)(\forall^{st}\theta)(\forall x)(\exists^{st}\varepsilon)[(|x| < b \wedge |\delta| < \varepsilon) \Rightarrow |f'(x) - \frac{\delta f}{\delta x}| < \theta]$$

$$\Leftrightarrow \qquad (\neg I)$$

$$(\forall^{st} b)(\forall^{st}\theta)(\exists^{st}_{fin} E)(\forall x)(\exists\varepsilon\in E)[|x|<b \wedge |\delta|<\varepsilon \Rightarrow |f'(x) - \frac{\delta f}{\delta x}| < \theta]$$

$$\Leftrightarrow \qquad (\text{replace } E \text{ by } \min E)$$

$$(\forall^{st} b)(\forall^{st}\theta)(\exists^{st}\varepsilon)(\forall x)[(|x| < b \wedge |\delta| < \varepsilon) \Rightarrow |f'(x) - \frac{\delta f}{\delta x}| < \theta].$$

When f' and f are standard (using $(\neg T)$ and T to remove all the st's) we have the standard definition:

$$f'(x) = \lim_{\delta\to 0} \frac{\delta f}{\delta x} \quad \text{uniformly on bounded sets.}$$

5. QUANTIFIER REDUCTION RULES

Consider a sentence of $L(^*V(S), st(\cdot))$ with only standard constants which has all its external quantifiers "outside", i.e., $(Q_1^{st} v_1) \cdots (Q_m^{st} v_m)\varphi$, where φ is internal and each Q_i^{st} is either $\forall^{st}$ or $\exists^{st}$ with bounds understood. Using (T) and its negative form we have

$$(T^+) \qquad (Q_1^{st} v_1) \cdots (Q_m^{st} v_m)\varphi \Leftrightarrow (Q_1 v_1) \cdots (Q_m v_m)\varphi.$$

Nelson has given a reduction algorithm which will take any external sentence and re-write it as an equivalent sentence with all external quantifiers on the left. Adding (T^+) at the last step gives a procedure to convert sentences of $L(^*V(S), st(\cdot))$ into equivalent (but perhaps quite complicated) sentences of $L^*(V(S))$ *provided the original sentence had only standard constants*. Nelson's reduction algorithm is also very useful for sentences with internal constants, but then (T^+) is not applicable at the last step.

We may re-write a formula as an equivalent formula with only the logical connectives $\neg$ and $\Rightarrow$. We replace $a \vee b$ by $\neg a \Rightarrow b$, $a \wedge b$ by $\neg(a \Rightarrow \neg b)$ and $a \Leftrightarrow b$ by $\neg((a \Rightarrow b) \Rightarrow \neg(b \Rightarrow a))$.

As a syntactical rule it is legal to apply (I) to formulas with free variables. Say $\varphi(x,y;v)$ has free variables $v = (v_1,\cdots,v_m)$. For any internal constants (I) yields

$$(\forall^{st}_{fin} Y)(\exists x)(\forall y \in Y)[\varphi(x,y;b)] \Leftrightarrow (\exists x)(\forall^{st} y)[\varphi(x,y;b)]$$

hence the replacement rule (i.e. replace either side with the other)

$$(I_0) \qquad (\forall^{st}_{fin} Y)(\exists x)(\forall y \in Y)[\varphi(x,y;v)] \Leftrightarrow (\exists x)(\forall^{st}_{y})[\varphi(x,y;v)]$$

results in equivalent formulas: once substitutions are made so they have truth values, those values are the same for both formulas.

Similar remarks apply to (S), but in addition, we only need (S) for internal formulas: the full theorem (3.3) follows from the reduction algorithm based on the syntactical rule

$$(S_0) \qquad (\forall^{st} x)(\exists^{st} y)[\varphi(x,y;v)] \Rightarrow (\exists^{st} y(\cdot))(\forall^{st} x)[\varphi(x,y(x);v)]$$

where bounds are understood on all quantifiers and $\varphi(x,y;v)$ is internal, but may contain free variables.

5.1 The (I_0)-(S_0) Reduction Rules

We say a formula Φ is in *reduced form* if

$$\Phi = (\forall^{st} u)(\exists^{st} v)[\varphi(u,v,w)]$$

where φ is internal and contains only bounded internal quantifiers and where $(\forall^{st} u)$ and $(\exists^{st} v)$ are abbreviations for bounded external quantifiers. If Φ, Φ_1 and Φ_2 are reduced, define the reduced forms of

$\neg\Phi$ *to be* $\qquad (\neg + (S_0))$

$$(\forall^{st} v(\cdot))(\exists^{st} u)(\neg\varphi(u,v(u),w)]$$

$\Phi_1 \Rightarrow \Phi_2$ *to be* $\qquad (\forall\exists + (S_0))$

$$(\forall^{st} u_2)(\forall^{st} v_1(\cdot))(\exists^{st} u_1)(\exists^{st} v_2)[\varphi_1(u_1,v_1(u_1),w_1) \Rightarrow \varphi_2(u_2,v_2,w_2)]$$

$(\forall x)\Phi$ *to be* $\qquad (\forall\forall + (I_0))$

$$(\forall^{st} u)(\exists^{st}_{fin} V)(\forall x)(\exists v \in V)[\varphi(u,v,w)]$$

$(\forall^{st}x)\Phi$ *to be* (already reduced)

$$(\forall^{st}x)(\forall^{st}u)(\exists^{st}v)[\varphi(u,v,w)]$$

$(\exists x)\Phi$ *to be the result of 3 reductions* $\neg(\forall x)\neg\Phi$

$(\exists^{st}x)\Phi$ *to be the result of 3 reductions* $\neg(\forall^{st}x)\neg\Phi$

Any formula of $L(*V(S),st(\cdot))$ with bounded internal and external quantifiers and only the connectives $\Rightarrow$ and $\neg$ may be put in its equivalent reduced form by application of the above rules working from the inside out. If either of the external quantifiers in the reduced form $(\forall^{st}u)(\exists^{st}v)\varphi$ is missing, we delete the associated quantifier in the rules above. For example, if φ is internal, φ is reduced, $(\forall^{st}x)[\varphi]$ is reduced, $(\exists^{st}x)[\varphi]$ is reduced. If $(\exists^{st}v)[\varphi]$ is reduced, the reduced form of $(\forall x)(\exists^{st}v)[\varphi]$ is $(\exists^{st}_{fin}V)(\forall x)(\exists v\in V)[\varphi]$, and so forth. Since every sentence of $L(*V(S),st(\cdot))$ may be written with bounded internal and external quantifiers and only the connectives $\neg$ and $\Rightarrow$, we have:

5.2 Theorem

Every sentence of $L(*V(S),st(\cdot))$ *is equivalent to a sentence in reduced form* $(\forall^{st}u)(\exists^{st}v)[\varphi(u,v;b)]$ *with* φ *internal, where one or both of the external quantifiers may be missing.*

In sections 6 and 7, we shall see that special reduced forms lacking one external quantifier yield additional information about the kinds of external sets such formulas define.

The general form of axiom (S) given above may be proved syntactically with only (S_0) and (I_0).

Another important application of the reduced form is

5.3 The Extension Principle

Let X *and* Y *be standard sets and let* $\Phi(x,y)$ *be a formula of* $L(*V(S),st(\cdot))$ *with only* x *and* y *free. Suppose that for each standard* x *in* X, *there exists* y *in* Y *such that* $\Phi(x,y)$ *holds. Then there is an internal* y *in* Y^X *such that* $\Phi(x,y(x))$ *holds for all* *standard* x *in* X, *that is*,

$$\text{(E)} \qquad (\forall x \in {}^{\sigma}X)(\exists y \in Y)[\Phi(x,y)] \Rightarrow (\exists y \in Y^X)(\forall x \in {}^{\sigma}X)[\Phi(x,y(x))]$$

Nelson (1980) proves this by writing Φ in reduced form and then syntactically manipulating both sides of the implication. The reader can give an ultrapower proof of the principle by replacing $^*\mathbb{N}$ in Lindstrøm's proof of II.1.4 with X.

An important special use of extension is in the construction of Loeb measure. Here is the first step. Suppose Ω is an internal set and $\Lambda_1, \Lambda_2, \cdots$ is an external countable sequence of internal subsets of Ω, $\Lambda_{(\cdot)} : {}^{\sigma}\mathbb{N} \longrightarrow {}^*P(\Omega)$. We want to extend $\Lambda_{(\cdot)}$ to an internal sequence $\Lambda'_{(\cdot)} : {}^*\mathbb{N} \longrightarrow {}^*P(\Omega)$. Nelson's (E) above says that we can do this provided we can find a formula $\Phi(m,\Theta)$ describing the standard values of Λ_m. Lindstrøm's extension principle II.1.4 says the internal $\Lambda'_{(\cdot)}$ exists. Since $\Lambda'_{(\cdot)}$ is an internal constant, $\Phi(m,\Theta) = [\Theta = \Lambda'_m]$ is such a formula. This seems like cheating relative to pure syntactical rules. It is, but it is correct cheating for bounded IST.

General Loeb sets can be constructed in bounded IST by using Souslin schemes as in Stroyan and Bayod (1986, section 2.2).

Suppose we had *not* included all internal constants from Lindstrøm's polysaturated $^*V(S)$ in the language of bounded IST. Then just by cardinality arguments there are only card(V(S)) formulas and at least $\text{card}^+(V(S))$ internal subsets of Ω. It wouldn't seem that all external sequences $\{\Lambda_m : m \in {}^{\sigma}\mathbb{N}\}$ would be definable in the smaller language. Indeed they are not. One can take a countable ultralimit which is not ω_1-saturated thus has external sequences $\{\Lambda_m^0\}$ with no internal extension. Countable ultralimits satisfy Nelson's axioms when only standard constants are included (see Stroyan and Luxemburg (1976, 7.5.5)), hence $\{\Lambda_m^0\}$ is not definable. Pure axiomatic IST says "definable external sequences have internal extensions". While that is a little more awkward, it amounts to nearly the same thing provided you "don't know" about undefinable external sets.

6. GENERALIZED TRANSFER AND IDEALIZATION

The transfer axiom cannot be applied to sentences with internal constants because $(\exists x)[x = \delta]$ is true, while $(\exists^{st}x)[x = \delta]$ is false when $\delta \approx 0$. Transfer (T) cannot be applied to $\varphi = (\exists^{st}v)[\psi(x,v)]$ because $(\forall^{st}x)(\exists^{st}y)[y = x]$ is true, while $(\forall x)(\exists^{st}y)[y = x]$ is not. However,

$$(T\forall) \qquad (\forall^{st}x)[\Phi(x)] \Leftrightarrow (\forall x)[\Phi(x)]$$

is true provided Φ is a monadic formula with only x free. (Bounded quantifiers are understood in the above abbreviations.) Monadic formulas are defined next. They are a very useful "topological" class.

6.1 Definition

A formula Φ of $L(^*V(S),st(\cdot))$ is called a $\forall^{st}$ *formula* if its reduced form does not contain a $\exists^{st}$ quantifier, that is,

$$\Phi \text{ has reduced form } (\forall^{st}u)[\varphi(u)]$$

with φ internal. (We allow the "empty" case where Φ has no external quantifiers.) A formula is called *monadic* if it is a $\forall^{st}$-formula and contains only standard constants.

Φ is called a $\exists^{st}$-*formula* if its reduced form does not contain a $\forall^{st}$ quantifier, that is,

$$\Phi \text{ has reduced form } (\exists^{st}v)[\varphi(v)].$$

Proof of **(T∀)**. Let $(\forall^{st}x)[\Phi(x)] = (\forall^{st}x)(\forall^{st}u)[\varphi(x,u)]$ be reduced. Fix any standard constants $b = u$. By ordinary transfer $(\forall^{st}x)[\varphi(x,b)] \Leftrightarrow (\forall x)[\varphi(x,b)]$, in other words we have shown $(\forall^{st}u)(\forall x)[\varphi(x,u)]$ and (T∀) follows by interchanging (∀∀).

A dual form of T with $\exists^{st}$ is also true.

$$(T\exists) \qquad (\exists^{st}x)[\Phi(x)] \Leftrightarrow (\exists x)[\Phi(x)]$$

provided Φ is a $\exists^{st}$-formula with only standard constants and only x free. The proof is left to the reader.◂

Idealization has two generalized forms:

$$(I\forall) \qquad (\forall^{st}_{fin}Y)(\exists x)(\forall y \in Y)[\Phi(x,y,v)] \Leftrightarrow (\exists x)(\forall^{st}y)[\Phi(x,y,v)]$$

for any $\forall^{st}$-formula Φ with free variables x,y,v, where bounds are understood on the quantifiers. In this case Φ may contain internal constants. The syntactical rule means that if internal constants are substituted for any free variables, $c = v$, then both sides have the same truth value. The dual form for a $\exists^{st}$-formula Ψ is:

$$(\mathrm{I}\exists) \qquad (\exists^{st}_{fin}Y)(\forall x)(\exists y \in Y)[\Psi(x,y,v) \Leftrightarrow (\forall x)(\exists^{st}y)[\Psi(x,y,v)].$$

Proof of **(I∀)**. Substitute any internal $c = v$ for the free variables and write the left side of the implication in reduced form:

$$(\forall^{st}_{fin}Y)(\exists x)(\forall y \in Y)[\Phi(x,y,c,u)]$$

$$\Leftrightarrow \quad (\forall^{st}_{fin}Y)(\exists x)(\forall^{st}u)(\forall y \in Y)[\varphi(x,y,c,u)] \qquad (\forall\forall)$$

$$\Leftrightarrow \quad (\forall^{st}_{fin}Y)(\forall^{st}_{fin}U)(\exists x)(\forall y \in U)(\forall y \in Y)[\varphi(x,y,c,u)] \qquad (\mathrm{I})$$

$$\Leftrightarrow \quad (\exists x)(\forall^{st}y)(\forall^{st}u)[\varphi(x,y,c,u)]. \qquad (\mathrm{I})◂$$

Monadic formulas are so named because they define monads (see 6.5). The more general class of $\forall^{st}$-formulas has two useful syntactical descriptions which we give next. This broadens the use of the axiom of the ideal point. If, in addition, such formulas only have standard constants, these characterizations give us a syntactical description of monads.

In our description of the reduction algorithm we only wanted the standard predicate appearing as bounds on external quantifiers. Now we want $st(\cdot)$ inside. We may write each formula in an equivalent form with the standard predicate $st(\cdot)$ inside and using only the connectives $\neg, \wedge, \vee$. For example, $a \Rightarrow b$ is replaced by $\neg a \vee b$,

$$(\forall x \in^{\sigma} B)[\cdots] \longleftrightarrow (\forall x \in B)[st(x) \Rightarrow \cdots] \longleftrightarrow (\forall x \in B)[\neg st(x) \vee \cdots]$$
$$(\exists x \in^{\sigma} B)[\cdots] \longleftrightarrow (\exists x \in B)[st(x) \wedge \cdots].$$

6.3 Definition

We say that the standard predicate *at most occurs negatively* in a formula Φ of $L({}^*V(S), st(\cdot))$ if Φ may be written in an equivalent form with $st(\cdot)$ and only the connectives $\neg, \wedge, \vee$ in which the expressions $st(t)$ only occur in the scope of an odd number of negations, $\neg$. (We allow the "empty" case where $st(\cdot)$ does not occur at all.)

The reduced form $(\forall^{st}x)[\varphi]$, with φ internal, has $st(\cdot)$ occurring negatively, $(\forall x \in B)[\neg st(x) \vee \varphi]$.

Every formula of predicate calculus may be written in an equivalent form with all the quantifiers on the left. This is called the prenex normal form. This result applies to the four quantifiers of $L({}^*V(S), st(\cdot))$, because they are abbreviations for ordinary quantifiers, $\forall, \exists$, with two kinds of bounds, $x \in B$ and $x \in^{\sigma} B$. Putting a formula in prenex form can be quite confusing in practice.

Since we want our syntactical rules to apply to formulas with free variables and since we sometimes wish to use formulas with $st(\cdot)$ inside, rather than as bounds on quantifiers, we have a technical problem to dispose of. The formula $\neg st(x)$ with free variable x cannot be converted into a bounded quantifier formula with $st(\cdot)$ only in quantifier bounds. We might try $(\forall^{st} y)[y \neq x]$ or $\neg(\exists^{st} y)[y = x]$, but what bound do we put on the quantifier? None exists. However, if the free variable is bounded in the sense that there is a standard B such that $x \in B$, then the formula has an external quantifier form:

$$[x \in B \wedge \neg st(x)] \Leftrightarrow (\forall y \in^{\sigma} B)[y \neq x].$$

The next two results are due to Benninghofen & Richter (1985).

6.4 Proposition

Let Φ *be a formula of* $L({}^*V(S), st(\cdot))$ *with only bounded free variables. Then the following are equivalent.*

a) Φ *is a* $\forall^{st}$*-formula.*

b) Φ *is equivalent to a formula with* $st(\cdot)$ *only in quantifier bounds whose prenex form does not contain* $\exists^{st}$ *(only* $\forall, \exists$ *and* $\forall^{st}$*).*

c) Φ *at most contains the standard predicate negatively.*

Proof. (a) $\Rightarrow$ (b) is clear and (b) $\Rightarrow$ (a) follows from repeated uses of (I) and interchange of like quantifiers. (b) $\Rightarrow$ (c) is also clear, just re-writing $(\forall x \in^{\sigma} B)[\cdots]$ as $(\forall x \in B)[\neg st(x) \vee \cdots]$.

(c) $\Rightarrow$ (a): First, replace quantified variables so that all quantifiers bind different variables. Next, replace each $st(t)$ by $(\exists v \in^{\sigma} B)[v = t]$, with still different variables for each new quantifier. Bounds on these quantifiers exist because of the bounds on the free variables of Φ. Bring the negations all the way in to the atomic formulas, leaving an equivalent form of Φ written with $(\forall)(\exists)(\forall^{st}), \wedge, \vee$ and positive and negative internal atomic formulas. The prenex form of this formula is obtained simply by moving all these quantifiers to the left.◂

6.5 Definition

A set $\mu \in V(^*S)$ is called a *monad* if there is a standard set U and a standard family $\mathcal{F}$ of subsets of U and that $\mu = \{x \in U : (\forall^{st} F \in \mathcal{F})[x \in F]\} = \cap^{\sigma}\mathcal{F}$.

The set of infinitesimal *real numbers is a monad where $\mathcal{F} = \{(-\varepsilon,\varepsilon): 0 < \varepsilon < {}^*\mathbb{R}\}$, so $o = \cap[(-\varepsilon,\varepsilon): 0 < \varepsilon \in {}^{\sigma}\mathbb{R}]$. The halo of *reals infinitely close to an unlimited *real, $o + h = \text{hal}(h) = \{x \in {}^*\mathbb{R} : x \approx h\}$, is *not* a monad (i.e., no standard $\mathcal{F}$ exists so that $o + h = \cap\, {}^{\sigma}\mathcal{F}$.) Neighborhood monads in topological spaces are used by Lindstrøm in his Chapter III.2.

6.6 The Syntactical Theorem on Monads

Let U *be a standard entity in* $V(^*S)$. *A set* $\mu \subseteq U$ *is a monad if and only if there exists a formula* $\Phi(x)$ *of* $L(^*V(S),\text{st}(\cdot))$ *with only* x *free, at most containing* $\text{st}(\cdot)$ *negatively, and containing only standard constants such that* $\mu = \{x \in U : \Phi(x)\}$. *In other words, monadic formulas define monads.*

Proof. We may assume $\Phi(x)$ is written in reduced form $(\forall^{st} v \in V)[\varphi(x,v)]$ by Proposition (6.4). The sets $U_v = \{x \in U: \varphi(x,v)\}$ for $v \in {}^{\sigma}V$ form a standard filter base where $\mu = \cap[U_v : v \in {}^{\sigma}V] = \{x \in U : \Phi(x)\}$.◂

7. PERMANENCE PRINCIPLES

Theorem 5.2 above has been proved in bounded IST, but, as it is purely syntactical, it is of course also valid in IST. It leads to the following classification of external formulas of IST:

a) Formulas whose reduced form contains no $\exists^{st}$ quantifiers (called $\forall^{st}$ formulas in the case of $L(^*V(S),\text{st}(\cdot))$).

b) Formulas whose reduced form contains no $\forall^{st}$ quantifiers (called $\exists^{st}$ formulas in the case of $L(^*V(S),\text{st}(\cdot))$).

c) Formulas whose reduced form contains $\forall^{st}$ as well as $\exists^{st}$ quantifiers.

We said that we often use external sets in IST as abbreviations of external formulas. Interactions between internal and external sets often give rise to permanence principles. Roughly speaking, permanence principles say that some phenomenon continues to hold past some point. The most important permanence principle, the so-called *Fehrele's principle*, asserts that an external set which can be defined by a $\forall^{st}$ formula (a *halo*) cannot be defined by an $\exists^{st}$ formula (a *galaxy*).

Before stating this principle we prove the following permanence principle: *any internal property that holds for all limited numbers also holds up to some unlimited number.* Our proof is based on a tautological but useful fact:

7.2 **Cauchy's Principle**

No external set is internal.

Proof of the permanence principle. We know (from a syntactical argument in Lindstrøm's article) that the set of limited numbers $\mathbb{G}$ is external (since if it was internal it would have a supremum and that sup could be neither limited nor unlimited). Assume that some internal property $\Phi(x)$ holds for all real limited x. The set $\{y \in \mathbb{R}^+ : \forall x(|x| \le y \Rightarrow \Phi(x))\}$ is internal and contains the external set $\mathbb{G}$. As the internal set may not be equal to $\mathbb{G}$ (by Cauchy's Principle), it contains an unlimited h and thus $\Phi(x)$ holds for all $|x| \le h$.◂

7.3 **Definition**

Let X be a standard set and $(A_x)_{x\in X}$ be an internal family of internal sets. Then $H = \bigcap_{x\in {}^\sigma X} A_x$ is called a *prehalo* and a *halo* if it is actually external. And $G = \bigcup_{x\in {}^\sigma X} A_x$ is called a *pregalaxy* and a *galaxy* if it is actually external.

Monads are special kinds of halos, obtained by assuming the family (A_x) to be standard. Of course there are halos which are not monads. One of those, which plays an important role in asymptotics, is the so-called *ε-microhalo* $\varepsilon\text{-}m$ defined as follows: Let $\varepsilon > 0$ be infinitesimal,

$$\varepsilon\text{-}m = \bigcap_{n\in {}^\sigma\mathbb{N}} [-\varepsilon^n, \varepsilon^n] = \{x,\ x \in \mathbb{R} \wedge \forall^{st} n\ |x| < \varepsilon^n\}.$$

Halos and galaxies are the two kinds of external sets corresponding respectively to the two first classes of external formulas in the classification above.

7.4 **Fehrele's Principle**

No halo is a galaxy.

Proof (van den Berg (1987)).

We shall prove that if G is a galaxy and H a halo such that $G \subseteq H$, then there exists an internal set I such that $G \subseteq I \subseteq H$. Thus, by the Cauchy Principle, G is strictly contained in H, $G \subset H$.

Let X, Y, (A_x), (B_y) be such that $G = \bigcup_{x \in {}^\sigma X} A_x$ and $H = \bigcap_{y \in {}^\sigma Y} B_y$ and let R be the *internal* relation on $X \times Y$ defined by

$$R(x,y) = [A_x \subseteq B_y].$$

Since $G \subseteq H$, we have $\forall^{st} x\ \forall^{st} y\ \ R(x,y)$. Moreover, for any standard finite sets $X_0 \subset X$ and $Y_0 \subset Y$ we have $R(x,y)$ for all $(x,y) \in X_0 \times Y_0$. It follows from the idealization axiom, applied to relation

$$F(Z,u) = [u \in Z \wedge Z \subseteq R \wedge Z \text{ is a cartesian product}]$$

that there exist internal sets X_1 and Y_1, ${}^\sigma X \subset X_1 \subseteq X$, ${}^\sigma Y \subset Y_1 \subseteq Y$ such that $R(x,y)$ holds for all $(x,y) \in X_1 \times Y_1$. Then $I = \bigcup_{x \in X_1} A_x$ (or $I = \bigcap_{y \in Y_1} B_y$) is an internal set such that $G \subset I \subset H$.◂

REFERENCES

Behrens, M. (1974). A local inverse function theorem, *Victoria Symposium on Nonstandard Analysis*, (eds. Hurd & Loeb), *Springer Lecture Notes in Mathematics* **369**.

Benninghofen, B. & Richter, M.M. (1985). A general theory of superinfinitesimals, *Fund. Math.* (to appear).

van den Berg (1987). *Nonstandard Asymptotic Analysis*, *Springer Lecture Notes in Mathematics* **1249**.

Lawler, G.F. (1980). A self-avoiding random walk, *Duke Math. J.* **47**, 655-693.

Nelson, E. (1977). Internal set theory, *Bull. Amer. Math. Soc.* **83**, 1165-1198.

Nelson, E. (1980). The syntax of nonstandard analysis, *Ann. Pure Appl. Logic* (to appear).

Robinson, A. (1973). Metamathematical problems, *J. Symbolic Logic* **38**, 500-515.

Stroyan, K.D. & Bayod, J.M. (1986). *Foundations of infinitesimal stochastic analysis*, North-Holland, Amsterdam.

Stroyan, K.D. & Luxemburg W.A.J. (1976). *Introduction to the Theory of Infinitesimals*, Academic Press, New York.

Some Introductions to Internal Set Theory

Cartier, P. (1982). Perturbations singulières des équations différentielles ordinaires et analyse nonstandard, *Astérisque* **92-93**.

Diener, F. (1983). *Cours d'Analyse Non-Standard*, Office des Publications Universitaires, Alger.

Diener, F. & Reeb G. (198?). *Analyse Non-Standard*, Hermann (Paris), to appear.

Diener, M. (1985). *Une Initiation aux Outils Non Standard Fondamentaux*, O.P.U.(Alger) et CNRS (Paris), 9-11.

Lutz, R. & Goze M. (1981). *Nonstandard analysis: a practical guide with applications*, *Springer Lecture Notes in Mathematics* **881**.

Nelson, E. (1977). Internal set theory, see references above.

Richter, M. (1982). *Ideal Punkte, Monaden und Nichtstandard Methoden*, Vieweg, Wiesbaden (1982).

Robert, A. (1985). *Analyse Non Standard*, Presses Polytechniques Romande, Lausanne.

Robert, A. (1988). *Nonstandard Analysis*, Wiley, New York & Chichester. (English translation of Robert (1985).)

Zvonkin, A.K. & Shubin, M.A. (1984). Nonstandard analysis and singular perturbations of ordinary differential equations, *Russian Math. Surveys* **39**(2), 69-131.

SOME ASYMPTOTIC RESULTS IN ORDINARY DIFFERENTIAL EQUATIONS

FRANCINE AND MARC DIENER

The theory of ordinary differential equations has benefited in the last few years from the use of nonstandard methods, especially the theory of singular perturbations.

The simplest type of singular peturbation is the equation

$$\varepsilon y' = f(x,y) \tag{1}$$

for small $\varepsilon > 0$. In classical mathematics, the fact that ε is assumed to be small is expressed by letting ε tend to 0. In nonstandard analysis, this is done by chosing ε infinitesimal, but fixed once for all. This very simple idea, due to G. Reeb, is very fruitful, especially when f depends on parameters.

As an example of nonstandard reasoning in this context, we will give here the proof of one of the central theorems of that theory: the existence of an expansion in powers of ε for the slow solutions of equation (1)(Theorem 3.1).

In the first section, we shall describe the main properties of the solutions of (1). In the classical approach, the point that is difficult to treat rigorously is the matching of the slow and the fast motions of the solutions. In nonstandard analysis, this is done by permanence principles. The second section is devoted to the description of the main tool: the ε-shadows expansions. These expansions are the nonstandard analogs of asymptotic expansions in powers of ε, in the sense that they are equivalent for standard objects (as are S-continuity and continuity of functions). They give good approximations of solutions, and are rather easy to compute once the crucial question of existence has been solved. This question will be dealt with in the third part, with Theorem 3.1.

Let us point out that most of the nonstandard results in ODE (see the bibliography) have been obtained using a version of nonstandard analysis called IST (*Internal Set Theory*), which is somewhat different from the one adopted in the remainder of this book. For a short introduction to IST, see the paper by K. Stroyan & F. Diener in this volume.

1. PROPERTIES OF THE TRAJECTORIES

Let $\varepsilon > 0$ be an infinitesimal constant, U a standard open subset of $\mathbb{R}^2$ and f: $U \to \mathbb{R}$ an internal C^∞ function. Assume that at all near standard points in U, f and f' are S-continuous and have limited values. Let $f_0: U \to \mathbb{R}$ be the shadow of f. Let $\overline{\mathbb{R}} = \mathbb{R} \cup \{-\infty\} \cup \{+\infty\}$; $x \approx \pm\infty$ means that x is unlimited and that $\pm x > 0$. Let $\mathcal{F}$ be the boundary of U in $\overline{\mathbb{R}}^2$.

1.1 Definition

We shall call the standard set $\mathcal{L} = \{(x,y) \in U\text{: } f_0(x,y) = 0\}$ the *slow curve*. We shall call the (external) set

$$\mathcal{G} = \{(x,y) \text{ near standard in U such that } f_0(x,y) \neq 0\}$$

the *fast galaxy*.

We shall denote by $\mathcal{H}$ the complement of $\mathcal{G}$ in the closure of U in $\overline{\mathbb{R}}^2$.

For all $M_0 = (x_0, y_0)$ the equation

$$y' = f(x,y)/\varepsilon \qquad (1)$$

has a solution $x \mapsto y(x)$, defined on some maximal interval $I \subset \mathbb{R}$, such that $y(x_0) = y_0$. The sets $\gamma^+ = \{(x,y(x)),\ x \in I,\ x \geq x_0\}$ and $\gamma^- = \{(x,y(x)),\ x \in I,\ x \leq x_0\}$ are called the *positive half-trajectory* (or sometimes just *trajectory*) and the *negative half-trajectory* of M_0 respectively.

As ε is infinitesimal, the shadow of any trajectory of (1) can be described in the following way: it is built up of segments of the slow-curve connected by segments of vertical lines.

To prove this, one has to study the transition behaviour of the solutions going from $\mathcal{G}$ to $\mathcal{H}$ and vice-versa. In a sense that will be made precise below, the trajectory of a point of $\mathcal{G}$ is quasi vertical (i.e. y' is unlimited) as long as it stays in $\mathcal{G}$; it has to leave $\mathcal{G}$ to enter $\mathcal{H}$

(Theorem 1.5) and is trapped in $\mathcal{R}$ as long as $°(f'_y)$ keeps its sign (Proposition 1.3).

1.2 **Definition**

Let $\gamma^+ = \{(x,y(x))\}$ be the positive half-trajectory of $M_0 = (x_0,y_0)$. We say that a point $\underline{M} = (\underline{x},\underline{y}) \in \bar{\mathbb{R}}^2$ is an *entrance point* of γ^+ in $\mathcal{R}$ if $\underline{M}$ is a standard point of $\mathcal{L} \cup \mathcal{F}$ and if there exists x_1 such that $(x_1,y(x_1)) \approx \underline{M}$ and such that for all $x \in [x_0,x_1]$,

$$(x,y(x)) \in \mathcal{R} \Rightarrow [(\xi,y(\xi)) \approx \underline{M} \text{ for every } \xi \in [x,x_1]].$$

One defines analogously an *exit point of* γ^- *from* $\mathcal{R}$.

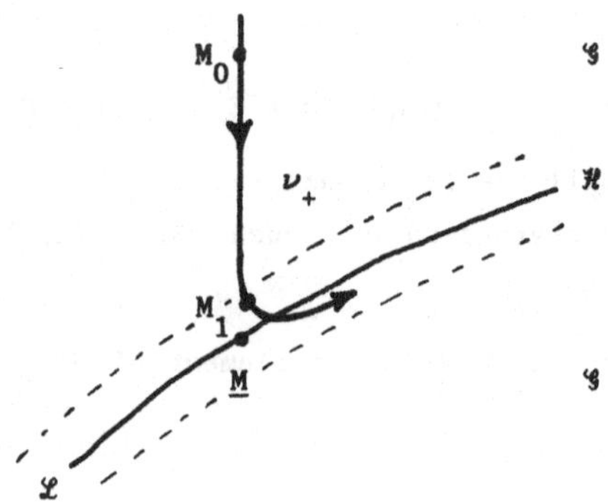

Figure 1. Entrance point $\underline{M}$ of the half-trajectory of $M_0 \in \mathcal{G}$.

1.3 **Proposition**

Any half-trajectory of any M_0 *in* $\mathcal{G}$ *has at most one entrance or exit point. If* $\underline{M}$ *is the entrance point and* $\underline{M} \in \mathcal{L}$ *then* $(f_0)'_x(\underline{M}) \le 0$. *If* $\underline{M}$ *is the exit point and* $\underline{M} \in \mathcal{L}$ *then* $(f_0)'_x(\underline{M}) \ge 0$.

Proof. Uniqueness: assume $\underline{M}' \approx (x'_1,y(x'_1))$ and $\underline{M}'' \approx (x''_1, y(x''_1))$ are two entrance points, $x_0 < x'_1 < x''_1$. As $\underline{M}''$ is an entrance point and $(x'_1, y(x'_1)) \approx \underline{M}' \in \mathcal{R}$, one has $\underline{M}' \approx (x'_1,y(x'_1)) \approx \underline{M}''$, and thus $\underline{M}' = \underline{M}''$. Uniqueness of the exit point follows analogously.

Let $\underline{M}$ be the entrance point of $\gamma^+ = \{(x,y(x)),\ x \ge x_0\}$. We have to prove that $(f_0)'_x(\underline{M}) \le 0$. Assume that $(f_0)'_x(\underline{M}) > 0$. As f_0 and $\underline{M}$ are standard, this implies that $(f_0)'_x(\underline{M}) \not\approx 0$ and thus there exists a standard neighbourhood V of $\underline{M}$ on which $(f_0)'_x(\underline{M}) > 0$ and $(f_0)'_x(\underline{M}) \not\approx 0$.

By hypothesis there exists x_1 such that $M_1 = (x_1,y(x_1)) \approx \underline{M}$. The set

$$\{x'\in[x_0,x_1]:\ (\xi,y(\xi)) \in \mathcal{H} \text{ for any } \xi \in [x'.x_1]\}$$

is not empty. It is external since $(x_0,y_0) \notin \mathcal{H}$ and it is contained in

$$\{x' \in [x_0,x_1]:\ (\xi,y(\xi)) \in V \text{ for any } \xi \in [x',x_1]\}.$$

By Cauchy's principle, the inclusion is strict. Thus there exists $M' = (x',y(x'))$ on γ^+ such that $M'\in\mathcal{G} \cap V$ and such that $(\xi,y(\xi))$ stays in V for all $\xi \in [x',x_1]$.

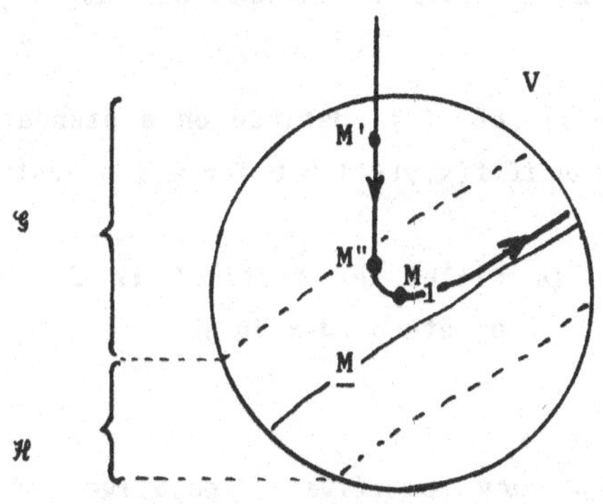

Figure 2

Let us consider the function $x \mapsto \varphi(x) = f_0^2(x,y(x))$ on $[x',x_1]$. Since $M' \in \mathcal{G}$ and $\varphi(\xi) \approx 0$ for all ξ such that $(\xi,y(\xi)) \in \mathcal{H}$, $\varphi(x') \not\approx 0$ (and >0). In particular, $\varphi(x_1) \approx 0$. But

$$\begin{aligned}\varphi'(x) &= \frac{d}{dx} f_0^2(x,y(x)) \\ &= 2[f_0^2(x,y(x))\ (f_0)'_y(x,y(x))/\varepsilon + f_0(x,y(x))(f_0)'_x(x,y(x)))]\end{aligned}$$

Thus $\varphi'(x)$ has the same sign as $(f_0)'_y(x,y(x))$ provided $(x,y(x)) \in \mathcal{G}$, because in that case $f_0^2(x,y)(x)) \not\approx 0$ and f_0 and f'_x are limited. So the set of all $x \geq x'$ for which φ is increasing contains all $x \geq x'$ such that $(x,y(x)) \in \mathscr{G}$, and then also an x'' which can be chosen in $[x',x_1]$ such that $M'' = (x'',y(x'')) \in \mathscr{H}$. So, on one hand $\varphi(x'') > \varphi(x')$ and thus $\varphi(x'') \not\approx 0$, and on the other hand $\varphi(x'') \approx 0$. This is a contradiction. Thus $(f_0)'_x(\underline{M}) \leq 0$.◂

Remark. This proposition is often used as an "external trajectory trap" as we shall see in the proof of Theorem 3.1. Suppose for example that a trajectory enters $\mathscr{H}$ at $\underline{M} = (\underline{x},\underline{y})$ and that the sign of $(f_0)'_y$ does not change between $\underline{x}$ and some $x' > \underline{x}$. It follows from the previous proposition that the trajectory is trapped in the halo of $\mathscr{H}$ until x'.

1.4 Definitions

A point M of $\mathscr{L}$ is called *non-critical* if $(f_0)'_y({}^oM)$ is non-zero.

A solution $x \mapsto y(x)$ of (1) defined on a standard interval $I \subset \mathbb{R}$ is called a *slow solution* if $f(x,y(x)) \approx 0$ for all nearstandard x in I.

A slow solution is called *non-critical* if $f(x,y(x))$ is a non-critical point of $\mathscr{L}$ for all near-standard x in I.

1.5 Theorem

For any half-trajectory, positive or negative, of any point $M_0 = (x_0,y_0)$ of $\mathscr{G}$ defined on a non infinitesimal interval I there exists an entrance point in $\mathscr{H}$ or an exit point $\underline{M} = (\underline{x},\underline{y})$ from $\mathscr{H}$ such that $\underline{x} \approx x_0$.

Proof. Let $\gamma^+ = \{x,y(x)), x \geq x_0, x \in I\}$ be the positive half-trajectory of M_0. The set $g = \{x \in I: x \geq x_0, \forall\xi \in [x,x_0], (\xi,y(\xi)) \in \mathscr{G}\}$ is contained in the halo of x_0. Indeed, if there is x' in g such that $x' \gg x_0$ then

$$|y(x')-y(x_0)| \geq |x'-x_0| \operatorname{Inf}\{|y'(\xi)|: \xi \in [x',x_0]\} \approx +\infty$$

which is impossible because $y(x')$ and $y(x_0)$ are limited. But I is not infinitesimal and thus it contains all $x \approx x_0$, $x \geq x_0$. Thus, as g is a pregalaxy, by Fehrele's or Cauchy's principle, there exists $x_2 \approx x_0$ such that $(x_2, y(x_2)) \notin \mathcal{G}$. As f is not infinitesimal on $\mathcal{G}$, one can assume, after having possibly replaced x_2 by another point of $[x_0,x_2]$, that $f(x,y(x))$ exists and does not vanish on $[x_0,x_2]$. Thus $y(x)$ is monotonic on this interval. Assume for example that y is increasing.

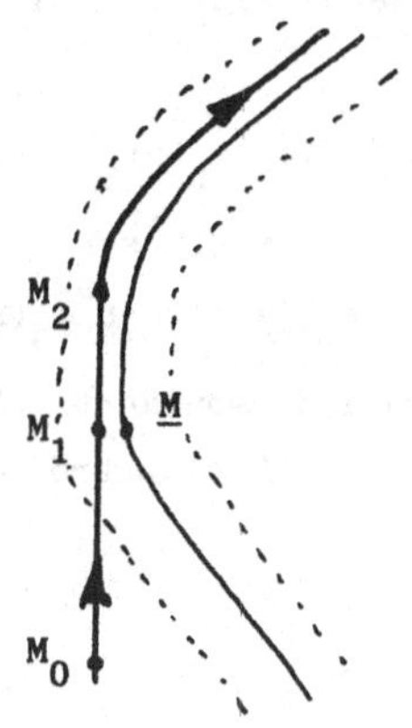

Figure 3

The intersection of the standard line segment $\{x_0\} \times [y_0, {}^{o}(y(x_2))]$ with $\mathcal{L} \cup \mathcal{F}$ is non empty since it contains $(x_0, {}^{o}(y(x_2)))$. Let $\underline{M}(\underline{x},\underline{y})$ be its element of minimal ordinate. Let us show that $\underline{M}$ is an entrance point of γ^+ in $\mathcal{H}$.

As y is monotonic from $[x_0,x_2]$ on $[y_0,y(x_2)]$, there exists $x_1 \in [x_0,x_2]$ such that $y(x_1) \approx \underline{y}$. Moreover, if for any $x \in [x_0,x_1]$, one has $(x,y(x)) \in \mathcal{H}$, then ${}^{o}(x,y(x)) \in \mathcal{L} \cup \mathcal{F}$ and thus by definition of $\underline{M}$, ${}^{o}y(x) = \underline{y}$. But as $y(x) \le y(x_1) \approx \underline{y}$, it follows that ${}^{o}y(x) = \underline{y}$. By monotonicity of y, this is still true for all $\xi \in [x,x_1]$.◂

2. ε-SHADOWS EXPANSIONS OF IMPLICIT FUNCTIONS

In section 1 we studied the shadows of the trajectories, which are the first approximation. In order to consider the following approximations we have to define the kind of expansions we are interested in. For solutions of equations (1) it is natural to introduce the so called *ε-shadows expansions* (van den Berg (1987), Diener F. (1983)).

2.1 Definition

Let k and n be standard integers, U a standard open subset of $\mathbb{R}^k$ and F: U → $\mathbb{R}$ an internal function. We say that F admits an *ε-shadows expansion to the* n^{th} *order* if there exist functions $F_1,\ldots,F_n$ and δ from U to $\mathbb{R}$ such that:

(i) the functions $F_1,\ldots,F_n$ are standard,

(ii) $\delta(x) \approx 0$ for all $x \in U$, near standard in U,

(iii) $F(x) = F_1(x) + \varepsilon F_2(x)+\ldots+\varepsilon^n F_n(x)$ for all $x \in U$.

A function will be called *regular to the* n^{th} *order* if it is C^∞ and if it and all its derivatives of standard order admit an ε-shadows expansion to the nth order.

2.2 Example

The number $F = 1/(1+\varepsilon)$ (corresponding to a constant function F) admits $\sum_{n\geq 0} (-1)^n\varepsilon^n$ as ε-shadows expansion, whereas $F = \sqrt{\varepsilon}$ has just an expansion to the order 0 but not to the order 1. More generally, if there exists a standard C^∞ function Φ such that $F(x) = \Phi(x,\varepsilon)$, then F has an ε-shadows expansion to any standard order n.

The number $F = \exp(1/\varepsilon)$ admits $\sum_{n\geq 0} 0\varepsilon^n$ as ε-shadows expansion, as the number 0 itself: indeed $\exp(1/\varepsilon)/\varepsilon^n \approx 0$ for all standard n. This example shows that two different functions may have the same expansion.

On the other hand if F(x) has an ε-shadows expansion, then this expansion is unique. Indeed, by external induction, it suffices to show that if for all $x \in U$, $\varepsilon^n(F_n(x)+\delta(x)) = \varepsilon^n(F'_n(x)+\delta'(x))$ then $F_n(x) = F'_n(x)$. This is true since F_n and F'_n are standard.

Let $d \geq 1$ and n be standard integers, U a standard open subset of $\mathbb{R}^d \times \mathbb{R}$ and F: U → $\mathbb{R}$ a $\mathscr{C}^\infty$ function regular to the nth order, with coefficients $(F_p)_{0\leq p\leq n}$. Let $(\underline{x},\underline{y})$ be standard and assume that $hal((\underline{x},\underline{y})) \subseteq U$ and $F(\underline{x},\underline{y}) \approx 0$. Let $L = \partial F/\partial y(\underline{x},\underline{y})$. Assume also that L is appreciable (i.e. limited and non-infinitesimal) and let L_0 be the shadow of L.

2.3 **Theorem**

There exists an internal, real valued, function $\bar{y}$ *defined on* $hal(\underline{x})$ *at least, regular to the* n^{th} *order, such that for all* $x \in hal(\underline{x})$, $\bar{y}(x) \in hal(\underline{y})$ *and* $F(x,\bar{y}(x)) = 0$. *Moreover, if* $(x,y) \in hal(\underline{x},\underline{y})$ *then* $F(x,y) = 0$ *if and only if* $y = \bar{y}(x)$.

Proof. Let $G: U \to \mathbb{R}^d \times \mathbb{R}$, $(x,y) \mapsto G(x,y) = (x,F(x,y))$. This is a $\mathcal{C}^\infty$ function, and $K := DG(\underline{x},\underline{y}) = (pr_x, \frac{\partial G}{\partial x}(x,y)+L\cdot): \mathbb{R}^d\times\mathbb{R} \to \mathbb{R}^d\times\mathbb{R}$ has a linear bijective shadow. By the Nearstandard Local Inverse Function Theorem (see the paper of Diener & Stroyan in this book) if follows that G admits a local inverse H defined on $hal(\underline{x},F(\underline{x},\underline{y})) = hal(\underline{x},0)$ at least, such that $G\circ H = Id$ and $H\circ G = Id$ on the domains of these functions. It suffices to let $\bar{y}(x) = pr_y\circ H(x,0)$. Let us show that $\bar{y}$ is regular to the nth order. We begin with a lemma.

Let $a_0(x,y_0) = F_0(x,y_0)$

and
$$a_r(x,y_1,\ldots,y_r) = F_r(x,y_0) + \sum_{\substack{k\geq 0 \\ m,i_1,\ldots,i_m\geq 1 \\ k+i_1+\ldots+i_m = r}} \frac{1}{m!}\frac{\partial^m F k}{\partial y^m}(x,y_0(x))\cdot y_{i_1}\cdots y_{i_m}$$

Lemma

For any standard sequence $(y_i(x))_{1\leq i\leq n}$ *of functions defined on some neighbourhood of* $\underline{x}$, *and all* $p \leq n$,

$$F(x,y_0(x)+\ldots+\varepsilon^p y_p(x)) = \sum_{r\leq p} \varepsilon^r a_r(x,y_1(x),\ldots,y_r(x)) + \varepsilon^{p+1}R_{p+1}(x)$$

where $\varepsilon^{p+1}R_{p+1}(x) \approx 0$, *and* $R_p(x) \in S^0$, *for each* $p \leq n$.

Proof. This follows from a straightforward computation, by expanding F in ε shadows to the order p, and then introducing a Taylor expansion of each (standard) function $F_0, F_1,\ldots,F_p$ in the neighbourhood of $(x,y_0(x))$.◂

Proof of Theorem 2.3 (concluded). We shall prove, by induction on $p \le n$ that $\bar{y}$ has an ε-shadows expansion to the nth order,

$$\bar{y}(x) = y_0(x) + \varepsilon y_1(x) + \ldots + \varepsilon^n y_n(x) + \varepsilon^n \varphi \text{ (where } \varphi \approx 0)$$

such that for all $p \le n$ we have $a_p(x, y_1(x), \ldots, y_p(x)) = 0$.

Let $p = 0$. As H is S-continuous, $\bar{y}$ $(= pr_y \circ H)$ is S-continuous. Let y_0 be its shadow. For all x in the domain of $\bar{y}$, we have $F_0(x, y_0(x)) \approx F(x, \bar{y}(x)) = 0$, and so $F_0(x, y_0(x)) = 0$ for standard x, and thus for all x, by transfer.

Now assume $\bar{y}$ has an expansion to the pth order $(0 < p < n)$ and $a_r(x, y_1(x), \ldots, y_r(x)) = 0$ for all $r \le p$. Then

$$\begin{aligned} 0 &= F(x,\bar{y}(x)) \\ &= F(x,\bar{y}(x)) - F(x, y_0(x) + \varepsilon y_1(x) + \ldots + \varepsilon^p y_p(x)) \\ &\qquad + F(x, y_0(x) + \varepsilon y_1(x) + \ldots + \varepsilon^p y_p(x)) \\ &= \frac{\partial F}{\partial y}(x,\eta)[\bar{y}(x) - (y_0(x) + \varepsilon y_1(x) + \ldots + \varepsilon^p y_p(x))] \\ &\qquad + \sum_{r \le p} \varepsilon^r a_r(x, y_1(x), \ldots, y_r(x)) + \varepsilon^{p+1} R_{p+1} \end{aligned}$$

with $\eta \approx y_0(x)$ between $\bar{y}(x)$ and $y_0(x) + \varepsilon y_1(x) + \ldots + \varepsilon^p y_p(x)$

$$= \frac{\partial F}{\partial y}(x,\eta)[\bar{y}(x) - (y_0(x) + \varepsilon y_1(x) + \ldots + \varepsilon^p y_p(x))] + \varepsilon^{p+1} R_{p+1}.$$

It follows that

$$\frac{\bar{y}(x) - (y_0(x) + \varepsilon y_1(x) + \ldots + \varepsilon^p y_p(x))}{\varepsilon^p} = \left[\frac{\partial F}{\partial y}(x,\eta)\right]^{-1} [R_{p+1}(x)]$$

$$\approx \left[\frac{\partial F_0}{\partial y}(x, y_0(x))\right]^{-1} [{}^{\circ}R_{p+1}(x)].$$

Thus $\bar{y}$ is regular to the $(p+1)^{th}$ order, and

$$y_{p+1}(x) = \left[\frac{\partial F_0}{\partial y}(x, y_0(x))\right]^{-1} [{}^{\circ}R_{p+1}(x)].$$

Finally:

$$\begin{aligned} 0 &= F(x,\bar{y}(x)) \\ &= F(x,\bar{y}(x)) - F(x, y_0(x) + \varepsilon y_1(x) + \ldots + \varepsilon^p y_p(x)) \\ &\qquad + F(x, y_0(x) + \varepsilon y_1(x) + \ldots + \varepsilon^p y_p(x)) \end{aligned}$$

$$= \sum_{r\le p+1} \varepsilon^r a_r(x,y_1(x),\dots,y_r(x)) + \varepsilon^{p+1}\delta$$

$$= \varepsilon^{p+1} a_{p+1}(x,y_1(x),\dots,y_{p+1}(x)) + \varepsilon^{p+1}\delta$$

from the induction hypothesis on the a_p. Thus

$$a_{p+1}(x,y_1(x),\dots,y_{p+1}(x)) = 0$$

as this function of x is standard.

Thus $\bar{y}$ has an expansion to the nth order. It follows immediately that the derivatives of $\bar{y}$ have also an expansion to the nth order, as can be seen by external induction on the order of derivation on $\bar{y}$, just by deriving the relation $F(x,\bar{y})) = 0$.◂

3. EXISTENCE OF AN EXPANSION

For a differential equation of type $y' = F(x,y,\varepsilon)$ where F is regular, any solution with initial conditions having an ε-shadows expansion admits an expansion and, in general, two solutions having different initial conditions have different expansions.

The situation is completely different for equation (1). In general solutions do not have an expansion even of order 0 because they are not S-continuous. But, as we shall see, any slow-solution admits an ε-shadows expansion, and moreover, two slow-solutions infinitely close at one point have the same expansion.

3.1 Theorem

Let $c > 0$ *be a standard integer,* $\varepsilon > 0$ *an infinitesimal,* U *a standard open set of* $\mathbb{R}^2$ *and let* $f\colon U \to \mathbb{R}$ *be regular to the* n^{th} *order. Let* I *be a standard open interval of* $\mathbb{R}$ *and* $\mathbf{y}\colon I \to \mathbb{R}^2$ *a non critical slow solution of*

$$\varepsilon^c y' = f(x,y) \tag{2}$$

Then **y** *has an* ε*-shadows expansion to the* n^{th} *order on* I.

Remark. It might be thought that, under the hypothesis of the theorem, **y** has not only an ε-shadows expansion but also an ε^c-shadows expansion. This is not true: for example, we can see that the solution $\mathbf{y}(x) = x + \varepsilon + \varepsilon^2 + K\exp(x/\varepsilon^2)$ of the equation $\varepsilon^2 y' = y - x - \varepsilon$ defined on $I = \mathbb{R}$ does not admit an ε^2-shadows expansion. Consider the following magnifying glass $\mathcal{M}$: $(x,y) \mapsto (x,y_1 = f(x,y)/\varepsilon)$ and denote by $y = \bar{y}(x,\varepsilon y_1)$ the implicit function defined by $\varepsilon y_1 = f(x,y)$. The mapping $\mathcal{M}$ transforms equation (2) into

$$\varepsilon^c y_1' = f_1(x,y) \tag{3}$$

where $f_1(x,y) = \varepsilon^{c-1} f_x'(x,\bar{y}(x,\varepsilon y_1)) + f_y'(x,\bar{y}(x,\varepsilon y_1))y_1$.

The equation of the slow curve $\mathcal{L}_1$ of (3) is

$$y_1 = \begin{cases} 0 & \text{if } c > 1 \\ (f_0)_x'(x,{}^\circ y(x))/(f_0)_y'(x,{}^\circ y(x)) & \text{if } c = 1 \end{cases}$$

Thus $\mathcal{L}_1$ is the graph of the function defined for all $x \in I$.

3.2 Lemma

The image by the magnifying glass $\mathcal{M}$ *of a non-critical slow solution* $y(x)$ *of* (2) *defined on* I *is, in* I, *a non-critical slow solution* $y_1(x)$ *of* (3).

Proof. Suppose not; i.e. assume that at some nearstandard point ξ in I, $f_1(\xi,y_1(\xi)) \not\approx 0$. We shall see that this implies the existence of some $\xi' \approx \xi$ such that $|y(\xi)-y(\xi')| \approx +\infty$ which is impossible since y is S-continuous.

Since y is a non-critical solution, we have $(f_0)_y'(x,{}^\circ y(x)) \not\approx 0$. Suppose for example that $(f_0)_y'(x,{}^\circ y(x)) > 0$ (repulsive case). It easily follows that $(f_1)_y'(x, {}^\circ y(x)) > 0$ (since y_1 is also repulsive).

Now apply Theorem 1.5 to the trajectory of (3) of the point $(\xi, y_1(\xi))$ belonging to the fast galaxy $\mathcal{G}_1$ of (2). This galaxy contains all points (x,y_1), with $x \in I$ and y_1 limited, that do not belong to the halo of $\mathcal{L}_1$, since the corresponding (x,y) are such that $f(x,y) \approx 0$. By Theorem 1.5 the positive half-trajectory of $(\xi,y(\xi))$ has an entrance point in the complement $\mathcal{H}_1$ of $\mathcal{G}_1$. This point can not be in $\mathcal{L}_1$, because of the sign of ${}^\circ(f_1)_y'$ and Proposition 1.3.

For the same reason the negative half-trajectory has an exit point from $\mathcal{H}_1$ which must be a point of $\mathcal{L}_1$ since $\mathcal{L}_1$ is the graph of a function. Thus there exists $x_1 \approx \xi$ and $x_1' \approx \xi$ such that $f(x,y_1(x_1)) \approx 0$ and $y_1(x_1')$ is unlimited.

Moreover, by Proposition 1.3 and from the sign of $°(f_1)'_y$, $y_1(x_y')$ stay unlimited for all $x \gg \xi$.

As $y_1(x) = f(x,y(x))/\varepsilon$ we have shown that $f(x,y(x))/\varepsilon$ is unlimited for all $x \gg \xi$. As ξ is nearstandard in I, which is open and standard, there actually exist points in I such that $x \gg \xi$. And for these,

$$|y(x) - y(\xi)| = \varepsilon^{1-c}\left|\int_\xi^x f(z,y(z))/\varepsilon\, dz\right| \approx +\infty$$

Thus $|y(x) - y(\xi)| \approx +\infty$ for all $x \gg \xi$. By Feherele's principle, there exists $\xi' \approx \xi$ such that $|y(\xi') - y(\xi)| \approx +\infty$. This ends the proof of the lemma.◂

Proof of Theorem 3.1. (concluded). We shall prove, by induction on $p \leq n$ that $\bar{y}$ admits an ε-expansion to the nth order on I.

It is true for n = 0 by hypothesis since **y** is a slow solution.

Assume that for all $p < n$, the fact that f admits an expansion to the pth order implies that **y** admits an expansion to the pth order. Assume that f has an expansion to the nth order.

Apply the induction hypothesis to equation (3) and its solution y_1. Since f is regular to the nth order, it follows by Theorem 2.1 that the implicit function $\bar{y}$ associated with the magnifying glass $\mathcal{M}$ is also regular to the nth order and thus that f_1 is regular to the nth order. By the lemma 3.2, y_1 is a non-critical slow solution of equation (3) on I. By induction y_1 is regular to the $(n-1)^{th}$ order and thus εy_1 is regular to the nth order. As $y = \bar{y}(x,\varepsilon y_1)$, y is itself regular to the nth order.◂

4. CONCLUSION: APPLICATION TO THE PROBLEM OF STREAMS

The problem of "streams" of ordinary differential equations gives a striking example of application of the previous results. We perceive two such streams on the drawing of the trajectories of

Liouville's equation

$$dY/dX = Y^2 - X.$$

This is a standard equation, which was introduced as an example of a non-solvable equation.

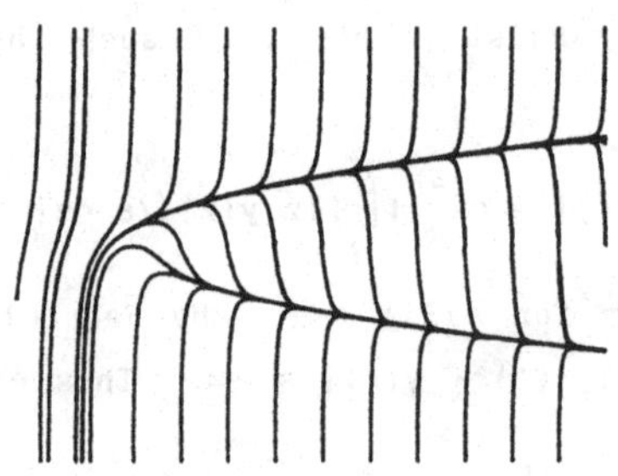

Figure 4

A stream consists of a gathering of the trajectories in a way that seems to defy the uniqueness theorem for solutions. The problem of streams, arising from the observation of computer plottings, has been treated by nonstandard means (see Diener F. (1986), Diener F. (198?), Diener F.& M. (198?), Diener M. (1985), Diener & Reeb (1987)), in the following way. Consider a rational differential equation $dY/dX = F(X,Y)$ with $F = P/Q$, and Q and P being non-zero polynomials. We turn such an equation to an equation of type (1) by using a "macroscope", that is a change of variable of type $x = \varepsilon^a X$, $x = \varepsilon^b Y$. The original equation becomes $dy/dx = f(x,y)/\varepsilon^c$. For a small number of values of the ratio b/a of the integers a and b, depending on F, one has $c > 0$ and f satisfying the hypothesis of theorem 3.1 It turns out that they are precisely the integers such that $dY/dX = F(X,Y)$ admits solutions asymptotic to $kX^{b/a}$ for some k, and those solutions are the streams that we observe. The expansion given by Theorem 3.1 can be pulled back to the (X,Y)'s as asymptotic expansions of the streams. An interesting fact is that these expansions are generally *not* convergent (in particular this is the case for the Liouville equation) but will give, with very few terms, excellent approximates of the "real" trajectories. Non-standard analysis provides also an insight in this amazing question (see van den Berg (1987)) but this is another story!

REFERENCES

Benoit, E. (1981). Tunnels et entonnoires, *C.R.Acad.Sc. Paris* **292**, 283-286.

Benoit, E. (1981). Relation entrée-sortie, *C.R.Acad.Sc. Paris* **293**, 293-296.

Benoit, E. (1983). Systèmes lents-rapides dans $\mathbb{R}^3$ et leurs canards, IIIe rencontre du Schnepfenried, **109-110:2**, *Asterisque*, 159-191.

Benoit, E. (1983). Canards et chaos dans $\mathbb{R}^3$, in *Outils et modèles mathématiques pour l'automatique, l'analyse des systèmes et le traitement du signal, Tome III*, Editions du CNRS, Paris, 335-340.

Benoit, E. (1985). Enlacements de canards *C.R.Acad.Sc. Paris* **300**, 225-230.

Benoit, E. (198?). Canards et enlacements; in *Mathématiques Finitaires et Analyse Non Standard, Journées SMF, Mai 1985*; to appear.

Benoit, E., Callot, J.L., Diener, F. & Diener, M. (1981). Chasse au canard, *Collectanea Mathematica* **31**, Fasc. 1-3, 37-119.

Benoit E., & Lobry, C. (1984). Les canards de $\mathbb{R}^3$, *C.R.Acad.Sc. Paris* **294**, 483-488.

van den Berg, I.P. (1987). *Nonstandard Asymptotic Analysis, Lecture Notes in Mathematics* **1249**, Springer.

Callot, J.L. (198?). Solutions visibles de l'équation de Schrödinger, *Mathématiques finitaires et Analyse Non Standard, Journées SMF, Mai 1985*; to appear.

Callot, J.L., Diener F. & Diener M. (1978). Le problème de la chasse au canards, *C.R. Acad.Sc. Paris* **286**, Série A, 1059-1061.

Callot, J.L. & Sari T. (1983). Stroboscopie infinitésimale et moyennisation dans les systèmes d'équations différentielles à solutions rapidement oscillantes; in *Outils et modèles mathématiques pour l'automatique, l'analyse des systèmes et le traitement du signal, Tome III*, Editions du CNRS, Paris, 345-353.

Cartier, P. (1981). Pertubations singulières des équations différentielles ordinaires et analyse non standard, Séminaire Bourbaki **580**.

Diener, F. (1978). Les équations $\varepsilon x''+(x^2-1)x'^{[s]}+x=0$, *Collectanea Mathematica* **24**:3, 217-247.

Diener, F. (1979). Famille d'équations à cycles limite unique, *C.R.Acad.Sc. Paris* **289**, Série A, 571-574.

Diener, F. (1983). Développements en ε-ombres; in *Outils et modèles mathématiques pour l'automatique, l'analyse des systèmes et le traitement du signal, Tome III*, Editions du CNRS, Paris, 315-328.

Diener, F. (1986). Sauts des solutions des équations $\varepsilon x''=\varphi(t,x,x')$, *SIAM Journal on Math. Anal.* **17**, 533-559.

Diener F. (1986). Propriétés asymptotiques des fleuves, *C.R.Acad.Sc. Paris* **302**, Série I, 55-58.

Diener F. (1987). Fleuves et variétés centrales; in *Actes du Colloque de Dijon , Singularités d'Equations Différentielles, Astérisque*, **150-151**.

Diener, F. (198?). Equations surquadratiques et disparition des sauts, *Cahiers Mathématiques de Paris X* (Preprint).

Diener, F., Canalis, M. & Gaetano, M. (198?). Calcul des valeurs à canards à l'aide de MACSYMA, *Mathématiques finitaires et Analyse Non Standard Journées SMF, Mai 1985*; to appear.

Diener, F. & Diener M. (1983). Sept formules relatives aux canards, *C.R.Acad.Sc. Paris* **297**, Série I, 557-588.

Diener, F. & Diener, M. (1985). A pedestrian proof of the Hopf-bifurcation theorem; in *Singularities & Dynamical Systems*, North-Holland, Amsterdam, 295-302.

Diener, F. & Diener, M. (198?). Fleuves 1-2-3: mode d'emploi, *Mathématiques finitaires et Analyse Non Standard, Journées SMF, Mai 1985*; to appear.

Diener, M. (1983). Canards et bifurcations; in *Outils et modèles mathématiques pour l'automatique, l'analyse des systèmes et le traitement du signal, Tome III*, Editions du CNRS, Paris, 315-328.

Diener, M. (1980). Deux nouveaux phénomènes-canards, *C.R.Acad.Sc. Paris* **290**, Série A, 541-544.

Diener, M. (1984). The unchained canard, or how fast/slow dynamical systems bifurcate, *The Mathematical Intelligencer* **6**, 38-49.

Diener, M. (1985). Determination et existence des fleuves en dimension deux, *C.R.Acad.Sc. Paris* **301**, Série I, 889-902.

Diener, M. (1986). Fleuves macroscopiques, *Publications Mathématiques de l'Université de Paris 6*, **82**, *Séminaire Choquet 1985/86*, 31-39.

Diener, M. & Reeb, G. (1987). Champs polynômiaux: nouvelles trajectoires remarquables, *Bull. Soc. Math. Belgique*, **38**, 131-150.

Lobry, C. & Wallet, G. (198?). La traversée de l'axe imaginaire n'a pas toujours lieu là où l'on croit l'observer, *Mathématiques finitaires et Analyse Non Standard, Journées SMF, Mai 1985* (to appear).

Lutz, R. (1983). L'intrusion de l'analyse non standard dans l'étude des perturbations singulières, *IIIe rencontre du Schnepfenried* **109-110:2**, *Asterisque*, 101-139.

Lutz, R. & Goze, M. (1982). *Non Standard Analysis: a Practical Guide with Applications, Lecture Notes in Mathematics* **881**, Springer, (1982).

Lutz, R. & Sari, T. (1981). Sur le comportement asymptotique des solutions dans un problème aux limites non linéaire. *C.R.Acad.Sc. Paris* **292**, 925-928.

Lutz, R. & Sari, T. (1982). Applications of non standard analysis to boundary value problems in singular thoery; in *Lecture Notes in Mathematics* **942**, Springer, 113-135.

Poston, T. & Diener, M. (1982). On the perfect delay convention; in *Chaos and Order in Nature*, (ed. H. Hacken), Springer.

Reeb, G. (1978). Equations différentielles et analyse non classique; in *Proceedings of the IV International Colloquium on Differential Geometry*, Santiago de Compostella, 1978.

Sanches-Pedreno, S. (198?). Sur certaines équations différentielles hautement non linéaires; in *Mathématiques finitaires et Analyse Non Standard, Journées SMF, Mai 1985* (to appear).

Sari, T. (1983). Sur la théorie asymptotique des oscillations non stationaires; in *IIIe rencontre due Schnepfenried* **109-110**:2, *Asterisque*, 141-158.

Troesch, A. (1982). Etude macroscopique de l'équation de Van der Pol; in *Lecture Notes in Mathematics* **942** 136-144.

Troesch, A. (1984). Etude macroscopique des systèmes différentiels, *Proc. London Math. Soc.* **48**, 207-237.

Troesch, A. (198?). Lorsque les canards naissent dans les tourbillons, *Mathématiques finitaires et Analyse Non Standard, Journées SMF, Mai 1985* (to appear).

Troesch, A. & Urlacher, E. (1978). Perturbations singulières et analyse non classique, *C.R.Acad.Sc. Paris* **286**, 1109-1111.

Troesch, A. & Urlacher, E. (1978). Perturbations singulières et analyse non standard. *C.R.Acad.Sc. Paris* **287**, 937-939.

Urlacher, E. (1984). Equations différentielles due type $\varepsilon x''+F(x')+x=0$, avec ε petit, *Proc. London Mathematical Soc.* **49**, 207-237.

Urlacher, E. (198?). Un système rapidement oscillant, *Mathématiques finitaires et Analyse Non Standard, Journèes SMF, Mai 1985* (to appear).

Wallet, G. (198?). Entrée-sortie dans un tourbillon, *Annales Fourier*, to appear.

Wallet, G. (198?). Dérive lente dans un champ de Liénard, *Mathématiques Finitaires et Analyse Non Standard, Journées SMF, Mai 1985*, to appear.

Zvonkin, A.K., Shubin, M.A. (1984). Non standard analysis and singular perturbations of ordinary differential equations, *Russian Math. Surveys* **39**:2, 69-131.

SUPERINFINITESIMALS AND INDUCTIVE LIMITS

KEITH D. STROYAN

Introduction. This article shows how to apply the ideas of Benninghofen & Richter (1985) to study two kinds of topological inductive limits whose monads do not usually have simple descriptions. The monads of these topologies do contain an easily computed core which we call the *limit infinitesimals*. Benninghofen and Richter's theory of product monads allows us to show that computations with limit infinitesimals extend to the full monad and consequently imply various new standard topological results.

The results of this paper are a summary of joint work with B. Benninghofen and M.M. Richter that will appear elsewhere in greater detail. We thank N. Cutland for his advice and for allowing us to publish his π-Transfer Theorem 2.3.

We shall work in Lindstrøm's polysaturated nonstandard model V(*S). We also need some of the syntactical ideas from the article of Diener & Stroyan in this volume. We shall use the notation and Syntactical Theorem on Monads from that article.

1. MONADS

If $\mathcal{U}$ is an entity of V(S), the set of standard elements of $\mathcal{U}$ is denoted ${}^{\sigma}U = \{U \in {}^*\mathcal{U} : st(U)\} = \{{}^*V : V \in \mathcal{U}\}$.

1.1 Definition

A nonempty set $\mu \in V({}^*S)$ is called a *monad* if and only if there is a standard family $\mathcal{U}$ of subsets of a standard set W such that

$$\mu = \cap{}^{\sigma}\mathcal{U} = \cap[{}^*U : U \in \mathcal{U}].$$

The monad $\mu \subseteq {}^*W$ is nonempty if and only if the corresponding standard family $\mathcal{U}$ is a filter base on W.

The infinitesimals in $^*\mathbb{R}$ form a monad, $o = \cap[^*(-\varepsilon,\varepsilon) : \varepsilon \in \mathbb{R}]$, but the points infinitely near an unlimited *real h, $o + h = \{x \in {}^*\mathbb{R} : x \approx h\}$, do not. Lindstrøm's article (esp. III.2) gives basic examples of monads. We shall concentrate our examples in spaces related to distributions (and their abstractions).

Let $\mathscr{C}_m^\infty = \mathscr{C}_m^\infty(\mathbb{R})$ denote the set of all infinitely differentiable real valued functions defined on $\mathbb{R}$, but carried on $[-m,m]$, (i.e., $f(x) = 0$, if $|x| > m$). The *uniform* $\mathscr{C}_m^\infty$*-topology* is given by the family of seminorms $p^k(f) = \sup|f^{(k)}(x)|$, where $f^{(k)}$ denotes the kth derivative. Thus if D_m is the monad of zero for this topology, we have

$$f \in D_m \iff \begin{cases} f(x) = 0\ , & \text{for } |x| > m \\ \text{and} \\ f^{(k)}(x) \approx 0, & \text{for all standard } k \in \mathbb{N} \text{ and all } x \in {}^*[-m,m]. \end{cases}$$

Let $\mathscr{D} = \mathscr{D}(\mathbb{R}) = \cup[\mathscr{C}_m^\infty(\mathbb{R}) : m \in \mathbb{N}]$ denote the set of smooth functions of compact support. A natural notion of infinitesimal on $\mathscr{D}$ is given by the union $D = \cup D_m$; i.e.,

$$f \in D \iff \begin{cases} f(x) = 0\ , & \text{for all unlimited } x \\ \text{and} \\ f^{(k)}(x) \approx 0, & \text{for all standard } k \text{ and limited } x \in {}^*\mathbb{R}. \end{cases}$$

Unfortunately, D is not a monad. We can, however, fatten D up as follows:

1.2 Definition

Let T be a non-empty subset of a standard entity $^*W \in V(^*S)$. The *discrete monad of* T is $\delta(T) = \cap[^*U : T \subseteq {}^*U \subseteq {}^*W]$.

Even the monad $\delta(D)$ on $^*\mathscr{D}(R)$ is not the monad of a vector topology on $\mathscr{D}(R)$. Nevertheless, we shall see that D does characterize enough of the behavior of the linear inductive limit topology on $\mathscr{D}$ to be useful in studying that topology. We have given this example to show that D_m is concrete – one can just use calculus on internal functions – as well as to raise two questions: 1) When do monads define a topology? and

2) When do monads define linear topologies? We will formulate these questions clearly and give answers using the tools developed in the next section.

2. PRODUCT MONADS

What could we mean by a "*standard family of monads?*" Our answer is: Suppose $\mathcal{F} : I \longrightarrow V_{n+1}$ is a standard function such that for each $i \in I$, $\mathcal{F}(i)$ is a filter. For each standard $i \in {}^{\sigma}I$ we have a nonempty monad $\mu(i) = \cap^{\sigma}\mathcal{F}(i) = \cap[{}^*F : F \in \mathcal{F}(i)]$. The family $\{\mu_i : i \in {}^{\sigma}I\}$ is a *standard family of monads in* V_n.

The product $\Pi\mathcal{F} = \{F: I \longrightarrow \cup\mathcal{F}(i) :$ *for all* i, $F(i) \in \mathcal{F}(i)\}$ is used to define monad-like sets at nonstandard indices.

2.1 Definition

Let $\{\mu_i : i \in {}^{\sigma}I\}$ be a standard family of monads. *The product- or π-monads* of this family are given for each $j \in {}^*I$ by

$${}^{\pi}\mu(j) = \{x : (\text{for all standard } F \in \Pi\mathcal{F})[x \in {}^*F(j)]\}.$$

Product monads are nonempty because the sets $F(j)$ have the finite intersection property, so saturation axiom (I) applies. When j is nonstandard, ${}^{\pi}\mu(j)$ need not be a monad. Even when the function $i \longrightarrow \mathcal{F}(i)$ is constant, ${}^{\pi}\mu(j)$ may be smaller than $\mu(i)$. For example, let $N = \{n \in {}^*\mathbb{N} : n$ is infinite$\}$. N is the monad of the Frechet filter $\mathcal{N} = \{M \subsetneq \mathbb{N} : \mathbb{N}\backslash M$ is finite$\}$. Consider the constant family of monads $N(h) = N$, for each standard $h \in {}^{\sigma}\mathbb{N}$. When $k \in {}^*\mathbb{N}\backslash{}^{\sigma}\mathbb{N}$ is infinite,

$${}^{\pi}N(k) = \{n \in {}^*\mathbb{N} : \text{for each standard } f : \mathbb{N} \longrightarrow \mathbb{N},\ n > {}^*f(k)\}$$

(as one may verify by using $f(k) = \min\{i \in \mathbb{N} : j > i \Rightarrow j \in F(k)\}$.) Numbers $n \in {}^{\pi}N(k)$ are 'superinfinite' compared to an infinite k because n is greater than every standard function at k, so $n > k^k > 2^k > k^2 > k$. For this reason we write

$$n \gg k$$

when $n \in {}^{\pi}N(k)$. (Reciprocals of superinfinite numbers are superinfinitesimal.)

In the standard spaces $\mathcal{X} = \mathcal{C}_m^\infty$ above we have neighborhood monads $\tau(f) = D_m + f = \{g \in {}^*\mathcal{C}_m^\infty : g = h + f,\ \text{some } h \in D_m\}$ for each standard f, because the topology is linear. We can take π-extensions of the family with respect to either f or m (or both).

Functions $f \in {}^\pi D_n$ for the family $\{D_m : m \in {}^\sigma\mathbb{N}\}$ are infinitesimal in $^*\mathcal{D}$ and yet may have unlimited carrier. For example, if k is infinite and $n >> k$ in $^*\mathbb{N}$, then

$$f(x) = \begin{cases} \frac{1}{n}\, e^{-1/(k^2-x^2)}, & |x| < k \\ 0, & |x| \geq k \end{cases}$$

belongs to ${}^\pi D_k$. In particular, this f is infinitesimal for the linear inductive limit topology on $^*\mathcal{D}$ and yet has unlimited carrier $[-k,k]$. Heuristically, this is permitted because the values of f are superinfinitesimal compared to k.

Benninghofen and Richter's (1985) main result is a transfer theorem that says suitable properties of a standard family of monads extend to the $^\pi$-monads of that family. The suitable class of formal properties that their $^\pi$-Transfer Theorem applies to is somewhat complicated. A key ingredient of the class of formulas is a monotone property that allows one to replace the finite set arising in Nelson's axiom (I) by its max or min. Nigel Cutland discovered the following useful special case of the $^\pi$-Transfer Theorem which is sufficient for our purposes. It is based on a simpler monotone property. Cutland's theorem relies on extending Lindstrøm's language $L^*(V(S))$ to substitution of external sets for free variables rather than using the language of bounded Internal Set Theory $L({}^*V(S), st(\cdot))$.

2.2 Definition

Let $\Phi(x,y)$ be a formula of $L^*(V(S))$ with only x and y free. Let $A : I \longrightarrow B$ be a standard function. We say $\Phi(x,A_i)$ is *externally increasing in* x *on* $^*V_n$ if for all $i \in {}^*I$ and all internal or external $X,Y \subseteq {}^*V_n$, $\Phi(X,A_i) \wedge X \subseteq Y \Rightarrow \Phi(Y,A_i)$. We say $\Phi(x,A_i)$ is *externally decreasing in* x *on* $^*V_n$ if for all $i \in {}^*I$ and all internal or external $X,Y \subseteq {}^*V_n$, $\Phi(X,A_i) \wedge X \supseteq Y \Rightarrow \Phi(Y,A_i)$. We say $\Phi(x,A_i)$ is *externally monotone on* $^*V_n$ it it is either increasing or decreasing.

2.3 Cutland's $^{\pi}$-Transfer Theorem

Let $\{\mu_i : i \in {}^{\sigma}I\}$ *be a standard family of monads in* $*V_n$ *and let* $A : I \to B$ *be a standard function. If* $\Phi(x,A_i)$ *is externally monotone, then* $(\forall\, i \in {}^{\sigma}I)[\Phi(\mu_i,A_i)] \Leftrightarrow (\forall\, j \in *I)[\Phi[{}^{\pi}\mu_j,A_j]$.

Proof. Suppose $\Phi(x,A_i)$ is externally increasing on $*V_n$ and assume that $(\forall\, i \in {}^{\sigma}I)[\Phi(\mu_i,A_i)]$ holds. Take any standard $i \in {}^{\sigma}I$ and $F \in {}^{\sigma}\Pi\mathcal{F}$, so $\mu_i \subseteq *F_i$. Since Φ is increasing $\Phi(*F_i,*A_i)$ holds and $(\forall\, F \in \Pi\mathcal{F})(\forall\, i \in I)\Phi(F_i,A_i)$ holds by ordinary transfer. Also, $(\forall\, G \in *\Pi\mathcal{F})(\forall\, i \in *I)\Phi(G_i,*A_i)$ holds by transfer.

Choose any $j \in *I$ and let $H \in *\Pi\mathcal{F}$ satisfy $H_j \subseteq {}^{\pi}\mu_j$. Such an H exists by saturation axiom (I). We know $\Phi(H_j,A_j)$ holds, so $\Phi({}^{\pi}\mu_j,A_j)$ follows from the increasing property.

Next, suppose $\Phi(x,A_i)$ is decreasing. We claim that $(\exists\, F \in \Pi\mathcal{F})(\forall\, i \in I)\Phi(F_i,A_i)$ holds. This is sufficient because if $F \in {}^{\sigma}\Pi\mathcal{F}$ satisfies the above then $(\forall\, i \in *I)\Phi(*F_i *A_i)$ holds by ordinary transfer and $(\forall\, i \in *I)\Phi({}^{\pi}\mu_i,*A_i)$ follows from ${}^{\pi}\mu_i \subseteq *F_i$ and the decreasing property of Φ.

The claim holds by the decreasing property and the hypothesis that $(\forall\, i \in {}^{\sigma}I)\Phi(\mu_i,A_i)$ holds. Fix any standard $i \in {}^{\sigma}I$. We will exhibit a standard F_i such that $\Phi(*F_i,*A_i)$ holds. Any $G \in *\mathcal{F}_i$ such that $G \subseteq \mu_i$ satisfies $\Phi(G,*A_i)$ by the decreasing property. Such G's exist by saturation axiom (I). The set of $G \in *\mathcal{F}_i$ such that $\Phi(G,*A_i)$ holds is standard, hence there is an $F \in {}^{\sigma}\mathcal{F}_i$ such that $\Phi(F,A_i)$ holds.◂

In general, a union of a family of monads is not a monad; however

2.4 Lemma

Let $\{\mu_i : i \in {}^{\sigma}I\}$ *be a standard family of monads. Then*

$$\delta(\cup[\mu_i : i \in {}^{\sigma}I]) = \cup[{}^{\pi}\mu_i : i \in *I].$$

Proof. Let $\{\mathcal{F}(i) : i \in I\}$ be the standard function associated with the monads μ_i and let $\Pi\mathcal{F}$ be their product. The formula

$$\psi(x) = (\exists\ j \in {}^*I)(\forall^{st}\ F \in \Pi\mathcal{F})[x \in F(j)]$$

defines a monad $P = \{x : \psi(x)\} = \cup[{}^{\pi}\mu_j : j \in {}^*I]$ by the Syntactical Theorem on Monads from the article of Diener & Stroyan. Since $N = \delta(\cup[\mu_i : i \in {}^{\sigma}I])$ is the smallest monad containing the standard family, $N \subseteq P$. Let U be any standard set in the filter associated with N, ${}^*U \supseteq N$. The property $\Phi(x) = x \subseteq {}^*U$ is decreasing and we know $\Phi(\mu_i)$ for all standard i. Cutland's ${}^{\pi}$-Transfer Theorem gives $(\forall\ i \in {}^*I)[{}^{\pi}\mu_i \subseteq {}^*U]$, or $P \subseteq {}^*U$. Since $N = \cap[{}^*U : {}^*U \supseteq N]$, we have $P \subseteq N$, thus $P = N$.◂

3. LIMIT SPACES AND TOPOLOGIES

3.1 Definition

Let ${}^*\mathbf{X}$ be a standard entity. A standard *limit space structure* on ${}^*\mathbf{X}$ is an assignment $\{\lambda(a): a \in {}^{\sigma}\mathbf{X}\}$ of a monad containing a, $a \in \lambda(a)$, for each standard $a \in {}^{\sigma}\mathbf{X}$. We denote the limit space $({}^*\mathbf{X},\lambda)$ or $({}^*\mathbf{X},\{\lambda(a): a \in {}^{\sigma}\mathbf{X}\})$ and call the $\lambda(a)$ *limit monads*.

If $\mathbf{X}$ is a (standard) topological space whose family of open sets is $\mathcal{T}$, then ${}^*\mathbf{X}$ is a standard limit space under the assignment of neighborhood monads of the topology given by $\tau(a) = \cap[{}^*U : a \in U \in \mathcal{T}]$. We shall denote topological spaces by $({}^*\mathbf{X},\tau)$ when τ is this monad family.

The way that topology can be formulated in terms of limit filters (in a standard model) is by giving a direct condition that guarantees the filters arise as neighborhoods of a topology. Let $({}^*\mathbf{X},\lambda)$ be a limit space with the associated standard filters, $\mathcal{L}(a) = \{U \subseteq \mathbf{X} : \lambda(a) \subseteq {}^*U\}$. We are supposing now that only $\mathbf{X}$ and $\{\mathcal{L}(a) : a \in \mathbf{X}\}$ are given and we work in a standard setting. Define a set U to be λ-open if for each $u \in U$, $U \in \mathcal{L}(u)$. The λ-open sets form a topology, $\mathcal{T}_{\lambda}$.

We want a condition that tells us whether the $\mathcal{T}_{\lambda}$-neighborhood filters are the same as the original $\mathcal{L}(a)$ filters. This happens if and only if the original system had a base of λ-open sets:

$$(\forall a \in \mathbf{X})(\forall V \in \mathcal{L}(a))(\exists U \in \mathcal{L}(a))[U \subseteq V \wedge (\forall u \in U)[U \in \mathcal{L}(u)]]$$

It is known that this property follows from the seemingly weaker property

$$(\forall a \in \mathbf{X})(\forall V \in \mathcal{L}(a))(\exists W \in \mathcal{L}(a))(\forall w \in W)[V \in \mathcal{L}(w)]$$

(To prove this let $U = \{x \in \mathbf{X} : V \in \mathcal{L}(x)\}$ and use the weaker property to show that for each $x \in U$, $U \in \mathcal{L}(x)$.)

Recall (Lindstrøm III.2.1) that a standard set U is $\mathcal{T}$-open if and only if $(\forall\, a \in {}^{\sigma}U)[\tau(a) \subseteq {}^{*}U]$. Theorem 3.2 is a generalization of this which says that limit monads define a topology if and only if they are 'superinfinitesimally open'.

3.2 Theorem

A standard limit space $({}^{*}\mathbf{X},\lambda)$ *is a topological space if and only if* $(\forall a \in {}^{\sigma}\mathbf{X})(\forall x \in \lambda(a))[{}^{\pi}\lambda(x) \subseteq \lambda(a)]$.

Proof. If the filters defining the $\lambda(a)$ are the neighborhood filters of a topology, then for standard open V and standard $a \in V$ we have $\lambda(a) \subseteq {}^{*}V$. The ${}^{\pi}$-Transfer Theorem implies $(\forall x \in {}^{*}V)({}^{\pi}\lambda(x) \subseteq {}^{*}V)$ and hence the assertion.

For the other direction we consider some standard $a \in \mathbf{X}$ and a standard neighborhood V of a. Let $W \in {}^{*}\mathcal{L}(a)$ satisfy $a \in W \subseteq \lambda(a)$. By assumption we have for all $w \in W$, ${}^{\pi}\lambda(w) \subseteq \lambda(a) \subseteq {}^{*}V$, which shows that V is a neighborhood of all points in W, that is, we have shown that

$$(\forall a \in \mathbf{X})(\forall V \in \mathcal{L}(a))(\exists W \in \mathcal{L}(a))(\forall w \in W)[V \in \mathcal{L}(w)]$$

This shows that the monads are topological (as explained above) and concludes the proof.◂

A limit structure always defines a finest compatible topology by: $\mathcal{T}_{\lambda} = \{U \subseteq \mathbf{X} : (\forall a \in {}^{\sigma}U)[\lambda(a) \subseteq {}^{*}U]\}$. We define the *neighborhood monads associated with* λ to be the neighborhood monads of $\mathcal{T}_{\lambda}$ given by $\tau_{\lambda}(a) = \cap\,[{}^{*}U : a \in U \in \mathcal{T}_{\lambda}]$, for standard $a \in {}^{\sigma}\mathbf{X}$. We always have $\tau_{\lambda}(a) \supseteq \lambda(a)$, but they are different from the original *limit monads*, $\tau_{\lambda}(a) \neq \lambda(a)$, when λ is not topological. The next result describes $\tau_{\lambda}(a)$ in terms of $*$-finite chains of ${}^{\pi}\lambda$-monads, but without reference to $\mathcal{T}_{\lambda}$.

3.3 Theorem

Let $({}^*\mathbf{X},\lambda)$ *be a limit space with the finest compatible topology as above. Then for each standard* $a \in {}^\sigma\mathbf{X}$, *the neighborhood monads associated with* λ *are given by*

$$\tau_\lambda(a) =$$
$$\{x\in{}^*\mathbf{X} : (\exists n\in{}^*\mathbb{N})(\exists y(\cdot)\in{}^*\mathbf{X}^{n+1})[a=y(0) \wedge y(n)=x \wedge (\forall k<n)[y(k+1)\in{}^\pi\lambda(y(k))]]\}$$

Proof. By the Syntactical Theorem on Monads and the formal definition of ${}^\pi\lambda$, we see that the set on the right hand side of the equation is a monad. Denote this monad by $\mu(a)$ for each standard a. One step chains, $y(0) = a \wedge y(1) = x$, show that $\lambda(a) \subseteq \mu(a)$, so μ is compatible with λ.

We can show that $({}^*\mathbf{X},\mu)$ is a topological space using (3.2). This requires a fairly technical application of ${}^\pi$-transfer or of syntactical reductions as in the article by Diener & Stroyan to reduce it to Cutland's ${}^\pi$-Transfer Theorem. We omit the proof. Theorem 5.4 is an important similar result.

3.4 Proposition

Standard uniform spaces $(\mathbf{E},\mathcal{U})$ *are in one-to-one correspondence with pairs* $({}^*\mathbf{E},\overset{u}{\approx})$, *where* $\overset{u}{\approx}$ *is a monad and an equivalence relation in* ${}^*\mathbf{E}\times{}^*\mathbf{E}$. *The correspondence is:* $\mathcal{U} = \{U \subseteq \mathbf{E}\times\mathbf{E} : {}^*U \supseteq (\overset{u}{\approx})\}$ *and* $(\overset{u}{\approx}) = \bigcap {}^\sigma\mathcal{U}$.

A monadic equivalence relation $\overset{u}{\approx}$ *induces a topology on* $\mathbf{E}$ *with neighborhood monads,* $\tau_u(a) = \{x \in {}^*\mathbf{E} : x \overset{u}{\approx} a\}$ *for* $a \in {}^\sigma\mathbf{E}$.

Proof. Omitted. The last part may be shown easily with (3.2).◂

4. LIMIT VECTOR SPACES

We will be mainly interested in vector spaces with a notion of infinitesimal vector. A monad at zero $\lambda(0)$ is given. In this case a limit structure can be obtained by adding $\lambda(0)$ to standard points.

The ${}^\pi$-monad of the constant family $\{\lambda(0) : z \in {}^\sigma\mathbf{E}\}$ is denoted $\Big|_z^\pi \lambda(0)[z] = {}^\pi\lambda[z]$.

4.1 Proposition

Let $(\mathbb{E},+)$ *be an abelian group and let* $0 \in \lambda(0)$ *be a monad containing the identity. Consider the limit structure* $\{\lambda(a) = a+\lambda(0) : a \in {}^{\sigma}\mathbb{E}\}$. *The neighborhood monads associated with* λ *satisfy:*

$$\tau_\lambda(a) = a + \tau_\lambda(0) =$$

$$\left\{x\in{}^*\mathbb{E}\text{: } (\exists n\in{}^*\mathbb{N})(\exists h(\cdot)\in{}^*\mathbb{E}^n)\left[x=a+\sum_{k=1}^{n} h(k) \wedge (\forall k\le n)\left[h(k)\in\prod_{z}^{\pi}\Big|\lambda(0)[z=\sum_{i\Delta k} h(i)]\right]\right]\right\}$$

Proof. This is simply Theorem (3.3) with the additive structure moving the limit monads.◂

Notice in (4.1) that we do not claim that $({}^*\mathbb{E},\tau_\lambda)$ is a topological group. Theorem (4.3) gives additional conditions required for a linear topology. In some noteworthy cases (such as bounded-weak-star topologies) τ_λ will satisfy these additional requirements (by the Banach-Dieudonne theorem.)

Next, let $\mathbb{E}$ be a topological vector space over the field $\mathbb{K}$ (= $\mathbb{R}$ or $\mathbb{C}$). A monad $\mu(0)$ containing zero arising from a linear topology on $\mathbb{E}$ makes addition and scalar multiplication jointly continuous. One conventional infinitesimal formulation of these continuity conditions is $\mu(0)+\mu(0) \subseteq \mu(0)$, $\mathcal{O}\mu(0) \subseteq \mu(0)$ and $o\cdot\mu(0) \subseteq \mu(0)$, where $\mathcal{O}$ and o denote the *limited* and *infinitesimal scalars* of ${}^*\mathbb{K}$, respectively.

4.2 Definition

Let S be a monad in an internal vector space $\mathbb{E}$ over ${}^*\mathbb{K}$. The *hyperconvex* (strictly speaking, the hyper-absolutely-convex) *hull* of S, ΓS, is the set of all $*$-finite (absolutely) convex combinations of elements of S,

$$\Gamma S = \{\ y : (\exists n \in{}^*\mathbb{N})(\exists x \in{}^*\mathbb{E}^n)(\exists \alpha \in{}^*\mathbb{K}^n)[\Phi(y,n,\alpha,x)]\}$$

where

$$\Phi(y,n,\alpha,x) = [\ \sum_{i=1}^{n} |\alpha_i| \le 1 \wedge y = \sum_{i=1}^{n} \alpha_i x_i \wedge (\forall i)[\ x_i \in S]].$$

A monad S is *hyperconvex* if $S = \Gamma S$.

This is simply the internal condition of (absolute) convexity applied to a (possibly) external set. The monad of a locally convex topology is hyperconvex. For monads, S, $\Gamma(\Gamma(S)) = \Gamma(S)$ and $\Gamma(S)$ is a monad.

4.3 Proposition

Let $\mathbb{E}$ *be a vector space over* $\mathbb{K}$ (= $\mathbb{R}$ or $\mathbb{C}$) *and let* $\mu(0)$ *be a monad in* $^*\mathbb{E}$ *containing* 0. *Then the limit space* $(^*\mathbb{E}, \{a+\mu(0) : a \in {}^\sigma\mathbb{E}\})$ *is a topological vector space over* $\mathbb{K}$ *if the following conditions hold:*

(a) $\mu(0) + \mu(0) \subseteq \mu(0)$,

(b) $o \cdot {}^\sigma\mathbb{E} \subseteq \mu(0)$,

(c) $o \cdot \mu(0) \subseteq \mu(0)$,

(d) ${}^\sigma\mathbb{K} \cdot \mu(0) \subseteq \mu(0)$.

This topology is locally convex if condition (e) *also holds.*

(e) $\mu(0)$ *is hyperconvex.*

Proof. We shall only mention one ingredient in our proof of the first part, because it is similar to other standard infinitesimal analysis in chapter 10 of Stroyan & Luxemburg (1976). Suppose $\mu(0)$ satisfies conditions (a)-(d). Define an equivalence relation on $^*\mathbb{E}$ by $x \overset{\mu}{\approx} y$ if and only if $x-y \in \mu(0)$. The relation $\overset{\mu}{\approx}$ is a monad in $^*\mathbb{E} \times {}^*\mathbb{E}$ by the Syntactical Theorem on Monads. Calculations show that if $x \overset{\mu}{\approx} x'$, $y \overset{\mu}{\approx} y'$ and $\lambda' \approx \lambda \in {}^\sigma\mathbb{K}$, then $\lambda x+y \overset{\mu}{\approx} \lambda' x'+y'$, so the operations satisfy the infinitesimal continuity criterion. Finally, we have proposition (3.4) to tell us that this continuity is topological.◂

The second part of proposition (4.3) is simply the infinitesimal formulation of the condition that there is a basis for the neighborhood filter of the origin consisting of (absolutely) convex sets.

4.4 Definition

Let $(^*\mathbb{E},\mu)$ be a locally convex topological vector space. The *μ-finite points* of $^*\mathbb{E}$ are given by:

$$\text{fin}_\mu(^*\mathbb{E}) = \{x \in {}^*\mathbb{E} : q(x) \textit{ is limited for every standard } \mu\textit{-seminorm } q\}$$

The *μ-bounded points* of $^*\mathbb{E}$ are given by:

$$\text{bd}_\mu(^*\mathbb{E}) = \cup\{^*B : B \subseteq \mathbb{E} \;\&\; {}^*B \subseteq \text{fin}_\mu(^*\mathbb{E})\}$$

The bounded points are always a proper subset of the finite points unless μ is given by a single seminorm. Moreover, the μ-infinitesimals, $\mu(0)$, are a subset of the bounded points only in the single seminorm case. (See Stroyan & Luxemburg (1976), (10.1.24).)

4.5 Definition

Let $({}^*\mathbb{E},\mu)$ be an (F)-space (i.e., a complete metrizable locally convex space) with continuous dual $\mathbb{E}'$. The *weak-star topology* σ or $\sigma(\mathbb{E}',\mathbb{E})$ on $\mathbb{E}'$ is given by the neighborhood monad of zero,

$$\sigma(0) = \{x \in {}^*\mathbb{E}' : \langle e,x\rangle \approx 0 \text{ for each } e \in {}^\sigma\mathbb{E}\}$$

The *strong topology* β or $\beta(\mathbb{E}',\mathbb{E})$ on $\mathbb{E}'$ is given by the neighborhood monad of zero,

$$\beta(0) = \{x \in {}^*\mathbb{E}' : \langle e,x\rangle \approx 0 \text{ for each } e \in \mathrm{bd}_\mu({}^*\mathbb{E})\}$$

Since $\mathbb{E}$ is complete, the strongly bounded and weakly bounded subsets of $\mathbb{E}'$ coincide (Robertson & Robertson (1964), Prop.3, p.72), $\mathrm{bd}({}^*\mathbb{E}') = \mathrm{bd}_\beta({}^*\mathbb{E}') = \mathrm{bd}_\sigma({}^*\mathbb{E}')$ so we shall simply refer to these points as the bounded points of ${}^*\mathbb{E}'$.

5. LIMIT INFINITESIMALS

Let $\mathbb{F}$ be a vector space over $\mathbb{K}$ ($= \mathbb{R}$ or $\mathbb{C}$) and suppose we have a standard sequence $\{B_m\}$ of (absolutely) convex subsets of $\mathbb{F}$ with $B_m+B_m \subseteq B_{m+1}$ and $\cup B_m = \mathbb{F}$. Let κ be the zero-monad of a (Hausdorff) locally convex vector topology on ${}^*\mathbb{F}$. For each m in ${}^\sigma\mathbb{N}$ define a monad $M_m = \kappa \cap {}^*B_m$. The sequence $\{M_m : m \in {}^\sigma\mathbb{N}\}$ has the following properties:

(1) $o \cdot {}^\sigma\mathbb{F} \subseteq \cup[M_m : m \in {}^\sigma\mathbb{N}] \subseteq {}^*\mathbb{F}$.

(2) M_m is hyperconvex, for m in ${}^\sigma\mathbb{N}$.

(3) $M_m+M_m \subseteq M_{m+1}$, for m in ${}^\sigma\mathbb{N}$.

This situation arises both in mixed topologies, bounded weak star topologies in particular, and in strict inductive limit topologies.

5.1 Definitions

We shall say that a set $M \subseteq {}^*\mathbb{F}$ is a *set of limit infinitesimals on* ${}^*\mathbb{F}$ if there is a sequence as above such that $M = \cup[M_m : m \in {}^\sigma\mathbb{N}]$.

If M is a set of limit infinitesimals on $^*\mathbb{F}$, the *limit infinitesimal relation* is given by $x \overset{M}{\approx} y$ if and only if $x-y \in M$.

The (locally convex) *limit monad* associated with a set of limit infinitesimals, M, is the hyperconvex hull of the discrete monad of M, $\mu = \Gamma[\delta(M)]$.

5.2 Limit Infinitesimals on ℓ^∞ and $\mathcal{C}^\infty$

When $\mathbb{F} = \ell^\infty$, the limit infinitesimals are characterized by

$$x \overset{M}{\approx} y \quad \textit{if and only if} \begin{cases} x(m) \approx y(m) & \textit{when} \quad m \in {}^\sigma\mathbb{N} \\ \textit{and} & \\ \|x-y\|_\infty & \textit{is limited.} \end{cases}$$

In this case $\kappa = \sigma(0)$, the weak-star monad of $\sigma(\ell^\infty, \ell^1)$, or $\kappa =$ the pointwise convergence monad. The sets $B_m = \{x : \|x\|_\infty \le m\}$. The limit monad μ is the bounded weak star monad.

A second example is given at the beginning of this article, $B_m = \mathcal{C}^\infty_m$. In this case the limit monad μ is the linear inductive limit monad of $\mathcal{D}$. (See (6.6).)

5.3 Lemma

Let $^*\mathbb{F}$ *have a set of limit infinitesimals,* M. *Let* $\mu = \Gamma[\delta(M)]$ *be the limit monad of* M. *The space* $(\mathbb{F}, \mu)$ *is the finest locally convex structure whose zero monad contains the limit infinitesimals,* $\mu = \mu(0) \supseteq M$.

Proof. We use Proposition (4.3) to prove the lemma. Notice that μ is a hyperconvex monad by the Syntactical Theorem on Monads.

We proved the following characterization of $\delta(M)$ in Lemma 2.2: $\delta(M) = \cup[{}^\pi M_n : n \in {}^*\mathbb{N}]$. Also, if ν_i and μ_i are standard families of monads satisfying $(\forall i \in {}^\sigma I)[\,\nu_i \subseteq \mu_i]$, then Cutland's $^\pi$-Transfer Theorem shows that $(\forall i \in {}^* I)[{}^\pi\nu_i \subseteq {}^\pi\mu_i]$. Let $F \in \Pi\mathcal{F}$, where $\mathcal{F}_i$ generates μ_i. We have $(\forall i \in {}^\sigma I)[\,\nu_i \subseteq {}^*F_i]$, so $(\forall F \in {}^\sigma\Pi\mathcal{F})(\forall i \in {}^*I)[\nu_i \subseteq {}^*F_i]$ $\equiv (\forall i \in {}^*I)[{}^\pi\nu_i \subseteq {}^\pi\mu_i]$.

We may show by induction that for all standard m,n in ${}^{\sigma}\mathbb{N}$, $mM_n \subseteq M_{m+n}$. By the π-Transfer Theorem on n, we have $(\forall m \in {}^{\sigma}\mathbb{N})(\forall n \in {}^{*}\mathbb{N})\ m^{\pi}M_n \subseteq \overset{\pi}{\underset{i}{|}} M_{m+i}[n]$. We have $\overset{\pi}{\underset{i}{|}} M_{m+i}[n] \subseteq {}^{\pi}M_{m+n}$, so $(\forall m \in {}^{\sigma}\mathbb{N})(\forall n \in {}^{*}\mathbb{N})\ m^{\pi}M_n \subseteq {}^{\pi}M_{m+n}$. Therefore, by our characterization of $\delta(M)$ above, ${}^{\sigma}\mathbb{N} \cdot \delta(M) \subseteq \delta(M)$. Now we can show condition (a) of Proposition (4.3), $\mu + \mu \subseteq \mu$, because two *-finite convex combinations may be written over the larger index set (by adding zeros) and

$$x+y = 2\left[\sum \tfrac{1}{2}\alpha_i x_i + \tfrac{1}{2}\beta_i y_i\right].$$

Notice that we have not shown that $\delta(M) + \delta(M) \subseteq \delta(M)$. This is false in general.

Condition (b) of Proposition (4.3) follows from the first condition on a sequence defining limit infinitesimals, so $o \cdot {}^{\sigma}\mathbb{F} \subseteq \mu$.

Condition (c) of Proposition (4.3) is immediate because an infinitesimal times a vector is a convex combination, $o \cdot \mu \subseteq \mu$.

We know that ${}^{\sigma}\mathbb{N} \cdot \delta(M) \subseteq \delta(M)$ and ${}^{\sigma}\mathbb{N} \cdot \mu \subseteq \mu$ follows easily from this. If b is a standard scalar, $b = m \cdot a + c$ where m is a standard integer and $|a|, |c| \le 1$. Therefore we see condition (d) of Proposition (4.3), ${}^{\sigma}\mathbb{K} \cdot \mu \subseteq \mu$.

Proposition (4.3) now proves our lemma (6.1).◂

5.4 **Lemma**

The limit monad $\mu = \Gamma[\delta(M)]$ *of* M *is equal to the set of all *-finite convex combinations of internal sequences with* $x_j \in {}^{\pi}M_j$, *for* $1 \le j \le n$; i.e.

$$\mu = \{x \in {}^{*}\mathbb{F}\colon (\exists n \in {}^{*}\mathbb{N})(\exists \alpha \in {}^{*}\mathbb{K}^n)(\exists y \in {}^{*}\mathbb{F}^n)[x = \sum_{j=1}^{n} \alpha_j y_j, \sum |\alpha_j| \le 1 \ \&\ (\forall j \le n) y_j \in {}^{\pi}M_j]\}.$$

Proof (Following a suggestion of Cutland). The set on the right side of the equality is a monad by the Syntatical Theorem on Monads. Denote it by ν. We will show that ν is hyperconvex. Once we know this, it follows that $\nu = \mu$. This is because ν contains each M_m and hence $\nu \supseteq M$, so we have $\nu \supseteq \mu$ the finest hyperconvex monad containing M. Clearly, ν

is a subset of μ because it consists of special convex combinations from $\delta(M) = \cup[{}^{\pi}M_n : n \in {}^*\mathbb{N}]$.

Now we prove that the monad ν is hyperconvex. Suppose $y^j = \sum \alpha_i^j x_i^j$ is an internal sequence of elements of ν with $\sum_i |\alpha_i^j| \le 1$ and $x_i^j \in {}^{\pi}M_i$ for $1 \le i \le n_j$, $1 \le j \le n$. We want to show that $z = \sum \beta_j y^j$ is in ν for any internal sequence with $\sum |\beta_j| \le 1$.

First, we show that the functions α_i^j and x_i^j may be chosen to be internal in both variables. This follows from general syntatical transformations. We know that $(\forall j)(\exists \alpha, x)[\ \Phi(y^j, \alpha, x)\ ]$ where Φ says y^j is a convex combination with α and x in ${}^{\pi}M_i$. Since we only have the external quantifiers in the condition $(\forall i)[x_i \in {}^{\pi}M_i]$ we may write Φ in the form $(\forall^{st} u)[\ \varphi(y^j, \alpha, x, u)\ ]$. Apply Nelson's axioms and ordinary rules of quantifiers to pull the external quantifier in front of $(\forall j)(\exists \alpha, x)$. Then use the internal axiom of choice to re-write these quantifiers as $(\exists \alpha^{(\cdot)}, x^{(\cdot)})(\forall j)$. Finally, push the external quantifiers back inside to obtain

$$(\exists \alpha^{(\cdot)}, x^{(\cdot)})(\forall j)[\ \Phi(y^j, \alpha^j, x^j)\ ]$$

This shows that the α_i^j and x_i^j may be chosen internally. We may also choose the i-lengths of the α and x sequences to all be the same length m by letting short sequences equal zero up to the maximum i-length.

Each filter $\mathcal{F}_i$ for M_i has a base of convex sets, so we may apply the axiom of the ideal point (I) to obtain an internal function $H \in \Pi\mathcal{F}$ such that for all standard $F \in \Pi\mathcal{F}$, $H \subseteq {}^*F$ and for each $i \le m$, H_i is hyperconvex and $x_i^j \in H_i$. Take β_j as above. The sum

$$z = \sum \beta_j y^j = \sum_j \sum_i \beta_j \alpha_i^j x_i^j = \sum_i \sum_j \beta_j \alpha_i^j x_i^j = \sum_i \gamma_i \sum_j \frac{\beta_j \alpha_i^j}{\gamma_i} x_i^j$$

where $\gamma_i = \sum_j |\beta_j \alpha_i^j|$ (and don't divide if $\gamma_i = 0$.) The inside sums satisfy

$$z_i = \sum_j \frac{\beta_j \alpha_i^j}{\gamma_i} x_i^j \in H_i \subseteq {}^{\pi}M_i$$

so we have expressed the *convex combination of elements of ν in the special form $z = \sum_i \gamma_i z_i \in \nu$ and thus ν is hyperconvex. This completes the proof.◂

5.5 **Proposition**

Let M *be limit infinitesimals on* $*\mathbb{F}$ *defined by* $M_m = \kappa \cap *B_m$. *Suppose that the monads* M_m *are strictly increasing,* $M_m \not\supseteq M_{m+1}$. *Then* M *is not a monad,* $M \neq \delta[M]$, *and* $(*\mathbb{F},\{a+\delta[M] : a \in {}^{\sigma}\mathbb{F}\})$ *is not a topological vector space, in particular,* $\delta[M] + \delta[M] \not\subseteq \delta[M]$.

Proof. See Benninghofen, Stroyan & Richter (1988).◂

5.6 **Corollary**

Let $\mathbb{F} = \mathbb{E}'$ *be the dual of an infinite-dimensional* (F)*-space and let* M *be the limit infinitesimals associated with the bounded-weak-star topology,* $M = \sigma(0)\cap bd(*\mathbb{E}')$. *Then* $(*\mathbb{F},\{a+\delta[M] : a \in {}^{\sigma}\mathbb{F}\})$ *is not a topological space.*

Proof. If $\mathbb{E}$ is infinite-dimensional, then $(\mathbb{F},\sigma)$ is not a normed space. By Theorem (10.1.24)c from Stroyan & Luxemburg (1976), $\sigma(0)\backslash bd(*\mathbb{F}) \neq \emptyset$. Thus we may choose a defining sequence so that M_m is increasing.

The finest topology τ that agrees with σ on bounded sets is locally convex by the Banach-Dieudonne theorem (see (6.1)). We know that $\tau(0) \supseteq M$, so $\tau(0) \supseteq \delta[M]$. If $\{a+\delta[M] : a \in {}^{\sigma}\mathbb{F}\}$ defines a topology on $*\mathbb{F}$, then $\tau(0) = \delta[M]$ defines a locally convex vector topology contrary to the lemma. This proves the corollary.◂

5.7 **Theorem**

Let $*\mathbb{F}$ *have limit infinitesimals* M. *The finest uniformity which contains the infinitesimal relation* $\overset{M}{\approx}$ *is* $\overset{\mu}{\approx}$ *for the locally convex limit monad* μ.

Proof. See Benninghofen, Stroyan & Richter (1988). This is a two-variable result, while the next result is a one variable result and consequently has a simpler proof.◂

We consider standard uniform spaces, $\mathbb{X}$, as given by a monadic infinitesimal relation, $\overset{u}{\approx}$, on $^*\mathbb{X}$ as in Proposition (3.4). If $^*\mathbb{F}$ has limit infinitesimals M and limit monad μ, then the monadic infinitesimal relation on $^*\mathbb{F}$ is given by $x \overset{\mu}{\approx} y$ if and only if $x-y \in \mu$. The standard infinitesimal translation of uniform continuity of a function $f : \mathbb{F} \longrightarrow \mathbb{X}$ is as follows (Stroyan & Luxemburg (1976), (8.4.22)). A standard function f is uniformly continuous in the classical "epsilon" sense if and only if for every $x,y \in {}^*\mathbb{F}$ if $x \overset{\mu}{\approx} y$ then $f(x) \overset{u}{\approx} f(y)$ in $^*\mathbb{X}$. Our criterion allows a test only on the finer limit infinitesimals.

5.8 **Theorem**

Let $^*\mathbb{F}$ *be a standard space with limit infinitesimals,* M, *and limit monad* μ. *Let* $(^*\mathbb{X},\overset{u}{\approx})$ *be a standard uniform space. A standard function* $f : \mathbb{F} \longrightarrow \mathbb{X}$ *is* μ*-u-uniformly continuous if and only if for every* $x,y \in {}^*\mathbb{F}$, *if* $x \overset{M}{\approx} y$, *then* $f(x) \overset{u}{\approx} f(y)$ *in* $^*\mathbb{X}$.

Proof. Since $x \overset{M}{\approx} y$ implies $x \overset{\mu}{\approx} y$, ordinary μ-u-uniform continuity implies the limit infinitesimal condition (using the remarks before the theorem).

Conversely, if $x-y \in \mu$, then (by 5.4) there is a *-finite sequence $z_i \in {}^{\pi}M_i$, for $1 \le i \le n$ and a sequence α_i with $\Sigma|\alpha_i| \le 1$ satisfying

$$x-y = \sum_{i=1}^{n} \alpha_i z_i .$$

Define a new sequence w_j by $w_0 = y$ and $w_j = y + \sum_{i=1}^{n} \alpha_i z_i$, so that $w_n = x$ and $w_j - w_{j-1} = \alpha_j z_j \in {}^{\pi}M_j$ for $1 \le j \le n$.

Let $\overset{u}{\approx}$ be given by the standard family of semimetrics $\{\rho : \rho \in R_u\}$. For each standard $\rho \in {}^{\sigma}R_u$ we will show that $\rho(f(x),f(y)) \approx 0$. We know that

$$(\forall m \in {}^{\sigma}\mathbb{N})(\forall u,v \in {}^*\mathbb{F})[u-v \in M_m \Rightarrow 2^m \rho(f(u),f(v)) \approx 0]$$

Since the formulas $\Phi(x,2^m\rho\circ f) = [u-v \in x \Rightarrow 2^m\rho(f(u),f(v)) < \varepsilon]$ are monotone decreasing in x for each standard ε, Cutland's $^{\pi}$-Transfer

Theorem gives

$$(\forall n \in {}^*\mathbb{N})(\forall u,v \in {}^*\mathbb{F})[u-v \in {}^{\pi}M_n \Rightarrow 2^n\rho(f(u),f(v)) \approx 0]$$

Thus, for all $j \le n$, $2^j\rho(f(w_j),f(w_{j-1})) \approx 0$. By the triangle inequality for ρ,

$$\rho(f(x),f(y)) \le \sum_{j=1}^{n} 2^j\ \rho(f(w_j),f(w_{j-1}))/2^j$$

$$\le \max_{1\le j\le n}\ [2^j\rho(f(w_j),f(w_{j-1}))]\sum 1/2^j$$

$$\approx 0$$

This proves the theorem.◂

Benninghofen, Stroyan & Richter (1988) use similar techniques to prove the following extension of the uniform continuity criterion.

5.9 The Relative Uniform Continuity Theorem

Let ${}^*\mathbb{F}$ *have limit infinitesimals* M *given by the sequence* $M_m = \kappa \cap {}^*B_m$. *Suppose that the finest topology that agrees with* κ *on the sets* B_m *is a locally convex vector topology, hence given by the limit monad* μ. *Let* $\mathbb{G}$ *be a* μ*-closed linear subspace of* $\mathbb{F}$ *and let* $f : \mathbb{G} \longrightarrow (\mathbb{X},\overset{u}{\approx})$ *be a standard map. Then* f *is* $\mathbb{G}$*-relatively uniformly* μ*-u-continuous if and only if whenever* $x,y \in {}^*\mathbb{G}$ *satisfy* $x \overset{M}{\approx} y$, *then* $f(x) \overset{u}{\approx} f(y)$.

Proof. See Benninghofen, Stroyan & Richter (1988).◂

5.10 The Convex Continuity Theorem

Let ${}^*\mathbb{F}$ *have limit infinitesimals* M *given by the sequence* $M_m = \kappa \cap {}^*B_m$. *Suppose that the finest topology on* $\mathbb{F}$ *that agrees with* κ *on the sets* $\{B_m\}$ *is a locally convex vector topology, hence given by the limit monad* μ. *Let* $\mathbb{G}$ *be either a* μ*-open or* μ*-closed subset of* $\mathbb{F}$ *and let* $f : \mathbb{G} \longrightarrow \mathbb{X}$ *be a standard function into a topological space*

$(\mathbb{X},\tau)$. *Then* f *is* μ-τ-*continuous on* $\mathbb{G}$ *(in the relative* μ-*topology) if and only if whenever* $x \in {}^*\mathbb{G}$ *and* $a \in {}^{\sigma}\mathbb{G}$ *satisfy* $x \overset{M}{\approx} a$, *then* $f(x) \in \tau(f(a))$.

Proof. See Benninghofen, Stroyan & Richter (1988).◂

Remark. The bounded-weak-star topology satisfies the hypotheses of Theorems 5.9 and 5.10 with limit infinitesimals $M = \sigma(0) \cap bd({}^*\mathbb{E}')$.

6. EXAMPLES

This section contains examples of various concrete spaces with limit infinitesimals.

6.1 Bounded Weak-Star Topologies

Let $({}^*\mathbb{E},\mu)$ be an (F)-space with continuous dual $\mathbb{E}'$. The *bounded weak-star topology* γ or $\gamma(\mathbb{E}',\mathbb{E})$ on $\mathbb{E}'$ is defined to be the finest topology on $\mathbb{E}'$ that agrees with the weak-star topology on bounded subsets of $\mathbb{E}'$.

The "Banach-Dieudonne" theorem (Dunford & Schwartz (1958), p.427, Schaeffer (1966), p.151 or Jarchow (1981), 9.4.1) asserts that γ is a locally convex topology. If we let $\mathbb{F} = \mathbb{E}'$, $\kappa = \sigma$, the weak-star topology (4.5) and B_m a basis for the bounded sets, then the limit infinitesimals satisfy $M = \sigma(0) \cap bd({}^*\mathbb{E}')$. Note that the dual of an (F)-space always has a fundamental sequence B_m of convex σ-closed bounded sets as in (5.1) above. The limit monad, $\Gamma\delta(M) = \gamma(0)$, is the bounded-weak-star monad.

Rubel & Ryff (1970) is a nice introduction to bounded weak-star topologies.

6.2 Limit Infinitesimals on ℓ^{∞}

When $\mathbb{E} = \ell^1$ and $\mathbb{E}' = \ell^{\infty}$, the limit infinitesimals are characterized in 5.2 above. This characterization makes continuity calculations for (ℓ^{∞},γ) easy. Applications to ℓ^{∞} are given in Stroyan (1983). Bounded weak star limit infinitesimals also have a simple description for H^{∞}.

6.3 Strong Duals of (FM)-Spaces

Suppose $\mathbb{E}$ is an (FM)-space, so that bounded, compact, and precompact points all coincide: $bd({}^*\mathbb{E}) = cpt({}^*\mathbb{E}) = pcpt({}^*\mathbb{E})$. Since $\mathbb{E}$ is metrizable, $\mathbb{E}$ is also an (HM)-space, so finite points are near standard: $fin({}^*\mathbb{E}) = ns({}^*\mathbb{E})$. (See Stroyan & Luxemburg (1976), section 10.6 or Henson & Moore (1973).)

The bounded-weak-star topology $\gamma(\mathbb{E}',\mathbb{E})$ is the topology of precompact convergence. (This is one version of the "Banach- Dieudonne" theorem.) The strong topology $\beta(\mathbb{E}',\mathbb{E})$ is the polar topology of $bd({}^*\mathbb{E})$. Since these sets coincide, we have $\gamma(\mathbb{E}',\mathbb{E}) = \beta(\mathbb{E}',\mathbb{E})$, when $\mathbb{E}$ is an (FM)-space.

Now we make a general observation about duals of (FM)-spaces. The limit infinitesimals on $\mathbb{E}'$ are given by

$$M = \sigma(0) \cap bd({}^*\mathbb{E}') = \beta(0) \cap bd({}^*\mathbb{E}'),$$

the bounded infinitesimals for either the weak or strong topology. In general the finite points of a space are easier to characterize, so it is of interest to note:

6.4 Proposition

If $\mathbb{E}$ *is an* (FM)*-space and* $x \in M$, *then* $\langle e,x\rangle \approx 0$, *for all* $e \in fin({}^*\mathbb{E})$.

Proof. This follows from the fact that $\mathbb{E}$ is an (HM)-space, every finite point of ${}^*\mathbb{E}$ is μ-near-standard. If $x \in M$ and $e \in fin({}^*\mathbb{E})$, then there is a standard $f \in {}^*\mathbb{E}$ such that $e-f \in \mu(0)$. Since $x \in \sigma(0)$, $\langle f,x\rangle \approx 0$. Since $\mathbb{E}$ is barrelled, $\mu(0) = bd({}^*\mathbb{E}')^a = \mu(0)^{aa}$, so $\langle e-f,x\rangle \approx 0$ because $x \in bd({}^*\mathbb{E}')$, thus $\langle e,x\rangle = \langle e-f,x\rangle + \langle f,x\rangle \approx 0$.◂

In various concrete examples the condition on x that $\langle e,x\rangle \approx 0$ for e in $fin({}^*\mathbb{E})$, together with other natural conditions, implies that $x \in M$. The following is an example.

6.5 Distributional Limit Infinitesimals

Let G be an open region in $\mathbb{R}^n$ and let $\mathcal{E}(G) = \mathcal{C}^\infty(G)$ be the (FM)-space of smooth functions defined on G. The space $\mathcal{E}'(G)$ consists of the distributions of compact support. (See Treves (1976), for example.)

By (6.4),

$$\mu \overset{M}{\approx} \nu \text{ in } {}^*\mathcal{E}'(G) \iff \begin{cases} \langle f,\mu\rangle \approx \langle f,\nu\rangle, & \text{for every } f \in \mathrm{fin}({}^*\mathcal{E}(G)), \\ \mu-\nu & \text{is supported in a standard compact subset of } G \\ \mu-\nu & \text{has standard order.} \end{cases}$$

6.6 Inductive Limit Infinitesimals

Strict inductive limits are related to limit infinitesimals in the following way. Suppose that $\{(B_m,\mu_m) : m \in \mathbb{N}\}$ is a sequence of locally convex spaces with $B_m \subseteq B_{m+1}$ and zero-monads $\mu_m(0) \subseteq \mu_{m+1}(0)$. Let $(\mathbb{F},\mu_\infty)$ denote the classical inductive limit of this sequence (see for example Jarchow (1981), 4.6). Define $M_m = \mu_m(0) = \mu_\infty(0) \cap {}^*B_m$. The limit infinitesimals given by $M = \cup M_m$ are related to the inductive limit by $\mu_\infty = \Gamma[\delta(M)] = \mu$.

A detailed description of inductive limits with infinitesimals can be found in Benninghofen (1982b). The distribution space $\mathcal{D}(\mathbb{R})$ given above is an example of a strict inductive limit.

7. COUNTEREXAMPLES

This section contains counterexamples to various natural, but false, generalizations of our results on continuity measured by limit infinitesimals.

7.1 Counterexample

Let $\mathbb{F} = \ell^\infty$ and take the limit infinitesimals of (5.2). Let m in ${}^*\mathbb{N}$ be an infinite natural number and let n be infinite compared to m, $n >> m$ or $n \in {}^\pi N_m$ (where N is the monad of infinite *natural numbers). Define an internal function $f : {}^*\ell^\infty \longrightarrow {}^*\mathbb{R}$ by $f(x) = x(n)/m$. If x is in M, then $x(n)$ is finite, so $f(x) \approx 0$. Since f is linear, we have the condition of (5.8), $x \overset{M}{\approx} y$ implies $f(x) \approx f(y)$. However, f is not uniformly S-continuous as we now show.

Define a vector x in ${}^*\ell^\infty$ by:

$$x(j) = \begin{cases} m, & \textit{if } j = n \\ 0, & \textit{otherwise.} \end{cases}$$

Certainly $f(x) = 1 \not\approx 0$, yet we claim that $x \in \mu$, so f is not bounded-weak-star continuous at zero. Notice that our claim that x is in μ shows that there are μ-infinitesimals of unlimited $*\ell^{\infty}$-norm, $\|x\|_{\infty} = m$.

Now we prove that x belongs to the limit monad μ. For each standard j, let N equal the monad of infinite natural numbers. We have

$$(\forall j \in {}^{\sigma}\mathbb{N})M_j = \{z : \|z\| \le j \ \&\ [i \in N \text{ or } z(i) \in o]\}$$

so that

$$(\forall k \in *\mathbb{N})^{\pi}M_k = \{z : \|z\| \le k \ \&\ [j \in {}^{\pi}N_k \text{ or } z(j) \in {}^{\pi}o_k]\}.$$

Since $x(j) = m\delta_j^n$ with $n \in {}^{\pi}N_m$ we see that $x \in {}^{\pi}M_m$. In (2.4) we saw that $\delta(M) = \cup[{}^{\pi}M_m : m \in *\mathbb{N}]$, hence $x \in \mu = \Gamma[\delta(M)]$.

7.2 Counterexample

In (7.1) we saw that the uniform continuity criterion does not apply to internal functions. The internal function in (7.1) is linear, so it also shows that the continuity criterion does not apply to internal functions.

The continuity criterion also does not apply to standard functions at a single point. Again we use ℓ^{∞} to construct our counterexample. Define $f : \ell^{\infty} \longrightarrow \mathbb{R}$ by $f(x) = x(1) \cdot \|x\|_{\infty}$. If $x \in M$, then $\|x\|_{\infty}$ is finite and $x(1) \approx 0$, so f is M-continuous at $a = 0$. The function f is not bounded-weak-star continuous at zero. Let $m \in *\mathbb{N}$ be infinite and let $n >> m$, $n \in {}^{\pi}N[m]$, be infinite compared with m. Then the sequence

$$x(k) = \begin{cases} \frac{1}{m}, & \textit{if } k = 1 \\ m, & \textit{if } k = n \\ 0, & \textit{otherwise} \end{cases}$$

satisfies $x \in \gamma(0)$ (see (7.1) and apply similar arguments), yet $f(x) = 1 \approx 0$.

7.3 Counterexample

The "Banach-Dieudonne" property is required in order to apply our continuity criterion (5.10). The following is an example of an everywhere M-continuous function which is μ-discontinuous at zero. In this example the finest topology that contains the limit structure a + M is finer than the locally convex one induced by the limit monad μ.

Let $\ell^1 = \ell^1(\mathbb{N})$ be the usual sequence space with norm $\|x\|_1 = \sum_{k=1}^{\infty} |x(k)|$. Let $\mathscr{C}_K(\mathbb{R}) = \mathscr{C}_K$ be the continuous functions of compact support with the uniform norm $\|f\|_\infty = \sup[|f(r)| : r \in \mathbb{R}]$. The space we consider is $\mathbb{F} = \ell^1 \times \mathscr{C}_K$ with the norm topology on the first factor and the inductive limit topology on the second (see section 1, but take zero derivatives). The limit infinitesimals are given by:

$$(x,f) \overset{M}{\approx} (y,g) \textit{ if and only if } \begin{cases} \|x-y\|_1 \approx 0 \\ \|f-g\|_\infty \approx 0 \\ (f-g) \textit{ has limited support.} \end{cases}$$

Define a norm $p(x) = \sum_{k=1}^{\infty} |x(k)/k|$ on ℓ^1. Then $p(x) \le \|x\|_1$ and the function $t : (\ell^1\backslash\{0\}) \longrightarrow \mathbb{R}$ given by $t(x) = \frac{1}{p(x)}$ is $\|\cdot\|_1$-continuous on the space minus zero.

Define a function $q : (0,\infty) \times \mathscr{C}_K \longrightarrow \mathbb{R}$ by $q(r,f) = \sup[|f(s)| : s \ge r]$, so that q is jointly continuous for the metric on $\mathbb{R}$ and the $\|\cdot\|_\infty$-norm on $\mathscr{C}_K$.

Our counterexample, $\theta : \mathbb{F} \longrightarrow \mathbb{R}$, is defined as follows:

$$\theta(x,f) = \begin{cases} 0, & \textit{if } \|x\|_1 = 0 \\ \dfrac{1}{p(x)} \cdot q\left[\dfrac{1}{\|x\|_1}, f\right], & \textit{otherwise.} \end{cases}$$

This function θ is continuous in the norm $(\|\cdot\|_1, \|\cdot\|_\infty)$ on the set $(\ell^1\backslash\{0\}) \times \mathscr{C}_K$ but is μ-discontinuous at zero. (See Benninghofen, Stroyan & Richter (1988).)

REFERENCES

Benninghofen, B. (1982a). *Nichtstandardmethoden und Monaden*, Diplomarbeit, RWTH Aachen.

Benninghofen, B. (1982b). *Infinitesimalien und Superinfinitesimalien*, Dissertation, RWTH Aachen.

Benninghofen, B. & Richter, M.M. (1985). *A General Theory of Superinfinitesimals*, to appear, Fund. Math.

Benninghofen, B., Stroyan, K.D. & Richter, M.M. (1988). Superinfinitesimals in topology and functional analysis, to appear in *Proc. Lond. Math. Soc.*

Dunford, N. & Schwartz, J. (1958). *Linear Operators, Part I*, Interscience, New York.

Henson, C.W. & Moore, L.C. Jr. (1973). Invariance of the nonstandard hulls of locally convex spaces, *Duke Math. J.* 40, 193-205.

Jarchow, H. (1981). *Locally Convex Spaces*, B.G. Teubner, Stuttgart.

Robertson, A.P. & Robertson, W. (1964). *Topological Vector Spaces*, Cambridge University Press, Cambridge.

Rubel, L.A. & Ryff, J.W. (1970). The bounded weak-star topology and the bounded analytic functions, *J. Funct. Anal.* **5**, 167-183.

Schaefer, H.H. (1966). *Topological Vector Spaces*, Macmillan, New York.

Stroyan, K.D. (1983). Myopic utility functions on sequential economies, *J. Math. Econ.* **11**, 267-276.

Stroyan, K.D. & Luxemburg, W.A.J. (1976). *Introduction to the Theory of Infinitesimals*, Academic Press Series on Pure & Appl. Math. **72**, New York.

Francois, T. (1967). *Topological Vector Spaces, Distributions and Kernels*, Academic Press Series on Pure and Appl. Math. **25**, New York.

THE NON-LINEAR BOLTZMANN EQUATION FAR FROM EQUILIBRIUM

LEIF ARKERYD

Abstract. This paper surveys the theory of Loeb solutions of the mild, space-dependent Boltzmann equation under natural initial conditions of finite mass, energy, and entropy. The existence theory for large initial data is presented, and compared with the classical theory. The limit of zero mean free path and infinite time are also discussed, including some new results.

1. INTRODUCTION

Consider the Boltzmann equation, that famous kinetic model of rarefied gases driven by binary collisons. Since a linearization would remove some of the interesting behaviour, we shall here retain the truly non-linear collision mechanism.

A fairly common characteristic of non-linear evolution equations is that the behaviour close to equilibrium is smooth and regular with nice asymptotic convergence, whereas initial values further away may result in wilder, possibly non-smooth behaviour. For space-dependent Boltzmann gases in full space or bounded containers, various contraction mapping estimates can be used to prove the existence of unique, smooth solutions converging to an equilibrium (Maxwellian) value with time, if the initial values are close enough to equilibrium. Such methods break down when the initial values are further away from equilibrium, and so do natural compactness arguments.

For another approach, let us first recall the physical background. From the point of view of physics, what happens at distances or within volumes below, say, the scale of elementary particle phenomena, is an artefact of the model with little experimental relevance. From this perspective, the question whether the model starts from an underlying set of rationally, really, or infinitesimally spaced points, should be decided purely on mathematical grounds.

The aim of this paper is to survey what is known about the Boltzmann gases on an underlying three-dimensional continuum filled with the denser set of nonstandard real points instead of the standard real ones. In such a setting a Loeb standard mild form of the actual Boltzmann equation can be solved for arbitrary initial mass distributions with finite initial energy and entropy. This is for space-dependent gases at the present state of the art in contrast to the classical situation, where the continuum is filled with only standard points. Moreover, Maxwellian behaviour can be proved in the small mean free path limit for these solutions. When unique classical solutions are known to exist, in many cases they coincide in the distribution sense with the present solutions. In the space-homogeneous situation, where classical solutions do exist for large data, convergence to a Maxwellian of those solutions when time tends to infinity, has in several cases first been proved using non-standard methods, only later to be followed by standard proofs. And just like all classical computational methods, whatever interesting physical quantities there are, such as moments, they correspond also in the present context to real valued integrals of the solutions multiplied by test functions. Finally, there are other important problems in gas kinetics for which this approach also seems promising.

We next give a short introduction to the Boltzmann equation. This is followed by the existence theory in an S- and Loeb integral context. The last section is devoted to the limiting Maxwellian behaviour.

2. THE BOLTZMANN EQUATION

Let F be the density of a gas in phase-space, which for the present we can think of as $\mathbb{R}^3 \times \mathbb{R}^3$. The amount of gas at time t in a region Y is then

$$\int_Y F(x,v,t)dxdv.$$

Boltzmann expressed the material derivative D_tF through a balance between the gain from molecules entering a region of collision and the loss from those leaving it, in the absence of exterior forces as

$$\partial_t F(x,v_1,t) + v_1.\nabla_x F(x,v_1,t) = QF(x,v_1,t)/\varepsilon.$$

Here ∇_x is the gradient with respect to the position $x \in \mathbb{R}^3$, ε is the mean free path (suppressed in this and the following section), and Q is the so called collision operator,

$$QF(x,v_1) = \int_{\mathbb{R}^3\times B} (F(x,v_1')F(x,v_2') - F(x,v_1)F(x,v_2))k(v_1,v_2,u)dv_2du.$$

Given two molecules of initial velocities (v_1,v_2), and initially separated in space, (v_1',v_2') denotes the velocities of the molecules after collision. Conservation of momentum and energy give

$$v_1 + v_2 = v_1' + v_2' \quad \text{and} \quad v_1^2 + v_2^2 = (v_1')^2 + (v_2')^2.$$

These equations are not sufficient to completely specify the collision. The further details of the collision process are described with help of the parameter set B,

$$B = \{u=(\theta,\varphi):\ 0 \le \theta \le \tfrac{1}{2}\pi,\ 0 \le \varphi \le 2\pi\}.$$

A general discussion of Q and B can be found in Cercignani (1975) and Truesdell & Muncaster (1980) for example . For inverse j^{th} power molecular forces, $j>1$,

$$k = |v_2-v_1|^{(j-5)/(j-1)}\,\beta(\theta),$$

where

$$\beta(\theta) \sim |\tfrac{1}{2}\pi-\theta|^{-(j+1)/(j-1)} \quad \text{as } \theta \uparrow \tfrac{1}{2}\pi.$$

For the convergence of

$$\int_B k(v_1,v_2,u)du,$$

we here assume a so called angular cut-off with

$$k(v_1,v_2,u) = k_1(|v_2-v_1|)\beta(\theta)$$

where β is measurable and bounded, and on the interior of $(0, \tfrac{1}{2}\pi)$ locally bounded from below by positive constants. We also take

$$0 < k_1(w) \le C(w^{-\gamma} + 1 + w^{\lambda}) \quad \text{for } w \ne 0, \text{ with } 0 \le \lambda < 2,\ 0 \le \gamma < 3,$$

where the condition $\gamma<3$ (corresponding to $j>2$ above) is used to make $k(v_1,v_2,u)$ locally integrable, and $\lambda<2$ to control $F(x,v_1)k(v_1,v_2,u)$ for large v_1.

In the discussion below we also consider a truncated version k^n of k. Set

$$\bar{k}^n(v_1,v_2,u) = k(v_1,v_2,u) \wedge n \quad \text{for } u \in B \text{ and } v_1^2 + v_2^2 \le n^2,$$

$$\bar{k}^n(v_1,v_2,u) = 0 \quad \text{otherwise},$$

$$\chi(s) = 0 \quad \text{for } s\le 0,$$
$$\chi(s) = 1 \quad \text{for } s\ge 1, \quad 0\le\chi\le 1, \quad \text{and } \chi \in C^\infty(\mathbb{R}),$$

and define

$$k^n(x,v_1,v_2,u) = \chi(n^3-|x|^2)\bar{k}^n(v_1,v_2,u).$$

A mild form of the corresponding truncated Boltzmann equation is

$$F(x+tv_1,v_1,t) = F_0(x,v_1) + \int_0^t Q_nF(x+sv_1,v_1,s)ds \tag{2.1}$$

with

$$Q_nF(x,v_1) = \int_{\mathbb{R}^3\times B} (\langle F(x,v_1')F(x,v_2')\rangle - \langle F(x,v_1)F(x,v_2)\rangle)k^n(x,v_1,v_2,u)dv_2du,$$

and

$$\langle F(x,v_1)F(x,v_2)\rangle = F(x,v_1)F(x,v_2) \qquad \text{if } |F(x,v_1)F(x,v_2)| \le n,$$
$$\langle F(x,v_1)F(x,v_2)\rangle = n \text{ sign } F(x,v_1)F(x,v_2) \qquad \text{otherwise.}$$

The equation (2.1) is easy to solve, since the truncated collision operator Q_n is locally Lipschitz continuous in L^∞.

2.1 **Proposition** (Arkeryd (1972))

For any initial value F_0 in the positive cone of $L^\infty(\mathbb{R}^6)$, there exists a unique, non-negative solution F of (2.1).

If

$$F_0,\ x^2F_0,\ v^2F_0,\ F_0\ln F_0 \in L^1(\mathbb{R}^6), \tag{2.2}$$

then the integrals

$$\int F(x,v,t)dxdv, \quad \int F(x,v,t)v^2dxdv, \quad \int F(x+vt,v,t)x^2dxdv$$

are conserved, and the integral

$$\int F(x,v,t)\ln F(x,v,t)dxdv$$

is non-increasing.

Remark. The notation ln stands for the natural logarithm. The condition $F_0 \ge 0$, $F_0 \in L^1(\mathbb{R}^6)$ expresses that F_0 is a density, and $v^2F_0 \in L^1(\mathbb{R}^6)$ comes from the physical interpretation of $\int v^2F_0dxdv$ as the total energy of the molecules. Also $-\int F_0 \ln F_0 dxdv$ can be interpreted as an entropy function.

In contrast to this case the Boltzmann equation without truncation is difficult to solve due to the lack of linearity, boundedness, and continuity of Q. This comes about in three different ways. First there are more v-factors in Q than in the left member of the Boltzmann equation. Secondly, at least when there is no cut-off in k, the two terms in Q are not well defined separately. Thirdly, Q as a function of x contains the square of a L^1-function. In the space-homogeneous case, where one looks at solutions F independent of x, the third problem, of course, disappears. Mainly for this reason existence and uniqueness have been easier to obtain when x is absent than otherwise.

3. LOEB SOLUTIONS TO THE BOLTZMANN EQUATION

By transfer the truncated existence result of Proposition 2.1 also holds in the nonstandard context; and condition (2.2) connects the proposition with S- and Loeb integrability techniques. We recall those concepts next, before using them to study the full Boltzmann equation.

Let λ denote the Lebesgue measure on $\mathbb{R}^n$, $\mathcal{A}$ the σ-algebra of Lebesgue measurable subsets of a measurable set $B \subseteq \mathbb{R}^n$, $X = {}^*B$, $A = {}^*\mathcal{A}$ (an algebra of sets), $m = {}^*\lambda$ (at least a finitely additive set function on A). To get a countably additive set function, define $L(m)(Y) = {}^\circ m(Y)$ for $Y \in \mathcal{A}$, and $\sigma(\mathcal{A})$ as the smallest σ-algebra of subsets of X with $\sigma(\mathcal{A}) \supseteq \mathcal{A}$. For $Y \in \sigma(\mathcal{A})$ set

$$L(m)(Y) = \inf L(m)(Y'),$$

where inf is taken over all $Y' \in \mathcal{A}$, $Y \subseteq Y'$. The *Loeb space* $(X, L(\mathcal{A}), L(m))$ which is the completion of $(X, \sigma(\mathcal{A}), L(m))$, turns out to be suitable extension of $(B, \mathcal{A}, \lambda)$ for our study of the Boltzmann equation.

To be able to use the new tool of nonstandard analysis, the *-mapping, we shall control the connection between the Loeb measure L(m) and *Lebesgue measure $m = {}^*\lambda$ using the concept of S-*integrability*. The function $f: {}^*\mathbb{R}^n \to {}^*\mathbb{R}$ is S-*integrable* on ${}^*\mathbb{R}^n$, if

(i) f is *Lebesgue measurable,

(ii) $\int_{{}^*\mathbb{R}^n} (|f| - |f| \wedge \omega) {}^*dx \approx 0 \quad (\omega \in {}^*\mathbb{N} \backslash \mathbb{N})$,

(iii) $\int_{{}^*\mathbb{R}^n} |f| \wedge \omega^{-1} {}^*dx \approx 0 \quad (\omega \in {}^*\mathbb{N} \backslash \mathbb{N})$.

Loosely (ii) means 'infinitesimal contribution from infinite function values', and (iii) means 'infinitesimal contribution from infinitesimal function values'. Under the technical requirement that the nonstandard model is countably saturated, S-integrability implies Loeb integrability (Loeb (1975)), and

$$^{\circ}\int_Y f dm = \int_Y {}^{\circ}f dL(m) \qquad \text{for } Y \in A.$$

For a more detailed description of Loeb spaces and S-integrability see the article by T. Lindstrøm (this volume).

Consider now the nonstandard solution f of the Boltzmann equation truncated at $n \in {}^*\mathbb{N}\backslash\mathbb{N}$, as given by Proposition 2.1 and transfer. In the case of bounded entropy and second moments, the corresponding solution f is S-integrable; thus $^{\circ}f$ is Loeb integrable. This is so because (iii) 'infinitesimal mass from infinitesimal function values' holds,

$$\int_{*\mathbb{R}^6} f(t) \wedge \omega^{-1} {}^*dxdv \le \int_{x^2+v^2\le\omega^{1/4}} \omega^{-1} {}^*dxdv +$$

$$+ (1+\omega^{1/4})^{-1} \int_{*\mathbb{R}^6} (1+x^2+v^2) f(t) {}^*dxdv \approx 0 \qquad (\omega \in {}^*\mathbb{N}\backslash\mathbb{N}),$$

together with (ii) 'infinitesimal mass from infinite function values', which can be proved as follows. We have by Proposition 2.1 and transfer

$$\int f \ln f {}^*dxdv \le \int {}^*F_0 \ln F_0 {}^*dxdv,$$

with the right member finite by assumption. Applying the elementary inequality $y\ln y \ge -z + y\ln z$ $(y\ge 0,\ z>0)$, to the case $y = f(x+vt,v,t)$, $z = \exp(-x^2-v^2)$ and using the previous inequality, we have with $\ln^+ x = 0 \vee \ln x$,

$$\int_{*\mathbb{R}^6} f \ln^+ f {}^*dxdv \le \int_{\mathbb{R}^6} F_0 \ln F_0 dxdv + \int_{\mathbb{R}^6} \exp(-x^2-v^2) dxdv +$$

$$+ \int_{\mathbb{R}^6} F_0 (x^2+v^2) dxdv = K.$$

From this we get for $\omega \in {}^*\mathbb{N}\backslash\mathbb{N}$

$$\int_{*\mathbb{R}^6} (f(t) - f(t)\wedge\omega) {}^*dxdv \le (\ln\omega)^{-1} \int_{*\mathbb{R}^6} f(t) \ln^+ f(t) {}^*dxdv \approx 0.$$

Below we shall also prove that in (2.1) $\langle f\otimes f\rangle$ can be substituted by $^{\circ}f\otimes{}^{\circ}f$, k^n by $^{\circ}k$, and $^*\mathbb{R}^3$ by $ns{}^*\mathbb{R}^3$, if Loeb integration is used. This leads to the following result.

3.1 Theorem (Arkeryd (1986a))

Let f *be the solution of the nonstandard, truncated Boltzmann equation (2.1) with* $n \in {}^*\mathbb{N}\backslash\mathbb{N}$, *and initial condition*

$$f(x,v,0) = {}^*F_0(x,v) \wedge n + n^{-1}\exp(-v^2-x^2),$$

where F_0 *satisfies (2.2). Then* ${}^\circ(1+v^2+x^2)f(x,v,t) \in \text{Loeb } L^1(ns^*\mathbb{R}^6)$ *for* $t \in ns^*\mathbb{R}_+$. *The function* ${}^\circ f$ *is a Loeb solution of the integrated Boltzmann equation; i.e. for Loeb a.e.* $(x,v) \in ns^*\mathbb{R}^6$ *and for* $t \in ns^*\mathbb{R}_+$

$${}^\circ f(x+tv_1,v_1,t) = F_0 \circ st(x,v_1) +$$

$$+ \int_0^t \int_{ns^*\mathbb{R}^3\times B} {}^\circ f(x+sv_1,v_1',s){}^\circ f(x+sv_1,v_2',s)k \circ st(v_1,v_2,u)L(dv_2duds) -$$

$$- \int_0^t \int_{ns^*\mathbb{R}^3\times B} {}^\circ f(x+sv_1,v_1,s){}^\circ f(x+sv_1,v_2,s)k \circ st(v_1,v_2,u)L(dv_2duds).$$

(Here B *is the previously defined set of collision parameters).*

The solution conserves mass and first moments in v *and* x-vt, *has globally bounded* H*-function, and satisfies*

$$\int_{ns^*\mathbb{R}^6} v^2\,{}^\circ f(x,v,t)Ldxdv \le \int_{\mathbb{R}^6} v^2F_0(x,v)dxdv,$$

$$\int_{ns^*\mathbb{R}^6} x^2\,{}^\circ f(x+tv,v,t)Ldxdv \le \int_{\mathbb{R}^6} x^2F_0(x,v)dxdv.$$

Remark. Corresponding results have been obtained in the space-periodic case (Arkeryd (1984)), and for bounded C^1-regions with reflection type boundary conditions and exterior forces (Elmroth (1984)). A wider class of non-negative, *Lebesgue measurable initial data can also be handled with the same method. The theorem also holds for radial cut-off in k. For forces of inverse j-th power type, $2<j\le 3$, and angular cut off, the solution concept of Theorem 3.1 suggested the first standard solution concept for which L^1 existence results could be proved in the standard space-homogenous case (Arkeryd (1986b)).

Proof of Theorem 3.1. We have already proved the first Loeb integrability statement of the theorem. The final statements about conservation and bounds are similarly proved and left as an exercise for the reader. To complete the proof it remains to show that given $t \in {}^*\mathbb{R}_+$ and finite,

$$
{}^\circ\!\int_0^t \int_{*\mathbb{R}^3} \int_{*B} \langle f(x+sv_1,v_1,s)f(x+sv_1,v_2,s)\rangle k^n(x+sv_1,v_1,v_2,u)\,{}^*dudv_2ds
$$

$$
= \int_0^t \int_{ns*\mathbb{R}^3} \int_{*B} {}^\circ f(x+sv_1,v_1,s)\,{}^\circ f(x+sv_1,v_2,s)k \circ st(v_1,v_2,u)L(dudv_2ds), \tag{3.1}
$$

and

$$
{}^\circ\!\int_0^t \int_{*\mathbb{R}^3} \int_{*B} \langle f(x+sv_1,v_1',s)f(x+sv_1,v_2',s)\rangle k^n(x+sv_1,v_1,v_2,u)\,{}^*dudv_2ds
$$

$$
= \int_0^t \int_{ns*\mathbb{R}^3} \int_{*B} {}^\circ f(x+sv_1,v_1',s)\,{}^\circ f(x+sv_1,v_2',s)k \circ st(v_1,v_2,u)L(dudv_2ds) \tag{3.2}
$$

for Loeb a.e. $(x,v_1) \in ns{}^*\mathbb{R}^6$.

To prove (3.1) we shall first show that:

(a) for Loeb a.e. $(x,v_1) \in ns{}^*\mathbb{R}^6$ the function

$$(v_2,u,s) \to f(x+sv_1,v_2,s)k^n(x+sv_1,v_1,v_2,u)$$

is S-integrable in ${}^*\mathbb{R}^3 \times {}^*B \times [0,t]$.

But after a change of variables, the positive, *measurable function $f.k^n$ can be estimated from above by the tensor product $f(x,v_2,s)k_1(v_1)\beta(\theta)$ of an S-integrable function and a weakly singular standard function. It follows that given $\bar{v}_1 \in ns{}^*\mathbb{R}^3$, $f.k^n$ is S-integrable with respect to (x,v_1,v_2,u,s) in

$${}^*\mathbb{R}^3 \times \{v_1\colon |v_1-\bar{v}_1|\le 1\} \times {}^*\mathbb{R}^3 \times {}^*B \times [0,t].$$

It then follows from the definition of S-integrability that (a) holds Loeb a.e. in ${}^*\mathbb{R}^3 \times \{v_1\colon |v_1-\bar{v}_1|\le 1\}$, hence, since $\bar{v}_1$ is arbitrary, (a) holds in general on $ns{}^*\mathbb{R}^6$.

As a second step in the proof of (3.1) we shall show that:

(b) for Loeb a.e. $(x,v_1) \in ns{}^*\mathbb{R}^6$ the function

$$(v_2,u,s) \to f(x+sv_1,v_1,s)f(x+sv_1,v_2,s)k^n(x+sv_1,v_1,v_2,u)$$

is S-integrable in ${}^*\mathbb{R}^3 \times {}^*B \times [0,t]$.

By (a) this holds, if for Loeb a.e. $(x,v_1) \in ns{}^*\mathbb{R}^6$ the function

$$(v_2,u,s) \to f(x+sv_1,v_1,s)$$

is uniformly finite on [0,t]. But we can take the ${}^*L^1$ derivative of (2.1) with respect to t (Mikusinski (1978) Ch. XIII), obtaining

$$D_s f(x+sv_1,v_1,s) = Q_n f(x+sv_1,v_1,s).$$

To this equation we add $h(x+sv_1,v_1,s)f(x+sv_1,v_1,s)$ with

$$h(x+sv_1,v_1,s) = \int_{*\mathbb{R}^3} \int_{*B} f(x+sv_1,v_2,s)k^n(x+sv_1,v_1,v_2,u)^*dudv_2.$$

We multiply the result with the integrating factor $\exp(\int_0^s h(\tau)^*d\tau)$, and take ${}^*L^1$ primitive functions, to get

$$f(x,+sv_1,v_1,s) = f(x,v_1,0)\exp(-\int_0^s h(x+\tau v_1,v_1,\tau)^*d\tau) + \tag{3.3}$$

$$+ \int_0^s \exp(-\int_\tau^s h(x+rv_1,v_1,r)^*dr)(Q_n f(x+\tau v_1,v_1,\tau)+hf(x+\tau v_1,v_1,\tau))^*d\tau$$

for *a.e. $s \in [0,t]$. By continuity the equality holds for all $s \in [0,t]$.

Notice that the last integrand in (3.3) is positive. For s=t, by the S-integrability of the left hand side, the integral in the right member stays uniformly bounded on ${}^*\mathbb{R}^6$-V, where the Loeb measure of V can be made arbitrarily small. We can also choose V so that the function $\exp(\int_0^t h(r)^*dr)$ is finite on $ns({}^*\mathbb{R}^6\setminus V)$. Then for $(x,v_1) \in ns({}^*\mathbb{R}^6\setminus V)$ the integral in the right member of (3.3) is uniformly finite on [0,t]. And so (b) holds, since the Loeb measure of V can be made arbitrarily small.

As a third step in the proof of (3.1) we notice that:

(c) for Loeb a.e. $(x,v_1) \in ns{}^*\mathbb{R}^6$

$$°(\langle f(x+sv_1,v_1,s)f(x+sv_1,v_2,s)\rangle k^n(x+sv_1,v_1,v_2,u))$$

$$= °f(x+sv_1,v_1,s)°f(x+sv_1,v_2,s)k \circ st(v_1,v_2,u)$$

for Loeb a.e. $(v_2,u,s) \in ns{}^*\mathbb{R}^3 \times {}^*B \times [0,t]$.

This is so because each of the three factors is finite Loeb a.e. in $ns{}^*\mathbb{R}^3 \times {}^*B \times [0,t]$.

From here (3.1) follows from (b) and (c) and the uniform finiteness of $f(x+sv_1,v_1,s)$ on $0\le s\le t$ from (b), if, for Loeb a.e. $(x,v_1)\in ns{*}\mathbb{R}^6$,

$$\lim_{m\to\infty} {}^{\circ}\int_0^t \int_{|v_2|>m} \int_{*B} f(x+sv_1,v_2,s)k^n(x+sv_1,v_1,v_2,u){*}dudv_2ds = 0.$$

But that is a consequence of

$$\lim_{m\to\infty} {}^{\circ}\int_{\mathcal{A}_m} f(x+sv_1,v_2,s)k^n(x+sv_1,v_1,v_2,u)d\sigma = 0,$$

where

$$\mathcal{A}_m = [0,t]\times {*}\mathbb{R}^3 \times \{v_1:\ |v_1-\bar{v}_1|\le 1\}\times\{v_2:\ |v_2|>m,\}\times {*}B,$$

and

$$d\sigma = {*}dudv_2dv_1dxds.$$

We now turn to the proof of (3.2). The rest of the proof holds for Loeb a.e. $(x,v_1)\in ns{*}\mathbb{R}^6$. First show that:

(d) $\quad {}^{\circ}\langle f(x+sv_1,v_1',s)f(x+sv_1,v_2',s)\rangle = {}^{\circ}f(x+sv_1,v_1',s){}^{\circ}f(x+sv_1,v_2',s)$

for Loeb a.e. $(v_2,u,s)\in ns{*}\mathbb{R}^3\times {*}B\times[0,t]$.

To prove this, let χ be the characteristic function of any set

$$\{(v_1,v_2):\ v_1^2+v_2^2\le m^2\}\qquad (m\in\mathbb{N}).$$

Since the integrals

$$\int_0^t\int \chi f(x+sv_1,v_j',s){*}dxdv_1dv_2duds \qquad (j=1,2)$$

are both finite, it follows that

$$\chi f(x+sv_1,v_1',s)\quad\text{and}\quad \chi f(x+sv_1,v_2',s)$$

are finite for Loeb a.e. $(v_2,u,s)\in ns{*}\mathbb{R}^3\times {*}B\times[0,t]$. This implies (d).

We next consider:

(e) the S-integrability in (v_2,u,s) of

$$\langle f(x+sv_1,v_1',s)f(x+sv_1,v_2',s)\rangle k^n(x+sv_1,v_1,v_2,u). \tag{3.4}$$

Since $\langle f(x+sv_1,v_1,s)f(x+sv_1,v_2,s)\rangle k^n(x+sv_1,v_1,v_2,u)$ is S-integrable, it is enough to prove for some finite $j>1$ that the function

(3.4) is S-integrable on the set

$$\Omega_j = \{(v_2,u,s) \in {}^*\mathbb{R}^3 x {}^*B x[0,t]: \langle f(x+sv_1,v_1',s)f(x+sv,v_2',s)\rangle \geq j\langle f(x+sv_1,v_1,s)f(x+sv_1,v_2,s)\rangle\}.$$

But on Ω_2

$$0 \leq \langle f(x+sv_1,v_1',s)f(x+sv_1,v_2',s)\rangle k^n(x+sv_1,v_1,v_2,s)$$
$$\leq 2(\langle f(x+sv_1,v_1',s)f(x+sv_1,v_2',s)\rangle -$$
$$- \langle f(x+sv_1,v_1,s)f(x+sv_1,v_2,s)\rangle)k^n(x+sv_1,v_1,v_2,u)),$$

and it suffices to consider the S-integrability of the right member on Ω_2. We only have to check (ii) and (iii) in the definition of S-integrability. Now for j>1 with evident notations

$$0 \leq \int_{\Omega_j} (\langle f'\otimes f'\rangle-\langle f\otimes f\rangle)k^n{}^*dv_2duds \tag{3.5}$$
$$\leq (\ln j)^{-1}\int_{\Omega_j} (\langle f'\otimes f'\rangle-\langle f\otimes f\rangle)\ln(f'\otimes f'/f\otimes f)k^n{}^*dv_2duds$$
$$= C/\ln j,$$

where C is finite a.e. in $ns{}^*\mathbb{R}^6$. That a majorization by a finite C is a.e. possible in (3.5) follows from an estimate of the third member in (3.5) integrated with respect to(x,v_1). By the usual proof of the H-theorem (Arkeryd (1972)) we have

$$\int_V (\langle f'\otimes f'\rangle-\langle f\otimes f\rangle)k^n\ln(f'\otimes f'/f\otimes f){}^*dxdv_1dv_2duds$$
$$\leq \int_{{}^*\mathbb{R}^6} (f_{t=0}\ln f_{t=0} + \exp(-x^2-v^2) + (x^2+v^2)f_{t=0}){}^*dxdv \qquad (\in ns{}^*\mathbb{R}_+),$$

where the integrand to the left is non-negative, the right member finite, and

$$V = {}^*(\mathbb{R}^9 \times B \times \mathbb{R}_+).$$

This implies (3.5).

To prove (ii) we let $\omega \in {}^*\mathbb{N}\backslash\mathbb{N}$ and set

$$\Omega = \{v_2,u,s)\in\Omega_2: \langle f'\otimes f'\rangle k^n > \omega\}.$$

It follows from (3.5) that

$$\int_{\Omega} {}^*dv_2 duds \approx 0.$$

Then

$$0 \le \int_{\Omega} \langle f \otimes f \rangle k^n {}^*dv_2 duds \le j_0^{-2} \approx 0 \quad \text{for some } j_0 \in {}^*\mathbb{N}\setminus\mathbb{N},$$

since $\langle f \otimes f \rangle k^n$ is S-integrable. Hence

$$\int_{\Omega-\Omega_{j_0}} \langle f' \otimes f' \rangle k^n {}^*dv_2 duds \approx 0.$$

As for the remaining part of Ω,

$$\int_{\Omega_{j_0}} (\langle f' \otimes f' \rangle - \langle f \otimes f \rangle) k^n {}^*dv_2 duds \approx 0$$

by (3.5). This proves (ii).

To prove (iii) we let $\omega \in {}^*\mathbb{N}\setminus\mathbb{N}$. It follows from the S-integrability of $\langle f \otimes f \rangle k^n$, that

$$0 \le \int_{\Omega_2} \langle f \otimes f \rangle k^n \wedge \omega^{-1} {}^*dv_2 duds \le j_0^{-2} \approx 0$$

for some $j_0 \in {}^*\mathbb{N}\setminus\mathbb{N}$.

Hence

$$\int_{\Omega_2-\Omega_{j_0}} \langle f' \otimes f' \rangle k^n \wedge \omega^{-1} {}^*dv_2 duds \approx 0.$$

As for the remaining part of Ω_2, by (3.5)

$$0 \le \int_{\Omega_{j_0}} \langle f' \otimes f' \rangle k^n \wedge \omega^{-1} {}^*dv_2 duds \le \int_{\Omega_{j_0}} \langle f' \otimes f' \rangle k^n {}^*dv_2 duds$$

$$< 2C/\ln j_0 \approx 0.$$

This proves (iii), and so the S-integrability (e).

Finally (3.2) follows from (d) and (e), if:

(f) $$\lim_{\nu\to\infty} {}^\circ\!\int_{A_\nu} \langle f' \otimes f' \rangle k^n {}^*dv_2 duds = 0,$$

where

$$A_\nu = \{(v_2, u, s) \in \Omega_2 : \ |v_2| \ge \nu\}.$$

But this holds by (3.5), since

$$\int_{A_\nu} \langle f'\otimes f'\rangle k^n {}^*dv_2 duds$$

$$\leq j \int_{A_\nu} (\langle f\otimes f\rangle)k^n {}^*dv_2 duds + 2\int_{\Omega_j} (\langle f'\otimes f'\rangle - \langle f\otimes f\rangle)k^n {}^*dv_2 duds,$$

and by (3.1)

$$\lim_{\nu\to\infty} {}^\circ\!\int_{A_\nu} \langle f\otimes f\rangle k^n {}^*dv_2 duds = 0.$$

This completes the proof of (f), thus of (3.2) and of Theorem 3.1.◂

The present Loeb solutions are - after division of the equation by (1+f) - equivalent to (generalised Young) measure valued standard solutions. When standard real valued solutions exist, they can be compared with the present Loeb solutions. In most cases if F is a standard solution, then F∘st is a Loeb solution of the corresponding type, and uniqueness of the standard solution implies uniqueness in the corresponding Loeb setting. When standard existence theorems can be proved, the idea of proof usually implies that our Loeb solution defines a distribution solution in the standard context, which then by regularity arguments is as smooth as could be hoped for. Let us illustrate this in the space-homogeneous case.

In the space-homogeneous, nonstandard case consider the following weak form of the truncated Boltzmann equation (2.1):

$$\int_{{}^*\mathbb{R}^3} f(v,t)g(v,t){}^*dv = \int_{{}^*\mathbb{R}^3} f(v,0)g(v,0){}^*dv + \tag{3.6}$$

$$+ \int_0^t \int_{{}^*\mathbb{R}^3} f(v,s)\partial_s g(v,s){}^*dvds + \int_0^t (Q_n f(s),g(s)){}^*ds,$$

where

$$(Q_n f(s),g(s))$$

$$= \int_{{}^*\mathbb{R}^3\times\mathbb{R}^3\times B} [g(v_1',s)-g(v_1,s)]\langle f\otimes f\rangle(v_1,v_2,s)k^n(v_1,v_2,u){}^*dv_1dv_2du.$$

The real-valued mapping L_t from the space $C_0(\mathbb{R}^3)$ of continuous functions on $\mathbb{R}^3$ with compact support defined by

$$L_tG(t) = {}^{\circ}\!\int_{*\mathbb{R}^3} f(v,t)*G(v)*dv$$

is linear and bounded,

$$|L_t(G)| \le {}^{\circ}\!\int_{*\mathbb{R}^3} f(v,t)*dv \sup_{\mathbb{R}^3}|G(v)|.$$

It follows that L_t defines a measure μ_t on $\mathbb{R}^3$. But by the energy bounds

$$\int_{|v|>j} |f(v,t)|*dv \approx 0$$

for $j \in *\mathbb{N}\backslash\mathbb{N}$, and so

$$\lim_{j\to\infty} \mu_t\{|v|>j\} = 0.$$

Let A_k be a sequence of measurable sets in $\mathbb{R}^3$, such that

$$\lim_{k\to\infty} \int_{A_k} dv = 0.$$

Again by the S-integrability of f(.,t)

$$\lim_{k\to\infty} \mu_t(A_k) = 0.$$

This implies that μ_t is absolutely continuous with respect to dv, and its Radon-Nikodym derivative is $d\mu_t = F_t dv$ with $F_t \in L^1_+(\mathbb{R}^3)$. This proves the first part of the following theorem.

3.2 Theorem

Suppose that $F_0 \ge 0$, *and that* F_0, v^2F_0, $F_0 \ln F_0 \in L^1(\mathbb{R}^3)$. *Then for* $n \in *\mathbb{N}\backslash\mathbb{N}$ *the solution* f *of* *(2.1) *with initial value*

$$f_0(v) = *F_0(v) \wedge n + n^{-1}\exp(-v^2)$$

defines a mapping

$$F : \mathbb{R}_+ \to L^1_+(\mathbb{R}^3)$$

through

$$\int_{\mathbb{R}^3} F(v,t)G(v)dv = {}^{\circ}\!\int_{*\mathbb{R}^3} f(v,t)*G(v)*dv \qquad (G \in C^1(\mathbb{R}^3))$$

If $\gamma < 1$ *in the definition of* k, *then the function* F *is a weak solution of the Boltzmann equation in the sense that*

$$\int_{\mathbb{R}^3} F_t(v)G(v,t)dv = \int_{\mathbb{R}^3} F_0(v)G(v,0)dv + \int_0^t \int_{\mathbb{R}^3} F_s(v)\partial_s G(v,s)dvds + + \int_0^t (QF_s, G(s))ds.$$

Remark. For the proof of the second part, notice that (3.6) holds rigorously under the assumptions of the theorem. From here the result follows by arguments close to those used in the proof of the first part. Using a suitably small nonstandard angular cut-off, we can obtain the same result for forces of infinite range and of inverse k-th power type, $k>3$.

4. THE MAXWELLIAN LIMIT

Theorem 3.1 holds for non-infinitesimal mean free paths, $°\varepsilon > 0$, whereas for the nonstandard solution f of (2.1), its standart part $°f$, is Loeb integrable also when $\varepsilon > 0$ is infinitesimal, $\varepsilon \approx 0$. We write $f = f_\varepsilon$ to stress the dependence of the solution on the mean free path in this section.

Any f_ε with $\varepsilon \approx 0$, is infinitely close to a local Maxwellian $a(x,t)\exp(-b(x,t)v^2+c(x,t).v)$. This can be seen in the following way. First a lemma which goes back to Gibbs is needed.

4.1 Lemma (Gibbs)

Consider the class $\mathscr{C}$ *of functions* $G \in L^1_+(\mathbb{R}^3)$ *which satisfy*

$$\int_{\mathbb{R}^3} G(v)dv = A, \qquad \int_{\mathbb{R}^3} G(v)vdv = B, \qquad \int_{\mathbb{R}^3} G(v)v^2dv \le C, \tag{4.1}$$

for some finite constants A,B,C. *Let* $E_0(v)$ *be the unique Maxwellian satisfying (4.1) with equality also in the third relation. Then*

$$\int_{\mathbb{R}^3} G(v)\ln G(v)dv \ge \int_{\mathbb{R}^3} E_0(v)\ln E_0(v)dv \qquad (G \in \mathscr{C}),$$

and the equality holds only for $G = E_0$.

If we multiply the Boltzmann equation with $\ln f_\varepsilon$, integrate, change variables, and use Gibbs' Lemma, we can rigorously obtain

$$0 \le \int_0^t \int (\langle f'_\varepsilon \otimes f'_\varepsilon\rangle - \langle f_\varepsilon \otimes f_\varepsilon\rangle) \ln(f'_\varepsilon \otimes f'_\varepsilon / f_\varepsilon \otimes f_\varepsilon) k^n {}^*dx dv_1 dv_2 du ds$$

$$= -4\varepsilon \int f_\varepsilon(t) \ln f_\varepsilon(t) {}^*dx dv + 4\varepsilon \int f_\varepsilon(0) \ln f_\varepsilon(0) {}^*dx dv \approx 0.$$

From this it follows that the integrand is infinitesimal.

4.2 Lemma

When $\varepsilon \approx 0$, *it holds for Loeb a.e.* $(x,t) \in ns{}^*\mathbb{R}^3 \times {}^*\mathbb{R}_+$, *that*

$$f_\varepsilon(x,v_1,t) f_\varepsilon(x,v_2,t) \approx f_\varepsilon(x,v'_1,t) f_\varepsilon(x,v'_2,t) \tag{4.2}$$

for Loeb a.e. $(v_1,v_2,u) \in ns{}^*\mathbb{R}^3 \times \mathbb{R}^3 \times B$.

Next, a careful analysis of (4.2) shows that f_ε is either Loeb a.e. infinitesimal or Loeb a.e. non-infinitesimal. We have (Arkeryd (1986a)):

4.3 Lemma

Let $g \in {}^*L^1_+(\mathbb{R}^3)$ *be given with*

$$\int g(v)(1+v^2) {}^*dv, \quad \text{and} \quad \int g(v) \ln g(v) {}^*dv$$

finite, and with

$$g(v_1)g(v_2) \approx g(v'_1)g(v'_2)$$

for Loeb a.e. $(v_1,v_2,u) \in ns{}^*\mathbb{R}^3 \times \mathbb{R}^3 \times B$. *Then either*

$$g(v) \approx 0 \quad \text{for Loeb a.e. } v \in ns{}^*\mathbb{R}^3,$$

or

$$°g(v) > 0 \quad \text{for Loeb a.e. } v \in ns{}^*\mathbb{R}^3.$$

Proof. We give a new proof of this, based on Fubini type arguments, and shorter than the original proof in Arkeryd (1986a).

Consider for simplicity a representative from the equivalence class g, still (somewhat incorrectly) denoted by g. The conditions of the lemma imply that g is S-integrable, hence that $°|g \otimes g - g' \otimes g'|$ is Loeb integrable. Either $g \approx 0$ for Loeb a.e. $v \in ns{}^*\mathbb{R}^3$, or there is a *measurable set A of finite diameter and positive Loeb measure in $ns{}^*\mathbb{R}^3$, where $°g > 0$. We can then choose this set A so that, moreover, g is finite on A with $C < g < 1/C$ for some $C \in \mathbb{R}_+$.

If $°g > 0$ Loeb a.e. on a sphere of positive Loeb measure, then the corresponding standard construction can be employed to prove that $°g > 0$ Loeb a.e. on $ns{*}\mathbb{R}^3$. Otherwise we partition A by four parallel planes at non-infinitesimal distances into five bounded sets of positive Loeb measure. Denote one of the 'end' sets by A_1, the middle one by A_2 and the other end set by A_3. The lemma follows, if we can construct a ball B' of positive Loeb measure in $ns{*}\mathbb{R}^3$, where $°g > 0$ for Loeb a.e. $v \in B'$. The rest of the proof is devoted to that construction, which is based on Fubini type arguments.

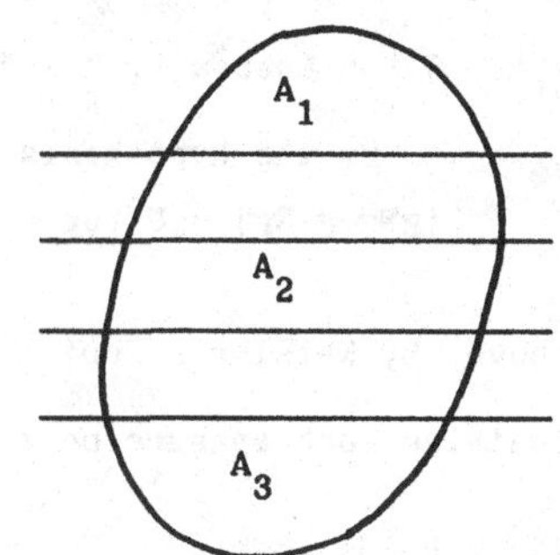

By Keisler's Fubini Theorem it follows that there is $\bar{v}_1 \in A_1$, and a line ℓ in ${*}\mathbb{R}^3$, such that $I = \ell \cap A_2$ has positive Loeb measure $|I|$ on the line, and such that for Loeb a.e. $v_2 \in I$

$$°|g(\bar{v}_1)g(v_2)-g(v_1')g(v_2')| = 0 \text{ for Loeb a.e. } u \in {*}B.$$

Now $°\int g{*}dv < \infty$, and so given $\varepsilon \in \mathbb{R}_+$, there is $\lambda_\varepsilon \in \mathbb{R}_+$, such that the Loeb measure of the set

$$\{v \in {*}\mathbb{R}^3:\ g(v) > \lambda_\varepsilon\},$$

is less than ε. Let μ denote the positive infimum for $v_2 \in I$ of the Loeb surface area of the sphere

$$S_{v_2} = \{v_2'(\bar{v}_1, v_2, u):\ \ u \in {*}B\}.$$

Take

$$\varepsilon_n = \mu|I|2^{-n}n^{-1},$$

and corresponding λ_{ε_n}. Remove those v_2 for which $g(v_2') > \lambda_\varepsilon$ on at least a n^{-1}-fraction of the sphere S_{v_2}. Starting from a large enough $n_1 \in \mathbb{N}$, do this for all $n \geq n_1$. Denote the remaining set of v_2 by I'. The Loeb measure of I' is positive and for $v_2 \in I'$, $g(v_2')$ is finite for Loeb a.e.

$u \in {}^*B_2$. For $v_2 \in I'$ define

$$T_{v_2} = \{v_1'(\bar{v}_1, v_2, u) \colon u \in {}^*B\}.$$

By construction, for $v_2 \in I'$ and Loeb a.e. $u \in {}^*B$, it holds that $g(v_2')$ is finite, $C < g(\bar{v}_1)$, $g(v_2) < 1/C$, $^\circ|g(\bar{v}_1)g(v_2) - g(v_1')g(v_2')| = 0$, and so $^\circ g(v_1') > 0$ for Loeb a.e. $u \in {}^*B$.

By the hypothesis of the lemma

$$^\circ|g\otimes g - g'\otimes g'| = 0 \text{ for Loeb a.e. } (v_1, v_2, u) \in A_3 \times \bigcup_{v_3 \in I'} T_{v_3} \times {}^*B.$$

As above, by Keisler's Fubini theorem, there is $\bar{\bar{v}}_1 \in A_3$, and a geodesic I" of positive Loeb measure on a sphere T_{v_3} $(v_3 \in I')$, such that for Loeb a.e. $v_2 \in I''$ it holds that

$$^\circ g(v_2) > 0 \text{ (thus } {}^\circ g(\bar{\bar{v}}_1)g(v_2) > 0),$$

and $$^\circ|g(\bar{\bar{v}}_1)g(v_2) - g(v_1')g(v_2')| = 0 \text{ for Loeb a.e. } u \in {}^*B.$$

But

$$\{v_1'(\bar{\bar{v}}_1, v_2, u) \colon v_2 \in I'', u \in {}^*B\}$$

contains a ball B' of positive Loeb measure. With

$$B_1 = \{(v_2, u) \colon v_1'(\bar{\bar{v}}_1, v_2, u) \in B'\},$$
$$B_2 = \{v_2'(\bar{\bar{v}}_1, v_2, u) \colon (v_2, u) \in B_2\},$$

a set of full Loeb measure on B' corresponds to a set of full Loeb measure on B_2 (B_3). Now there is a set of full Loeb measure on B_3, where $g(v_2')$ is finite. Since $\bar{\bar{g}}(v_1) > C$, $^\circ g(v_2) > 0$ for Loeb a.e. $v_2 \in I''$, and

$$^\circ|g(\bar{\bar{v}}_1)g(v_2) - g(v_1')g(v_2')| = 0 \text{ for Loeb a.e. } (v_2, u) \in I'' \times {}^*B,$$

we conclude that $^\circ g(v_1') > 0$ Loeb a.e. in B'. This completes the proof of the lemma. ◄

It is a straight-forward consequence of (4.2) and Lemma 4.3 that f_ε is almost a local Maxwellian (Arkeryd (1986a))

4.4 Theorem

For $\varepsilon \approx 0$ *and Loeb a.e.* $(x,t) \in ns{*}\mathbb{R}^3 \times R_+$, *there are*

$$a(x,t),\ b(x,t) \in \mathbb{R}_+,\ \textit{and}\ c(x,t) \in \mathbb{R}^3,$$

such that

$$f_\varepsilon(x,v,t) \approx a(x,t)\exp(-b(x,t)v^2 + c(x,t).v) \tag{4.3}$$

for Loeb a.e. $v \in ns{*}\mathbb{R}^3$.

Remarks. (i) For $\varphi \in {*}C_0^1(\mathbb{R}^3 \times \mathbb{R}_+)$

$$\int_{*\mathbb{R}^6}\int_0^t Q_n\, f_\varepsilon(x+sv_1,v_1,s)\varphi(x+sv_1,s){*}dsdxdv_1 = 0.$$

So if the maximum norm of φ in ${*}C_0^1$ is finite, then the function f_ε of Theorem 3.1 satisfies (also for infinitesimal $\varepsilon > 0$)

$$\int_{ns*\mathbb{R}^6} {}^\circ f_\varepsilon(x,v_1,t)^\circ\varphi(x,t)\ L(dxdv_1) = \int_{ns*\mathbb{R}^6} F_0 \circ st(x,v_1)^\circ\varphi(x,0)L(dxdv_1)$$

$$+ \int_{ns*\mathbb{R}^6\times[0,t]} {}^\circ f_\varepsilon(x,v_1,s)({}^\circ\partial_s\varphi(x,s) + {}^\circ v_1.\nabla_x\varphi(x,s))L(dxdv_1ds), \tag{4.4}$$

for $t \in ns{*}\mathbb{R}_+$.

Inserting (4.3) for $\varepsilon \approx 0$ into (4.4) with supp $\varphi \subseteq ns{*}\mathbb{R}^3 \times \mathbb{R}_+$, we obtain the first of the five compressible Euler equations. When the second and third v-moments of f_ε are S-integrable, then the other four compressible Euler equations also follow. In particular this is the case when the v^4-moment of f_ε is bounded for $t > 0$.

(ii) Consider the space-homogeneous case for cut-off hard potentials, with $F_0(1+|v|^{2+\delta}) \in L_+^1(\mathbb{R}^3)$ for some $\delta > 0$, and $F_0 \ln F_0 \in L^1(\mathbb{R}^3)$. Then (Arkeryd (1986a)) strong L^1-convergence to the Maxwellian when $t\to\infty$ for the corresponding standard solution follows from the relevant version of Theorem 4.4. There is also an L^p-version of this result (unpublished) by the present author.

REFERENCES

Arkeryd, L. (1972). An existence theorem for a modified space-inhomogeneous non-linear Boltzmann equation, *Bull. Amer. Math. Soc.* **78**, 610-614.

Arkeryd, L. (1984). Loeb solutions of the Boltzmann equation, *Arch. Rat. Mechs. Anal.* **86**, 85-97.

Arkeryd, L. (1986a). On the Boltzmann equation in unbounded space far from equilibrium, and the limit of zero mean free path, *Comm. Math. Phys.* **105**, 205-219.

Arkeryd, L. (1986b). On the Enskog equation in two space variables, *Transp. Theory & Stat. Phys.* **15**, 673-691.

Cercignani, C. (1975). *Theory and applications of the Boltzmann equation*, Academic Press, New York.

Elmroth, T. (1984). Loeb solutions of the Boltzmann equation with initial boundary values and external forces, Tech. Report, Dept. of Math., Goteborg.

Loeb, P. (1975). Conversion from nonstandard to standard measure spaces and applications in probability theory, *Trans. Amer. Math. Soc.* **211**, 113-122.

Mikusinski, J. (1978). *The Bochner integral*, Academic Press, New York.

Truesdell, C. & Muncaster, R.G. (1980). *Fundamentals of Maxwell's kinetic theory of a simple monatomic gas*, Academic Press, New York.

INDEX